Henry Stephens Randall

The Practical Shepherd

A complete treatise on the breeding, management and diseases of sheep. Sixth Edition

Henry Stephens Randall

The Practical Shepherd
A complete treatise on the breeding, management and diseases of sheep. Sixth Edition

ISBN/EAN: 9783337105709

Printed in Europe, USA, Canada, Australia, Japan

Cover: Foto ©berggeist007 / pixelio.de

More available books at **www.hansebooks.com**

THE
PRACTICAL SHEPHERD:

A COMPLETE TREATISE ON THE

BREEDING, MANAGEMENT AND DISEASES OF SHEEP.

BY

HENRY S. RANDALL, LL. D.,

AUTHOR OF "SHEEP HUSBANDRY IN THE SOUTH," "FINE-WOOL SHEEP HUSBANDRY," ETC., ETC.

WITH ILLUSTRATIONS.

SIXTH EDITION.

ROCHESTER, N. Y.:
D. D. T. MOORE, UNION BUILDINGS.
PHILADELPHIA: J. B. LIPPINCOTT & CO.
1863.

Entered according to Act of Congress, in the year 1863, by
D. D. T. MOORE,
In the Clerk's Office of the District Court of the United States for the Northern District of New York.

ROCHESTER, N. Y.:
STEREOTYPED BY JAMES LENNOX,
62 BUFFALO STREET.

INTRODUCTION.

An attempt has been made in the following pages to give an impartial history of all the most valuable varieties and families of sheep in the United States,— to explain the principles of breeding on which their improvement rests, and to describe their proper treatment in health and sickness, under the different climatic and other circumstances to which they are necessarily subjected in a country as extensive as our own.

Many of the topics of this work have been ably discussed, and are constantly being ably discussed in our Agricultural periodicals; but it is now eighteen years since the publication of the last elaborate American work which treats on them connectedly and with any considerable degree of fullness. It is fifteen years since the appearance of my own Sheep Husbandry in the South, which was confined to a portion of these subjects, and, in many instances, as the title would imply, to views and statements intended for local rather than general information.

In the mean time, a great change — almost an entire revolution — has taken place in the character of American sheep, and in the systems of American sheep husbandry. The fine-wool families which existed here in 1845 have, under a train of circumstances which will be found recorded in this volume, mostly passed away; and they have been succeeded by a new family, developed in our own country, which calls for essentially different standards of breeding and modes of practical treatment.

Our improved English, or, as they are often termed, mutton breeds of sheep, instead of being now confined to a few small, scattering flocks, have spread into every portion of our country, represent a large amount of agricultural capital, and throughout regions of considerable extent are more profitable than sheep kept specially for wool growing purposes. Some of the most valuable families of them were wholly unknown in this country — indeed, had scarcely been brought into general notice in England — fifteen years ago. And, finally, our advanced agricultural condition has created a new set of agricultural circumstances and interests which materially affect, and, in turn, are materially affected by, sheep husbandry, — so that their reciprocal relations must be understood to lead to the highest measure of success in almost any department of farming.

In view of these facts, a new work on American Sheep Husbandry brought down to the requirements of the present day — that is, embodying the results of the experience which sheep breeders have obtained down to the present time—is obviously called for. And the need is more urgent at a period when a great existing war has so raised the price of wool that multitudes are embarking in its production who have comparatively little knowledge of sheep or their management. This work is intended to be minute and explicit enough in regard to every detail of that management to meet the wants of the merest beginner.

I would gladly have seen this labor performed by another. But, during the past year, repeated public and private intimations have continued to reach me from breeders, agricultural editors, etc., scattered through various States of this Union, and representing personal interests the most diversified and even contrary, that my preparation of such a work was considered desirable. In complying with the wishes thus expressed, I can only bring to my task experience, and a disposition to state facts with accuracy and candor. As has been remarked in another portion of this volume, I have owned and been familiar with flocks of sheep from my infancy, and have had the direct and practical charge and management of them, in considerable numbers, for a period exceeding thirty years. During that time I have bred all the varieties of the Merino which have been introduced into our country, and several of the leading families of English sheep. But not having bred the latter extensively, or very recently, I have thought it would be more satisfactory, in most instances, to employ descriptions of them drawn from standard English writers, and from their actual breeders in the United States. Had I contemplated writing this work long enough in advance to make a collection of materials specially intended for it, I should also have taken pleasure in drawing out the opinions of the eminent and highly successful breeders of English sheep in the Canadas. My inquiries might even have extended to England. But the "Practical Shepherd" was commenced as soon as the writing of it was determined on, and the earlier Chapters, treating on Breeds, were in print before I could have sought in an appropriate mode and obtained the desired information from foreign lands.

When called upon to give the opinions of others in regard to points with which I am unacquainted, or less acquainted, I have chosen generally to quote their language,—and in all instances to mention their names. Disguised compilation is one of the pettiest forms of literary theft; and it deprives the reader of his fair and proper privilege of deciding for himself on the competence of the authority to which he is called upon to give credit. On various subjects, and more especially on the subject of those ovine diseases which are as yet unknown in the United States, these pages will be found enriched with the descriptions and the opinions of eminent foreign agricultural writers and veterinarians. For the invaluable privilege of thus availing myself of their knowledge, I, as well as the readers of this volume, owe them sincere acknowledgements.

I was at some loss whether or not it would be expedient for me to give descriptions of an extended list of diseases and remedies, the former of which have not appeared, or, at least, have not been recognized in our country. But judging from their increase thus far, and judging from their analogies derivable from the history of diseases in other domestic animals, and in man, we have strong reasons to apprehend that as our country grows older, and our systems of husbandry more artificial, the same causes will be generated or developed here which now produce many of the diseases of Europe It is already found, for example, that as we treat our English sheep according to English modes, maladies long known in England, but not previously known here, and not yet known among our other breeds of sheep, make their appearance among them. And some of the fellest ovine maladies of Europe are liable, at any time, to be introduced here by contagion. On the whole, I judged that it would be erring, if at all, on the safer side, to give descriptions drawn from the best existing sources of veterinary information of the symptoms and treatment of all the maladies unknown in this country which have thus far been recognized and classified in Europe.

I have quoted somewhat freely from my own previous works on Sheep. I could discover no objection to this, where my opinions remain unchanged; and where they are changed, omissions and, in a few cases, slight alterations have been made to conform the quoted statements to them. If occasional discrepancies are discoverable between my present and former views, I have only to say, in explanation, that further experience or further reflection has led me to change my conclusions.

A general history and description of all the breeds of sheep have not been attempted in this volume. Those desirous of such information are referred to Mr. Youatt's Work on Sheep. This unwearied investigator and copious writer exhausted this field of research — and he really left nothing, in what may be termed the literature of Sheep Husbandry, to be performed by another. Those who have followed him in the same field, have only repeated him; and these compilers have generally been as destitute of his grace as of his erudition.

I have alluded to all the distinct breeds of sheep which have, so far as my knowledge extends, been introduced into the United States, but I have particularly described only those leading and valuable ones which now employ the attention of enlightened agriculturists. And even in respect to these, no historic investigations have been indulged in which do not appear to me to have a direct bearing on the modes and means of their preservation or improvement. The province of this work embraces purely practical concerns, and history and disquisition are pertinent only so far as they throw a direct and instructive light on those concerns.

One of the greatest and most insuperable difficulties which I have experienced in the prosecution of my labors arises from the want of an established and systematic nomenclature to express the various divisions of species. The designations, species, race, kind, stock, breed, variety, family, etc., have been applied almost indiscriminately to the same divisions, as if the words were understood to be synonymous. Even Mr. Youatt falls into this loose and careless use of language. But unfortunately a confusion of terms can not but produce a corresponding confusion of ideas, on a subject not without intricacy, and in reference to distinctions or lines of demarkation which are frequently faint, and nearly always irregular and abounding in exceptions. The breeder who aspires to be an improver, ought to have clear ideas on this subject. Called upon early in the progress of this work, and without much previous consideration, to devise a uniform mode of classification in the premises, I adopted and have made use of the following:

The term breed is applied to those extensive and permanent groups of sheep which are believed to have had, respectively, a common origin — which exhibit certain common leading characteristics — and which transmit those characteristics with uniformity to their progeny. Examples of Breeds, are the Merino of Spain, including its pure blood descendants, wherever found; the Fat-Rumped Sheep of Asia, the Long-Wooled Sheep of England, and the Short-Wooled Sheep of England. The term Variety is applied to different national branches of the same breed, such as the Saxon, French and American varieties of the parent Spanish Merino. The term Family is used to designate those branches of a breed or variety found in the same country, which exhibit permanent, but ordinarily lesser differences than varieties. Thus the different kinds of Downs and the Rylands are families of the English Short-Wooled sheep; the Cotswolds and the Liecesters are families of the English Long-Wooled sheep; the Infantados and Paulars are families of both the Spanish and American Merinos. The term sub-family is occasionally used to designate a minor group, bearing about the same relation to a family that a family does to a variety. No satisfactory term was found to characterize the smallest and initial group of all, — those closely related animals, to which, among human beings, we apply the designation of a family, when we use that word in its most restricted sense. Perhaps I have sometimes, awkwardly enough, spoken of them as animals of the same individual blood, or as possessing the same strain of individual blood.

The system of classification above described, answers very well when applied to the Merino. This breed exhibits all the enumerated classes in permanent, distinct forms, each to a certain extent isolated from the others by separate breeding, for a considerable period, and totally isolated from all other and outside groups of sheep by perfect purity of blood. But this classification is wholly unsatisfactory when applied to the British breeds of sheep. I will not consume space to explain a fact, the causes of which will be so obvious to the observing reader.

I return my sincere thanks to the following gentlemen for valuable aid in collecting materials for this work — none the less valuable because, in many instances, they were contributed in a form which required no special mention in my pages. I arrange the names alphabetically to avoid making a distinction where, in most cases, none exists: — A. B. Allen, Lewis F. Allen, George Campbell, N. L. Chaffee, Edmund Clapp, Prosper Elithorp, George Geddes, James Geddes, W. F. Greer, James S. Grennell, Edwin Hammond, Benjamin P. Johnson, Geo. Livermore, R. A. Loveland, Daniel Needham, Theodore C. Peters, Virtulan Rich, William R. Sanford, Nelson A. Saxton, Homer L. D. Sweet, Samuel Thorne, and M. W. C. Wright.

HENRY S. RANDALL.

CORTLAND VILLAGE, N. Y.,
September, 1863.

CONTENTS.

CHAPTER I.
FINE-WOOLED BREEDS OF SHEEP.

The Spanish, French, Saxon, and Silesian Merinos,......... Page 13

CHAPTER II.
INTRODUCTION OF FINE-WOOLED SHEEP INTO THE UNITED STATES.

Early Importations of Spanish, French and Saxon Merinos,..... 23

CHAPTER III.
AMERICAN MERINOS ESTABLISHED AS A VARIETY.

The Mixed Leonese or Jarvis Merinos—The Infantado or Atwood Merinos—The Paular or Rich Merinos—Other Merinos,...... 27

CHAPTER IV.
LATER IMPORTATIONS OF FINE-WOOLED SHEEP INTO THE UNITED STATES.

French and Silesian Merinos Introduced,....................... 35

CHAPTER V.
BRITISH AND OTHER LONG AND MIDDLE-WOOLED SHEEP IN THE UNITED STATES.

Leicesters, Cotswolds, Lincolns, New Oxfordshires, Black-Faced Scotch, Cheviot, Fat-Rumped, Broad-Tailed, Persian and Chinese Sheep,.. 43

CHAPTER VI.
BRITISH SHORT-WOOLED SHEEP, ETC., IN THE UNITED STATES.

The South Downs, Hampshire Downs, Shropshire Downs and Oxfordshire Downs,........................... 55

CHAPTER VII.

THE POINTS TO BE REGARDED IN FINE-WOOLED SHEEP.

Carcass — Skin — Folds or Wrinkles — Fleece — Fineness — Evenness — Trueness and Soundness — Pliancy and Softness — Style and Length of Wool, ... 68

CHAPTER VIII.

THE SAME SUBJECT CONTINUED.

Yolk — Chemical Analysis of Yolk — Its Uses — Proper Amount and Consistency of it — Its Color — Coloring Sheep Artificially — Artificial Propagation and Preservation of Yolk, ... 77

CHAPTER IX.

ADAPTATION OF BREEDS TO DIFFERENT SITUATIONS.

Markets — Climate — Vegetation — Soils — Number of Sheep to be Kept — Associated Branches of Husbandry, ... 82

CHAPTER X.

PROSPECTS AND PROFITS OF WOOL AND MUTTON PRODUCTION IN THE UNITED STATES.

Page, ... 91

CHAPTER XI.

PRINCIPLES AND PRACTICE OF BREEDING.

Page, ... 101

CHAPTER XII.

BREEDING IN-AND-IN.

Page, ... 116

CHAPTER XIII.

CROSS-BREEDING.

Cross-Breeding the Merino and Coarse Breeds — Crossing Different Families of Merinos — Crossing Between English Breeds and Families — Recapitulation, ... 124

CHAPTER XIV.

SPRING MANAGEMENT.

Catching and Handling — Turning Out to Grass — Tagging — Burs — Lambing — Proper Place for Lambing — Mechanical Assistance in Lambing — Inverted Womb — Management of New-Born Lambs — Artificial Feeding — Chilled Lambs — Constipation — Cutting Teeth — Pinning — Diarrhea or Purging, ... 139

CHAPTER XV.

SPRING MANAGEMENT CONTINUED.

Congenital Goitre — Imperfectly Developed Lambs — Rheumatism — Treatment of the Ewe after Lambing — Closed Teats — Uneasiness — Inflamed Udder — Drying off — Disowning Lambs — Foster Lambs — Docking Lambs — Castration,............ 152

CHAPTER XVI.

SUMMER MANAGEMENT.

Mode of Washing Sheep — Utility of Washing Considered — Cutting the Hoofs — Time between Washing and Shearing — Shearing — Stubble Shearing and Trimming — Shearing Lambs and Shearing Sheep semi-annually — Doing up Wool — Frauds in Doing up Wool — Storing Wool — Place for Selling Wool — Wool Depots and Commission Stores — Sacking Wool,..... 163

CHAPTER XVII.

SUMMER MANAGEMENT — CONTINUED.

Drafting and Selection — Registration — Marking and Numbering — Storms after Shearing — Sun-Scald — Ticks — Shortening Horns — Maggots — Confining Rams — Training Rams — Fences — Salt — Tar, Sulphur, Alum, &c. — Water in Pastures — Shade in Pastures — Housing Sheep in Summer — Pampering,...... 179

CHAPTER XVIII.

FALL MANAGEMENT.

Weaning and Fall Feeding Lambs — Sheltering Lambs in Fall — Fall Feeding and Sheltering Breeding Ewes — Selecting Ewes for the Ram — Coupling — Period of Gestation — Management of Rams during Coupling — Dividing Flocks for Winter,.... 198

CHAPTER XIX.

WINTER MANAGEMENT.

Winter Shelter — Temporary Sheds — Hay Barns with Open Sheds — Sheep Barns or Stables — Cleaning out Stables in Winter — Yards — Littering Yards — Confining Sheep in Yards and to Dry Feed,... 211

CHAPTER XX.

WINTER MANAGEMENT — CONTINUED.

Hay Racks — Water for Sheep in Winter — Amount of Food Consumed by Sheep in Winter — Value of Different Fodders — Nutritive Equivalents — Mixed Feeds — Fattening Sheep in Winter — Regularity in Feeding — Salt,..................... 230

CHAPTER XXI.
PRAIRIE SHEEP HUSBANDRY.

Prairie Management in Summer—Lambing—Folds and Dogs—Stables—Herding—Washing—Shearing—Storing and Selling Wool—Ticks—Prairie Diseases—Salt—Weaning Lambs—Prairie Management in Winter—Winter Feed—Sheds or Stables—Water—Location of Sheep Establishment,........ 248

CHAPTER XXII.
ANATOMY AND DISEASES OF SHEEP.—THE HEAD.

Comparatively small Number of American Sheep Diseases—Low Type of American Sheep Diseases—Anatomy of the Sheep—The Skeleton—The Skull—The Horns and their Diseases—The Teeth—Swelled Head—Sore Face—Swelled Lips—Inflammation of the Eye,................................. 261

CHAPTER XXIII.
ANATOMY AND DISEASES OF THE SHEEP'S HEAD, CONTINUED.

Section of Sheep's Head—Grub in the Head—Hydatid on the Brain—Water on the Brain—Apoplexy—Inflammation of the Brain—Tetanus or Locked-Jaw—Epilepsy—Palsy—Rabies,... 273

CHAPTER XXIV.
DISEASES OF THE DIGESTIVE ORGANS.

Blain—Obstructions of the Gullet—The Stomachs and their Diseases—External and Internal Appearance of the Stomachs—The Mode of Administering Medicines into the Stomachs of Sheep—Hoove—Poisons—Inflammation of the Rumen, or Paunch—Obstruction of the Maniplus—Acute Dropsy, or Red-Water—Enteritis, or Inflammation of the Coats of the Intestines—Diarrhea—Dysentery—Constipation—Colic, or Stretches—Braxy, or Inflammation of the Bowels—Worms—Pining,.. 291

CHAPTER XXV.
DISEASES OF THE CIRCULATORY AND THE RESPIRATORY SYSTEMS.

The Pulse—Place and Mode of Bleeding—Fever—Inflammatory Fever—Malignant Inflammatory Fever—Typhus Fever—Catarrh—Malignant Epizootic Catarrh—Pneumonia, or Inflammation of the Lungs—Pleuritis or Pleurisy—Consumption,.. 314

CHAPTER XXVI.
DISEASES OF THE GENERATIVE AND URINARY ORGANS.

Abortion—Inversion of the Womb—Garget—Parturient, or Puerperal Fever—Cystitis, or Inflammation of the Bladder,.. 329

CONTENTS.

CHAPTER XXVII.
DISEASES OF THE SKIN.

The Scab — Erysipelatous Scab — Wild fire and Ignis Sacer — Other Cutaneous Eruptions — Small Pox, or Variola Ovina,.. 338

CHAPTER XXVIII.
DISEASES OF THE LOCOMOTIVE ORGANS.

Fractures — Rheumatism — Disease of the Biflex Canal — Gravel — Travel-Sore — Lameness from Frozen Mud — Fouls — Hoof-Rot,.... 354

CHAPTER XXIX.
OTHER DISEASES, WOUNDS, ETC.

The Rot — Scrofula — Hereditary Diseases — Cuts — Lacerated and Contused Wounds — Punctured Wounds — Dog Bites — Poisoned Wounds — Sprains — Bruises — Abscess,..................... 372

CHAPTER XXX.
LIST OF MEDICINES.

Page,... 383

CHAPTER XXXI.
THE DOG IN ITS CONNECTION WITH SHEEP.

The Injuries inflicted by Dogs on Sheep -- The Sheep Dog — The Spanish Sheep Dog — The Hungarian Sheep Dog — The French Sheep Dog — The Mexican Sheep Dog — The South American Sheep Dog — Other Large Races of Sheep Dogs — The English Sheep Dog — The Scotch Sheep Dog, or Colley — Accustoming Sheep to Dogs,........................... 393

APPENDICES.

A.— Origin of the Improved Infantados,........................ 412
B.— Origin of the Improved Paulars,........................... 416
C.— English Experiments in Feeding Sheep,..................... 418
D.— Sheep and Product of Wool in United States,............... 425
E.— Starting a Sheep Establishment in the New Western States, 427
F.— Climate of Texas,... 428
G.— Proportion of Meat to Wool in Sheep of Different Ages, Sexes and Sizes, ... 433
H.— The American Merinos at the International Exhibition of 1863,.. 438
LIST OF ILLUSTRATIONS,....................................... 440
INDEX,... 441

THE PRACTICAL SHEPHERD.

CHAPTER I.

FINE-WOOLED BREEDS OF SHEEP.

THE SPANISH, FRENCH, SAXON AND SILESIAN MERINOS.

THE SPANISH MERINO.—From a period anterior to the Christian era, fine-wooled Sheep abounded in Spain, and they were, or gradually ripened into, a breed distinct in its characteristics from all other breeds in the world. It was, however, divided into provincial varieties which exhibited considerable differences; and these were subdivided into great permanent cabanas or flocks which being kept distinct from each other and subjected to special courses of breeding, assumed the character of separate families varying somewhat, but in a lesser degree, from each other.

The first division recognized in Spain was into Transhumantes or traveling flocks and Estantes or stationary flocks. The first were regarded as the most valuable and were owned by the king and some of the principal nobles and clergy. They were pastured in winter on the plains of Southern Spain, and driven in spring (commencing the journey in April,) to the fresh green herbage of the mountains in Northern Spain. They began their return early in October. The route, each way, averaged about four hundred miles and was completed in six weeks. Through inclosed regions and where the feed was scarce, they often traveled from fifteen to twenty miles a day. The lambs were dropped early in January. Nearly half of them, and sometimes in seasons of bad pasturage, three-fourths of them were destroyed as soon as yeaned, and those which were preserved were usually suckled by two ewes. This was intended for the benefit of both lambs and ewes. The latter were thought to produce more wool than when each suckled a lamb. The lambs were little over three months old when the spring migration commenced, and about nine months old when the autumnal one commenced. Thus every year of its life the migratory Merino performed a journey of eight hundred miles, and passed nearly a fourth of the entire time on the road. It received neither shelter nor artificial food. Such a training constantly weeded out of the flock the old, the feeble and the weak in constitution, and developed among those which remained capabilities for enduring exertion and hardship to an extraordinary degree.

Some of the most esteemed families of migratory Merinos are thus mentioned by Lasteyrie:—"The Escurial breed is supposed to possess the finest wool of all the migratory sheep. The Gaudeloupe have the most perfect form, and are likewise celebrated for the quantity and quality of their wool. The Paulars bear much wool of a fine quality; but they have a more evident enlargement behind the ears, and a greater degree of throatiness, and their lambs have a coarse, hairy appearance, which is succeeded by excellent wool. The lambs of the Infantados have the same hairy coat when young. The Negretti are the largest and strongest of all the Spanish traveling sheep."

Vague and unsatisfactory as is this description, it is perhaps the best contemporaneous one extant, of that period near the opening of the present century when the flocks of Spain had reached their highest point of excellence — and before invasion and civil war had led to their sale into foreign countries and their almost general destruction or dispersion at home. I am inclined to think that the small pains taken by Lasteyrie and his contemporaries to point out the distinctions between the best Spanish families,— the "Leonesa" as they were collectively called—resulted from the fact, that the foreign breeders of that day, and the Spaniards themselves, attached but little importance to those distinctions in respect to value—though in respect to breeding they were rigorously preserved.

To furnish the reader with some data for comparison between the several Spanish families and their American descendants, I select the following facts from a table prepared by Petri, an intelligent and highly trustworthy writer, who visited Spain near the beginning of this century on purpose to examine its Sheep; and I add some measurements of

American Merinos made of Sheep in no wise extraordinary in their forms.*

NAMES OF FLOCKS.	Weight, including wool.	Length from mouth to horns.	Length from horns to shoulders.	Length from shoulders to tail.	The whole length.	Circumference of the belly.	Height of the fore legs.	Height of hind legs.	Distance of hip bones apart.
	lbs.	in.	ft. in.	ft. in.	ft. in.	ft. in.	ft. in.	in.	in.
NEGRETTI.									
Ram	97	9½	1 7	2 2	4 6¼	4 1½	1 3	10	6
Ewe	67	8½	1 5	2 1	4 2½	4 1½	1 1	9½	4½
INFANTADO.									
Ram	100½	10	1 6	2 3	4 7	4 2	1 0	9	6
Ewe	70	9	1 5½	2 1	4 3½	3 11	1 0	8½	5½
GUADELOUPE.									
Ram	97½	9	1 6	2 2	4 5	4 5½	1 0	8	6
Ewe	69	9	1 2	2 1	3 11	3 9		10½	6½	4
ESTANTES OF SIERRA DE SOMO.									
Ram	96½	9½	1 6	2 0	4 3¼	4 2½	1 0	8	6
Ewe	62½	9	1 2	2 1	4 0	3 10		11	7	5
SMALL ESTANTES.									
Ram	42	7½	1 3	1 9	3 7½	3 2		10	6½	3
Ewe	30	7	1 .1	1 6	3 2	2 10		8	6	3
AMERICAN MERINO.									
Ram	122	9	10	2 4	3 11	4 4½		11	9	9
Ewe	114	9½	10	2 4	3 11½	4 1½		11	9	8
Ewe	122	9	10	2 5	4 0	4 3		9	9	8
Ewe	100	9	11	2 3	3 11	4 0¾		8½	8	8

These weights and measures, except those of the American sheep, are Austrian. The Austrian pound is equal to 1.037 pounds avoirdupois; the Austrian foot to 1.234 English feet.

The fleece of the Spanish Merino was level on the surface and so dense that, like that of its American descendant, it opposed a firm resistance when grasped by the hand, instead of yielding under the fingers like fur, hair, or the thin wool of other races of sheep. The wool was shorter than that of the improved American Merino and particularly so on the belly, legs and head. It was very even in quality, both as between different sheep and on different parts of the same sheep. The most celebrated flocks, with the exception of the Escurial, were dark colored externally—about as dark as the present Merino sheep in our own Middle and Western States, which are not housed in summer. The wool was rendered moist to

---

* They were taken from my flock, and the measurements, &c., made in December, 1861. The ewes were a little over average size, but the ram was quite small. His usual weight immediately after shearing is but 100 pounds. I selected him more particularly to exhibit *another* contrast, with the Spanish Sheep. His unwashed fleece of a single year's growth has reached 21 lbs. and averages about 20 lbs. "21 per cent.," as he is called, was bred by Edwin Hammond, Esq., of Middlebury, Vt.

the feel, brilliant and heavy, by yolk, but it did not exhibit this in viscid or indurated masses within, or in a black, pitchy coating without. It opened with a fine, flashing luster, and with a yellowish tinge which deepened toward its outer ends.

Livingston gives the weight of the unwashed Spanish fleeces at 8½ lbs. in the ram and 5 lbs. in the ewe. Youatt places the weight of the ram's fleece half a pound lower. The King of England's flock of Negretti's, about one hundred in number, which were picked sheep and included some wethers (but no rams,) yielded, during five years, an annual average of a little over 3½ lbs. of brook-washed wool per head, and each fleece afterwards lost about a pound in scouring.*

Youatt measured the diameter of the wool of the various flocks first introduced from Spain into England. I judge from his statements that 1-750 part of an inch may be assumed as about the average diameter or fineness of the good Spanish wool of that period. The same ingenious investigator discovered that conformation of the fibers which causes the felting property. It is produced by "serrations," as he terms them,—tooth-like projections on the wool, all pointing in a direction from the root to the point, and so inconceivably minute that 2560 of them occur in the space of an inch of the fiber. They are more numerous in proportion to the fineness of the wool, and on their number, regularity and sharpness depends the perfection of the felting property. In this respect the finest grades of Merino wool exceed all others. The following cuts give the magnified appearance of a fine specimen of Spanish wool, viewed both as an opaque and transparent object.

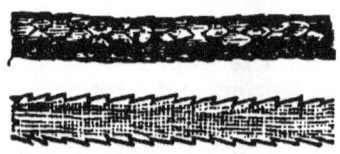

These tooth-like processes are still finer on choice specimens of Saxon wool; on that of the coarse-wooled varieties of sheep they are comparatively few, blunt and irregular.

The best flocks of Spain, as already mentioned, were lost to that country during the Peninsular war. In answer to an application for information from T. S. Humrickhouse, Esq., of

---

* See Sir Joseph Banks' five annual reports, from 1798 to 1802, in respect to this flock. The number of wethers is not given by him.

Ohio, made with a view to importations and directed to the Spanish Minister in Washington, in 1852, that functionary caused inquiry to be made in relation to the existing condition of the flocks of Spain. The statements sent back, in 1854, appear to have been derived from the Spanish "General Association of Wool Growers." The substance of them is condensed into the following paragraph:

"Although it is certain that, in the war of Independence, a great number of the said flocks, [the choice Transhumantes of Estremadura and Leon, such as the Infantado, Paular, Guadeloupe, Negretti, Escurial, Montarco, etc.,] were destroyed, and others diminished and divided, it is equally certain that they still exist in their majority and with the same good qualities which formerly made them so desirable and necessary. If, therefore, as it appears from the communication which has given rise to this report, the wool growers of the United States should have a desire and want to purchase fine sheep, they may come sure they will not be disappointed."

Then follows an extended list of flocks with the names of their owners.* The Escurial, the Negretti and the Arriza, are the only ones admitted to have been lost.

Conceding to these statements the merit of entire candor, they simply show that the Spaniards place a very different estimate on their present sheep from that placed on them by American breeders. The late John A. Taintor, Esq., o, Connecticut, who seven times visited Europe to buy sheep, carefully examined the flocks of Spain with an earnest wish to find superior animals in them for importation to the United States. He wrote to me in 1862, that the Spanish sheep "were so small, neglected and miserable, that he would not take one of them as a present."† In 1860 a gentleman of Estremadura, whose flock Mr. Taintor could not visit when in Spain, sent him a number of fleeces *as samples;* and one of these Mr. Taintor forwarded to me. It weighed, in the dirt, 5 lbs. 11 oz. The wool was about as long as ordinary American Merino wool, was not very even in quality, and was scarcely middling in point of fineness! Mr. William Chamberlain, of Red Hook, New York, the well known

---

* Scarcely any of these are the ancient owners, or those who held the flocks when the war "of Independence" commenced.

† See his letter to me in my Report on Fine-Wool Husbandry in Transactions of N. Y. State Agricultural Society for 1861. (The Report was made early in 1862 and will hereafter be cited as of that year.)

importer of Silesian Merinos, informs me that he imported about thirty Merinos from Spain, a few years since, and that after seeing them and shearing them he quietly sold them in the ensuing autumn to the butcher! William R. Sanford, of Orwell, Vermont, a Merino sheep breeder of great judgment and experience, visited the flocks of Spain, France and Germany, in 1851, in behalf of himself, Mr. Hammond and some other gentlemen of the same State, to ascertain whether fine-wooled sheep superior to those of the United States could be found in Europe. He thus wrote to me in respect to the sheep of Spain:

* * * "On arriving at Madrid I found that most of those who owned sheep to any amount lived in the city, and through our Minister I got introductions to them. From what I could learn from them in regard to the form, weight of fleece, etc., of their sheep, I became satisfied that they had none of much value. They finally admitted that they were not as good as formerly, and that they were going to Germany for bucks to improve them. I concluded, however, I would go and see for myself. It is about 200 miles from Madrid to the plains of Estremadura, where they winter their sheep. On examining the flocks, I found they had no fixed character. Occasionally there would be a fair looking sheep. At first they pretended that their sheep were pure and the best in the world. But when they found that I understood the history of their flocks, and what I wanted, they admitted they were not as good as the former ones, and they gave as a reason that they had no standard flocks to resort to as they had before the French invasion,—at which time those standard flocks were all broken up, those which were not eaten, being sold and mixed with the common sheep of the country, which were a very inferior kind. I did not see a sheep in Spain that I would pay freight on to this country. I do not believe they have any that are of pure blood."

I have conversed with several other American sheep breeders who have visited the Spanish flocks within the last fifteen years, and all of them substantially concur in the opinions above expressed.

THE FRENCH MERINO.—After several successful smaller experiments in acclimating the Spanish Merino in France, about 300 of them were imported under royal auspices to that country in 1786. Gilbert, a French writer of reputation, in a report made to the National Institute of France, ten years afterwards, thus speaks of them:

"The stock from which the flock of Rambouillet was derived, was composed of individuals beautiful beyond any that had ever before been brought from Spain; but having been chosen from a great number of flocks, in different parts of the kingdom, they were distinguished by very striking local differences, which formed a medley disagreeable to the eye, but immaterial as it affected their quality. These characteristic differences have melted into each other, by their successive alliances, and from thence has resulted a race which perhaps resembles none of those which composed the primitive stock, but which certainly does not yield in any circumstance to the most beautiful in point of size, form and strength, or in the fineness, length, softness, strength and abundance of fleece. * * * The comparison I have made with the most scrupulous attention, between this wool and the highest priced of that drawn from Spain, authorizes me to declare that of Rambouillet superior."

Lasteyrie thus gives their weight of fleeces, unwashed, through a series of years:—In 1796, 6 lbs. 9 oz.; 1797, 8 lbs.; 1798, 7 lbs.; 1799, 8 lbs.; 1800, 8 lbs.; 1801, 9 lbs. 1 oz. In 1802, he says:—"The medium weight of full grown nursing ewes' fleeces was 8 lbs. 7 oz.; of the ewes of three years old, which had no lambs, 9 lbs. 13 oz.; and two-tenths [grade] ewes, 10 lbs. 8 oz."

Mr. Trimmer, an English flock-master and writer of experience, thus described them in 1827:

"The sheep, in size, are certainly the largest pure Merinos I have ever seen. The wool is of various qualities, many sheep carrying very fine fleeces, others middling, and some rather indifferent; but the whole is much improved from the quality of the original Spanish Merinos. In carcass and appearance I hesitate not to say they are the most unsightly flock of the kind I ever met with. The Spaniards entertained an opinion that a looseness of skin under the throat, and other parts, contributed to the increase of fleece. This system the French have so much enlarged on that they have produced, in this flock, individuals with dewlaps almost down to the knees, and folds of skin on the neck, like frills, covering nearly the head. Several of these animals seem to possess pelts of such looseness of size that one skin would nearly hold the carcasses of two such sheep. The pelts are particularly thick, which is unusual in the Merino sheep. The rams' fleeces were stated at 14 lbs., and the ewes' 10 lbs., in the grease. By washing they would be reduced half, thus giving 7 and 5 lbs. each."

But the royal flock was already beginning to be outstripped by private ones in size of carcass and weight of fleece, and now there are a very few choice flocks in France which are said to average 14 lbs. of unwashed wool to the fleece in ewes, and from 20 lbs. to 24 lbs. in rams, the ewes weighing 150 lbs. and the rams 200 lbs.

THE SAXON MERINO.—In 1765, three hundred Merinos were introduced from Spain into Saxony. They, too, were a royal importation, and were placed in government establishments. It is understood they were selected principally if not exclusively from the Escurial cabana.

The course of breeding and management generally adopted in that country tended to develop a very high quality of wool at the expense of its quantity and at the expense of both carcass and constitution. The sheep were not only housed during the winter, but at night, during all rainy weather, and generally from the noonday sun in summer. They were not even allowed to run on wet grass. Their food was accurately portioned out to them in quantity and in varying courses; their stable arrangements were systematic and included a multitude of careful manipulations; at yeaning time they received (and came to require) about as much care as human patients.

When introduced into the United States (1824,) the Saxon lacked from a fifth to a quarter of the weight of the parent Spanish stock in the country, and the latter were materially smaller then than now. Their forms indicated a far feebler constitution than those of the Spanish sheep. They were slimmer, finer boned, taller in proportion, and thinner in the head and neck,— and shorter, thinner, finer and evener in the fleece. The wool had no hardened yolk internally or externally; was white externally; and opened white instead of having the buff tinge of the unwashed Spanish wool. It was from an inch to an inch and a half long on the back and sides and shorter on the head, legs and belly. Medium specimens of it measured about 1-840 parts of an inch in diameter. The washed fleeces on an average weighed from $1\frac{1}{2}$ lbs. to 2 lbs. in ewes, and from 2 lbs. to 3 lbs. in rams. There has been a regeneration and improvement of this variety in various parts of Germany, but an account of these changes would possess little interest for the mass of practical American breeders.

THE SILESIAN MERINO.—Prussian Silesia has numerous flocks of sheep descended from the Electoral and other Saxon flocks. These require no separate mention here. An importation of a different family of Merinos has been made from that country to the United States, and they have acquired, here, the distinctive appellation of Silesian Merinos. These will be described when an account is given of the importations of foreign fine-wooled sheep into the United States.

CHAPTER II.

INTRODUCTION OF FINE-WOOLED SHEEP INTO THE UNITED STATES.

EARLY IMPORTATIONS OF SPANISH, FRENCH AND SAXON MERINOS.

SPANISH MERINOS INTRODUCED.—Wm. Foster, of Boston, Massachusetts, imported three Merino sheep from Spain into that city in 1793. They were given to a friend, who killed them for mutton! In 1801 M. Dupont de Nemours, and a French banker named Delessert, sent four ram lambs to the United States. All perished on the passage but one, which was used for several years in New York, and subsequently founded some excellent grade flocks for his owner, E. I. Dupont, near Wilmington, Delaware. He was of fine form, weighed 138 lbs., and yielded 8½ lbs. of brook-washed wool,— the heaviest fleece borne by any of the early imported Merinos of which I have seen any account.* The same year, Seth Adams, of Zanesville, Ohio, imported into Boston a pair of Spanish sheep which had been brought from Spain into France. I know nothing of their later history. In 1802, Mr. Livingston, American Minister in France, sent home two pairs of French Merinos, purchased from the Government flock at Chalons. The rams appear from their recorded weights to have been larger than Spanish rams, but a picture of one of them which is extant exhibits no difference of form, and I have always learned from those who saw them, that they bore no resemblance to the modern French Merinos. Mr. Livingston subsequently imported a French ram from the Rambouillet flock. This eminent public benefactor was too much engrossed in a multitude of great undertakings to give

---

* As Dupont de Nemours was the head of the Commission appointed by the French Government to select in Spain the flocks of Merinos given up by the latter by the Treaty of Basle, I conjecture that this ram was from the original Spanish, and not from the French stock.

that close individual attention to his sheep which is necessary to marked success in breeding. But his statements show that he improved them considerably.

The following table in respect to his sheep in 1810, I take from a manuscript letter of his, not before published. As the weights given both of carcasses and fleeces considerably exceed those of the previous year (published in his Essay on Sheep, p. 186,) it is probable that the sheep had been highly kept. The wool was unwashed.

Stock rams.   *Weight.*   *Weight of fleece.*
One, 6 years old, ...146 lbs.   9 lbs.......imported from Rambouillé.
 "   2 years old, ...146 lbs.   9 lbs....... raised here.
 "   1 year old, ...145 lbs.   11 lbs. 11 oz. raised here.

Ewes.   *Average weight of fleece.*
Common (268)........................3 lbs. 10 oz.
Half-breed, or first cross,...........5 lbs. 1 oz.
Three-fourths, or second cross,......5 lbs, 3 oz., heaviest fleece, 8 lbs.
Seven-eighths, or third cross,.......5 lbs. 6 oz.   do.   8 lbs. 4 oz.
Full-blood, ...5 lbs. 13 oz.   do.   8 lbs. 12 oz.

His half-blood wool sold for 75 cents; three-fourths for $1.25; seven-eighths for $1.50; full-blood for $2.00. He sold four full-blood ram lambs for $4,000; fourteen fifteen-sixteenths blood do. for $3,500; twenty seven-eighths blood do. for $2,000; thirty three-fourths blood do. for $900. He says if the lambs had been a year old they would have sold 50 per cent. higher.*

Later in the year 1802 Col. Humphreys, the American Minister in Spain, brought home with him 21 rams and 70 ewes bought for him in that country. I find no definite early statistics of the flock, though in manuscript letters of Col. H. seen by me, he states that they constantly improved in weight of fleece and in carcass. He mentions as worthy of note that a ram raised on his farm yielded 7 lbs. 5 oz. of washed wool. The reputation of his flock, handed down by tradition, is an excellent one. Various facts which I cannot occupy space to give in detail, have led me to the undoubting conclusion that it was entirely from the Infantado cabana or family, and that it was selected from the best sheep of that family.

A gentleman of Philadelphia imported two pair of *black* Merinos in 1803, and Mr. Muller, a small number from Hesse Cassel, in 1807.† In 1809, and 1810 Mr. Jarvis, American

---

* This letter will appear entire in the Transactions of the New York State Agricultural Society for 1862.

† These crossed with Col. Humphreys' sheep, in the flock of Mr. Wm. Caldwell of Philadelphia, were the origin of the formerly highly celebrated flocks of Wells & Dickinson, of Ohio.

Consul at Lisbon, Portugal, taking advantage of the offers of the Spanish Junto to sell the confiscated flocks of certain Spanish nobles, bought and shipped to different ports in the United States, about three thousand eight hundred and fifty sheep. About one thousand three hundred of these were Aqueirres, two hundred Escurials and two hundred Montarcos. The remainder consisted of Paulars and Negrettis—mostly of the former.*

Mr. Jarvis very unfortunately crossed his own flock with the Saxons, when the latter were introduced, but he discovered his error in time to correct it, and bred a pure Spanish flock to the period of his death. But he mixed his different Spanish families together, consisting of about half Paulars, a quarter Aqueirres, and the other fourth Escurials, Negrettis and Montarcos.† He stated to me that the average weight of fleece in his full-blood Merino flock, before his Saxon cross, was about 4 lbs.§ This I suppose included rams' and wethers' fleeces. The subsequent history of these sheep will again be referred to. From 3,000 to 5,000 Spanish Merinos were imported into the United States by other persons in 1809, 1810, and 1811.

The earlier importations had attracted little notice until the commencement of our commercial difficulties with England and France, in 1807. When the embargo was imposed, that year, wool rose to $1 a pound. In 1809 and 1810 Mr. Livingston sold his full-blood wool, *unwashed*, for $2 a pound. During the war of 1812, it rose to $2.50 a pound. Many of the imported Merino rams sold for $1,000 apiece, and we have seen that Mr. Livingston sold ram lambs of his own raising at that price. Ewes sometimes sold for equal sums. The Peace of Ghent (1815,) re-opened commerce and overthrew our infant manufactories. Such a revulsion ensued that before the close of the year full-blood Merino sheep were sold for $1 a head! Wool did not materially rally in price for the nine succeeding years, and during that period most of the full-blood flocks of the country were broken up or adulterated in blood.

---

* See Mr. Jarvis' letter to me, in 1841, in New York Agricultural Society's Transactions of that year.

‡ See his letter to me on this subject in 1844, published that year in the Albany Cultivator and New York Agriculturist.

§ Mr. Jarvis gives the facts more precisely in a letter to L. A. Morrell, published in American Shepherd, p. 300. He says:—From 1811 to 1826, when I began to cross with the Saxonies, my average weight of wool was 3 lbs. 14 oz. to 4 lbs. 2 oz.—varying according to keep. The weight of the bucks was from 5¼ lbs. to 6¼ lbs. in good stock case, all washed on the sheeps' backs."

SAXON MERINOS INTRODUCED.—The woolen tariff enacted in 1824, gave a new impulse to the production of fine-wool, and during that and the four succeeding years Saxon Merinos were imported in large numbers into the United States. A detailed history of these importations was embodied in a report on sheep which I made to the New York State Agricultural Society in 1838,* the facts in regard to the Saxons being furnished to me by another member of the committee, Henry D. Grove, the leading German importer and breeder of that variety of sheep in our country. That history having been republished in the "American Shepherd," in "Sheep Husbandry in the South," and in various other publications, it is scarcely necessary to take up space here with its curious particulars concerning a variety now pretty generally discarded in our country. Suffice it to say, that the most enormous frauds were practiced; grade sheep were mixed with nearly every importation; and these miserable animals brought along with them scab and hoof-rot, those dire scourges of the ovine race.

The great discrimination made in favor of fine-wool by the tariff of 1828, excited a mania for its production, and every producer strove to obtain the finest, almost regardless of every other consideration. Size, weight of fleece, and constitution were totally overlooked. Yet the grower was feeding on hope. Fine-wool did not rise to a high price until after the middle of 1830, and neither then nor at any subsequent period did the average price of Saxon exceed that of Spanish wool by more than ten cents a pound—while at least a third more of the latter could be obtained from the same number of sheep,† or the same amount of feed. When we consider this fact, and consider the superiority of the Spanish sheep in every other particular except fineness of wool, we cannot sufficiently wonder that from 1824 to 1840 the Saxons should have received universal preference, have sold for vastly higher prices, and that those who owned Spanish sheep, should have in almost every instance made haste to cross them with their small and comparatively worthless competitors.

In about 1840, however, a reaction commenced, and the tariff of 1846, (which established an even ad valorem duty of

---

* Published in Albany Cultivator, March, 1838, and partially in the New York State Agricultural Society's Transactions, 1841.
† Mr. Grove's flock of picked breeding sheep—not excelled probably in the United States among *pure* bloods, for weight of fleece—yielded an average of 2 lbs. 11 oz. per head of washed wool in 1840, and he published this product as a proof of the superior value of his favorite variety. See his letter to me, Transactions New York State Agricultural Society, 1841, p. 833.

30 per centum on all wools and on cloths,) completed the overthrow of the Saxons.

SAXON RAM.

The cut of the Saxon ram above given, is copied from an engraving from a drawing by Mr. Charles L. Fleischmann, formerly draughtsman for the Patent Office. The engraving was published in the Patent Office Report of 1847. Mr. Fleischmann states that it is an accurate representation of the best ram of Von Thaer (son of the celebrated Albert Von Thaer,) made by its owner's permission at Moeglin, in 1844–'45. The flocks of Von Thaer are among the best and most highly improved in Germany. The drawing was made in the beginning of the month of August while the fleece was yet short.

CHAPTER III.

AMERICAN MERINOS ESTABLISHED AS A VARIETY.

THE MIXED LEONESE OR JARVIS MERINOS — THE INFANTADO OR ATWOOD MERINOS — THE PAULAR OR RICH MERINOS — OTHER MERINOS.

THE MIXED LEONESE OR JARVIS MERINOS.—The origin of Mr. Jarvis' flock has been given. Their pedigrees rested on his own direct statements; and his integrity and veracity were never challenged by friend or foe. As has been seen, he mixed five families of Spanish sheep, the Paulars considerably predominating in numbers,— but his son writes me that for the purpose of "accommodating the manufacturers" he bred "in the contrary direction" from the type of the darker colored and yolkier families.* The appearance of his sheep when I first saw them, something over twenty years since, I thought plainly indicated that he had "accommodated the manufacturers" by chiefly using rams of his Escurial family or which bore a large proportion of that blood. They were lighter colored than the original Spanish sheep of other families and their wool was finer. It was entirely free from hardened yolk, or "gum," internally and externally, and opened on a rosy skin with a style and brilliancy which resembled the Saxon. It was longish, for those times, on the back and sides, but shorter on the belly, and did not cover the head and legs anything like as well as those parts are covered in the improved sheep of the present day. It was of fair medium thickness on the best animals. The form was perhaps rather more compact than that of the original Spanish sheep, but altogether it bore a close resemblance to them. I think that prior to 1840, Mr. Jarvis had begun to breed back toward the other strains of blood in his flock. At about that period small and choice lots of breeding ewes were

---

* See Charles Jarvis' letter to me in my report on "Fine-Wool Sheep Husbandry," 1862.

occasionally obtained from him which yielded from 4 lbs. to 4½ lbs. of washed wool per head. These sheep long enjoyed great celebrity, and are now represented in the pedigrees of many excellent pure bred flocks; but as a distinct family, they have mostly been merged in the two next to be described.

THE INFANTADO OR ATWOOD MERINO.—In 1813, Stephen Atwood, of Woodbury, Connecticut, bought a ewe of Col. David Humphreys for $120. He bred this ewe and her descendants to rams in his neighborhood which he knew to be of pure Humphreys' blood, until about 1830, after which period he uniformly used rams from his own flock. This is the distinct and positive statement of a man of conceded good character, and has been persisted in from a period long before the asserted facts would have had any effect on the reputation of his flock. From 1815 to 1824, and indeed down to a much later period, the pedigrees of "old-fashioned Merinos," as they were then termed, received very little respect or attention; and, in fact, I am not aware that Mr. Humphreys' importation enjoyed any especial credit over several other of the principal importations, until its reputation was reflected back on it by Mr. Atwood's own flock. Mr. Atwood, moreover, is a purely practical man; has been specially and almost exclusively devoted to his sheep; and has always acted as his own shepherd. We have no right, then, to doubt either his sincerity or his accuracy.

In 1840, his sheep were not far from the size and form of Mr. Jarvis'—though I think they were inclined to be a little flatter in the ribs, and perhaps a little deeper chested. Their wool was short, fine, even, well crimped, brilliant, generally thick, and very dark colored externally for that day. Some of them (particularly among the rams,) had a black external coat of hardened yolk, which was sticky in warm weather and formed a stiff crust in cold weather. The inside yolk was abundant, and generally colorless. The wool was still shorter on the belly, and as with the Jarvis sheep, did not very well cover the legs and head. Few of them had any below the knees and hocks. Their skins were mellow, loose and of a rich pink color. The rams had a pendulous dew-lap and some of them neck-folds, or "wrinkles," of moderate size. They rarely exhibited them on other parts of the body, and the "broad tail" and deep pendulous flank of the present day, were unknown in both sexes. The ewes generally had dew-laps of greater or lesser width, sometimes dividing into two parts under the jaw, so as to form a triangular cavity or "pouch" between; and there was on most of them a horizontal fold of skin running across the lower portion of the bosom or front of the brisket,—which was known as "the cross," and which modern breeders have developed into that pendulous mass now sometimes termed "the apron."

When the Spanish Merinos came again into credit, this flock became a public favorite and colonies from it were rapidly scattered throughout the United States, and particularly in the State of New York. Some of these deteriorated, but most of them continued to improve. The great and leading improver of the family has been Edwin Hammond, of Middlebury, Vermont. He made three considerable purchases of Mr. Atwood's sheep between the beginning of 1844 and the close of 1846—in the two last, getting the average of the flock, i. e., a proportionate number of each quality.* By a perfect understanding and exquisite management of his materials, this great breeder has effected quite as marked an improvement in the American Merino, as Mr. Bakewell effected among the long-wooled sheep of England. He has converted the thin, light-boned, smallish, and imperfectly covered sheep above described, into large, round, low, strong-boned sheep—models of compactness, and not a few of them almost perfect models of beauty, for fine-wooled sheep. I examined the flock nearly a week in February, 1863. They were in very high condition, though the ewes were fed only hay. Two of these weighed about 140 lbs. each. Numbers would have reached from 110 lbs. to 125 lbs. One of the two largest ewes had yielded a fleece of 17½ lbs., and the other 14¾ lbs. of unwashed wool. The whole flock, usually about 200 in number, with the due proportion of young and old and including, say, two per cent. of grown rams, and no wethers, yields an average of about 10 lbs. of unwashed wool per head. The ram, "Sweepstakes," given as the frontispiece of this volume, bred and now owned by Mr. Hammond, has yielded a single year's fleece of unwashed wool weighing 27 lbs. His weight in full fleece is about 140 lbs. Rams producing from 20 lbs. to 24 lbs. are not unusual in the flock.

Mr. Hammond's sheep exhibit no hardened yolk within the wool and but little externally: in nearly all of them the curves of the wool can be traced to its outer tips. They are

---

* In one case he bought the entire lot of ewe lambs of a year; in another, one-third of the old ewes—Mr. Atwood selecting the first and third, and Mr. Hammond the second of each trio. He had partners in some of his purchases, but there is no occasion to name them here.

dark colored because they have abundance of liquid "circulating" yolk, and because they (like all the leading breeding flocks of Vermont,) are housed, not only in winter, but from summer rain storms. The great weight is made up not by the extra amount of yolk, but by the extra length and thickness of *every part* of the fleece. In many instances it is nearly as long and thick on the belly, legs,* forehead, cheeks, etc., as on the back and sides. The wool opens freely and with a good luster and style. It is of a high medium quality and remarkably even. Mr. Hammond is intentionally breeding it back to the buff tinge of the original Spanish wool. He has not specially cultivated folds in the skin. Sweepstakes has more of these than most of his predecessors and has much increased them in the flock. Some of his best ewes are nearly without them, though all perhaps have dew-laps and the "cross" on the brisket. In every respect this eminent breeder has directed his whole attention to solid value, and has never sacrificed a particle of it to attain either points of no value or of less value. He has bred exclusively from Mr. Atwood's stock, sire and dam; and since the rams originally purchased of Mr. Atwood by himself and associates, has only used rams of his own flock. The marked extent of his in-and-in breeding, will be adverted to in the Chapter which I shall devote to the general subject of in-and-in breeding. But this has not developed any delicacy of constitution in his flock. They are every way stronger and more robust sheep than their predecessors of 25 years ago, bring forth larger and stronger lambs, and are far better breeders and nurses.

There are in Vermont and other States a large body of spirited and intelligent breeders whose flocks were founded mainly or exclusively on sheep purchased of Mr. Hammond. Not a few of them have bred with distinguished success. It would be justly considered invidious to mention the flocks of a portion of them, without mentioning all of equal merit. This I am unable to do, both because I am unprovided with a full list of them, and because the prescribed limits of this work do not admit of it. I have aimed to do justice to all of this improved family of sheep at once, in describing the flock of its distinguished founder.

THE PAULAR OR RICH MERINOS.—These sheep were originally purchased in 1823, by Hon. Charles Rich, M. C.,

---

* I do not mean to be understood that it is thus long below the knees and hocks, though it is generally quite as long as *it ought to be* on the shanks.

and Leonard Bedell, of Shoreham, Vermont, of Andrew Cock, of Flushing, Long Island. Cock purchased all of the original stock and part of the individual sheep sold to them, of the importers. Their Spanish pedigree, the authenticity of which

MERINO EWE.

was attested by a Consular certificate, (undoubtedly Mr. Jarvis', but that fact is not now remembered,) showed them to be Paulars.* They have been bred by John T. Rich, son of the preceding, and his sons John T. and Virtulan Rich, on the old

---

* Cock delivered this certified pedigree to Bedell. Letters of the late John T. Rich, Esq., son of one of the purchasers, and of the late Hon. S. H. Jennison, ex-Governor of Vermont, were published in 1844, stating that they had seen this document; and both gentlemen remembered the ewes in the flock certified to be of the original importation. Gov. Jennison says he saw them often between 1824 and 1830. They were very old and toothless. The Hon. Effingham Lawrence, who resided in the same town with Cock, and who was himself a distinguished importer and breeder of Merinos, as well as an old-school gentleman, highly eminent for social position and integrity, wrote to me in 1844:—"Andrew Cock * * was my near neighbor. We were intimate and commenced laying the foundations of our Merino flocks about the same time. I was present when he purchased most of his sheep, which was in 1811. He first purchased two ewes at $1,100 per head. They were very fine, and of the Escurial flock imported by Richard Crowninshield. His next purchase was 30 of the Paular breed at from $50 to $100 per head. He continued to purchase of the different importations until he run them up to about eighty, always selecting them with great care. This was the foundation of A. Cock's flock, nor did he ever purchase any but pure blooded sheep to my knowledge or belief. Andrew Cock was an attentive breeder; saw well to his business; and was of unimpeachable character. His certificate of the kind and purity of blood I should implicitly rely on. I recollect of his selling sheep to Leonard Bedell, of Vermont." Much other testimony sustaining the pedigree might be given.

homestead in Shoreham, down to the present day, without the least admixture of other blood than pure Spanish, and with very little crossing with other Spanish or American families.

These sheep, in 1840, were heavy, short-legged, broad animals, full in the quarters, strong-boned, with thick, short necks and thick coarse heads. The ewes had deep and sometimes plaited dew-laps and folds of moderate size about the neck. The rams had larger ones. They were darker externally than the Jarvis sheep, but not so much so as the Atwood sheep—indicating that their wool contained more yolk than the former and less than the latter. The wool was longer than that of either of the other families, very thick and covered them better on the belly, legs and head. But it was inferior in fineness, evenness and style. It was quite coarse on the thigh, and hairs were occasionally seen on the neck folds. The lambs were often covered with hair when yeaned, and their legs and ears were marked by patches of tan color which subsequently disappeared except on the ears, where it continued to show faintly. They were better nurses and hardier than either of the other families. I have remarked in a former publication that "they were precisely the negligent farmer's sheep." They encountered short keep, careless treatment of all kinds, exposure to autumnal storms and winter gales, with a degree of impunity which was unexampled. Their lambs came big, bony and strong, and did not suffer much if they were dropped in a snow bank.

In 1842 and 1843 this flock was bred to a Jarvis ram—peculiarly dark, thick and heavy fleeced and compact in form for one of his family—the object of Mr. Rich being to avoid breeding in-and-in and to improve the quality of his wool. For the same object, and to increase the yolkness of the wool, a dip or two of Atwood blood has been since taken; but it has always been made a point to *breed back* after taking these crosses, so as essentially to preserve the blood and distinctive characteristics of the original family. The Messrs. Rich have succeeded in all these objects and have kept up well with the rapid current of modern improvement. Their sheep are not so large nor do they yield so much wool per head as the improved Infantados, but they possess symmetrical forms which are remarkable for compactness. The body is shortish, and very thick, with their ancient good fore and hind quarters; and their heads, though thick and short, have lost their coarseness. Their fleeces are even and good. But that merit which gives them their great popularity in Vermont and elsewhere, is their adaptation to thin, scant herbage, and to their qualities as "working flocks." They demand no extra care or keep to develop their qualities, are always lively and alert; and though gentle and perfectly free from restlessness of temperament, they are ready to rove far and near to obtain their food. And for all they consume they make the most ample returns. While they will pay for care, they will thrive with but little care. In a word, they remain, par excellence, the negligent farmer's sheep.

The ewe, the portrait of which is given on page 31, is a three year old of this family, and is one of a small number of equal appearance and excellence, which I bought of the Messrs. Rich a year since. Her second fleece, when she was not so large as a high-kept yearling, and when she had not been housed before autumn, weighed 10 lbs. unwashed. Having bred both these and the Infantados for years, and being now about equally interested in both the improved families, I trust I can speak of them with impartiality; and I may here add that I also described Mr. Jarvis' sheep on ample personal experience.*

OTHER MERINO FAMILIES.—There were in 1840, a few small Merino flocks descended from pure Spanish importations, and derived from other sources than the foregoing, scattered very thinly through the States lying west of New England. Like the best Infantados and Paulars of that day, some of them averaged about 4¼ lbs. of washed wool to the fleece. I have been unable to obtain any authentic portraits of known Infantados or Paulars of that period. The drawing from which the cut given on the following page was taken, was made in 1840, by Francis Rotch, Esq., of Morris, (then called Louisville,) N. Y., one of the most eminent and skillful cattle and sheep breeders in the United States, and remarkable then as since for the accuracy and spirit of his drawings of animals. The cut is a ewe of his own flock of thirty breeding ewes, which had been selected with much care from different flocks in New England; and this one was then regarded as a model. She is rounder in the rib, broader and rounder in the thigh and fuller in the brisket than was common among the Merinos of that day. The illustration will show the changes which

---

* The account which I have given of the characteristics, &c., of these families 20 years ago, was submitted, in substantially the same form, to some of the most prominent present breeders of each variety, including Mr. Hammond and Mr. Rich, preparatory to its publication in my Report on Fine-Wool Husbandry in 1862, and it received their unanimous concurrence. See that Report, p. 58.

have taken place in American Merino sheep during the last twenty-three years.

MERINO EWE.

Other persons in New York, (including myself,) and several in Pennsylvania, Ohio, and perhaps some other States, owned pure Spanish flocks, not differing essentially in quality from those of Connecticut and Vermont. But while some flock-masters in New England, and particularly in Vermont, made ram breeding a specialty, those of the Middle and Western States generally devoted their attention to wool-growing, and soon began to draw their rams from the former sources. The consequence has been that they neither preserved nor established distinct families, among their early sheep; and those that now have pure and distinct families of the improved American Merinos (and their number greatly exceeds that of the breeders of pure sheep in New England,) have generally obtained the origin of their flocks, within the last fifteen or twenty years, from Vermont, or from Mr. Atwood's flock in Connecticut. Consequently, there is not within my knowledge any other *separate* families that require a special description.

CHAPTER IV.

LATER IMPORTATIONS OF FINE-WOOLED SHEEP INTO THE UNITED STATES.

FRENCH AND SILESIAN MERINOS INTRODUCED.

FRENCH MERINOS INTRODUCED.—The first importation of French Merinos into the United States, since they have assumed those characteristics which constitute them a separate variety, was made in 1840, by D. C. Collins, of Hartford, Conn. He purchased fourteen ewes and two rams from the royal flock at Rambouillet, which were esteemed of such choice quality that one of the rams ("Grandee") and several of the ewes "could only be procured after they had been used in the national flock as far as it could be done with advantage." Grandee, says A. B. Allen, then Editor of the American Agriculturist, who attended Mr. Collins' sheep-shearing in 1843, was 3 feet 8½ inches long from the setting on of the horns to the end of the rump; his height over the rump and shoulders was 2 feet 5 inches, and his weight in good fair condition about 150 lbs. The ewes were proportionably large. At three years old, in France, Grandee produced a fleece of 14 lbs. of unwashed wool. His fleece was suffered to grow from 1839 to 1841, two years, and weighed 26 lbs. 3 oz. clean unwashed wool. One year's fleece in 1842 weighed 12¾ lbs. In 1843 the ewes yielded an average of 6 lbs. 9 oz. of unwashed wool. Mr. Allen commended their constitutions and longevity; stated that they had large loose skins full of folds, especially about the neck and below it on the shoulders, and not unfrequently over the whole body; and that they were well covered with wool on every part down to the hoofs. Their fleeces opened of a brilliant creamy color, on a skin of rich pink, and was soft, glossy, wavy, and very even over the whole body. It was exceedingly close and compact, and had a yolk free from gum and easily liberated by washing.*

---

* See Am. Agriculturist, vol. 2, p. 98. I mostly use Mr. Allen's language.

The late Mr. Taintor, of Hartford, Connecticut, commenced importing French Merinos in 1846, and continued it through several succeeding years. He selected mostly from private flocks like those of M. Cugnot and M. Gilbert, which had been bred much larger and heavier fleeced than the royal one. Having made some inquiries of him, in 1862, in relation to the sheep of his importations, he referred me to John D. Patterson of Westfield, New York, who had purchased very extensively of him and who owned as good animals as had ever been imported. That gentleman wrote to me:

"In answer to your inquiry as to the weight of fleece of the French sheep and their live weight, I can only reply by giving the result of my own flock. My French rams have generally sheared from 18 to 24 pounds of an even year's growth, and unwashed, but some of them, with high keeping and light use, have sheared more, and my yearling rams have generally sheared from 15 to 22 pounds each. My breeding and yearling ewes have never averaged as low as 15 pounds each, unwashed, taking the entire flock. Some of them have sheared over 20 pounds each, but these were exceptions, being large and in high condition. The live weight of any animal of course depends very much upon its *condition*. My yearling ewes usually range from 90 to 130 pounds each, and the grown ewes from 130 to 170 pounds each, and I have had some that weighed over 200 pounds each; but these would be above the average size and in high flesh. My yearling rams usually weigh from 120 to 180 pounds each, and my grown rams from 180 to 250 pounds each—some of them have weighed over 300 pounds each, but these were unusually large and in high flesh and in full fleece. I have had ram lambs weigh 120 pounds at seven months old, but they were more thrifty, fleshy and larger than usual at that age."

I have seen many sheep of Mr. Taintor's importation and their direct descendants. A large portion of them possessed good forms considering their great size. Their wool was not so fine as Mr. Collins', but of a fair medium quality and pretty even. Their fleeces were very light colored externally, compared with those of any American family, owing undoubtedly to their relative deficiency in yolk and to the more soluble character of their yolk. Unless housed with care from both summer and winter storms, they were about as destitute of yolk before washing as a considerable class of American Merinos are after it. Under common treatment, then, their fleeces are greatly lighter in proportion to bulk than those of the latter, and correspondingly unprofitable in a market where no adequate discrimination is made between clean and dirty wools.

"The only really weak point of the best French Merino as a *pure wool* producing animal, is the want of that hardiness which adapts it to our changeable climate and to our systems of husbandry. In this particular it is to the American Merino what the great pampered Short-Horn of England is to the little, hardy, black cattle of the Scotch Highlands—what the high-fed carriage horse, sixteen hands high, groomed and attended in a wainscoted stable, is to the Sheltie that feeds among the moors and mosses, and defies the tempests of the Orkneys. The French sheep has not only been highly kept and housed from storm and rain and dew for generations, but it has been bred away from the normal type of its race. The Dishley sheep of Mr. Bakewell are not a more artificial variety, and all highly artificial varieties become comparatively delicate in constitution."*

The French Merino, if well selected, has always proved profitable in this country, where the French, or an equally fostering system of management, has been faithfully kept up— but by far the largest portion of buyers have not kept up such a system, and consequently their sheep have rapidly deteriorated. Where the rams have been worked hard and exposed to rough vicissitudes of weather, they have frequently perished before the close of the first year. These facts account for that reaction which has taken place against this variety in the minds of many of our farmers. And the tide of prejudice has been enormously swelled by the impositions of a class of importers. It creates a smile to recall to memory the great, gaunt, shaggy monsters, with hair on their necks and thighs projecting three or four inches beyond the wool— mongrels probably of the second or third cross between French Merinos and some long-wooled and huge-bodied variety of mutton sheep—which were picked up in France and hawked about this country by greedy speculators, who knew that, at that time, *size* and "*wrinkles*" would sell any thing!

I regret that Mr. Patterson's absence in California has prevented me from obtaining original drawings of some of

---

* I quote this paragraph from my Report on Fine-Wool Husbandry, 1862, because Mr. Taintor, the Messrs. Allen, and several other distinguished breeders and advocates of French Sheep, wrote to me expressing their entire satisfaction with my description of that breed in the Report; and the above quotation may therefore be set down as *res adjudicata*.

these sheep in time for this volume. I have not known where else to look for pure and favorable specimens of the variety.

Colonies of French Sheep have been planted in the mild climate of the South, in California, and in other situations the most favorable to them. I cannot but hope that they will yet acclimatize into a valuable variety for portions of our country. They are good mothers. They often raise twins. As a fine-wool mutton sheep they should stand unrivaled.

SILESIAN MERINO RAM.

INTRODUCTION OF SILESIAN MERINOS.—The following account of the introduction of this variety and of its characteristics, is contained in a letter from the principal importer, William Chamberlain, of Red Hook, New York. He wrote to me in January, 1862:

"Your favor dated 24th ult. is received, and it gives me pleasure to furnish the required information in regard to my flock of Silesian sheep, with full liberty to make such use of the facts as you please.

"1st. I have made importations for myself and George Campbell of Silesian sheep, as follows:

In the year 1851, say		40 ewes	and	15 bucks.
do.	1853, do.	27 do.		4 do.
do.	1854, do.	111 do.		13 do.
do.	1856, do.	34 do.		2 do.
		212		34 do.

"In 1854 I visited Silesia and made the purchases myself.

"2d. The sheep were bred by Louis Fischer, of Wirchenblatt, Silesia, except a few which were bred by his near neighbor, Baron Weidebach, who used Fischer's breeders.

"3d. Their origin is Spain. In 1811 Ferdinand Fischer, the father of Louis Fischer, the present owner of the flock, visited Spain himself and purchased one hundred of the best ewes he could find of the Infantado flocks, and four bucks from the Negretti flock, and took them home with him to Silesia, and up to the present day they have not been crossed with any other flocks or blood, but they have been crossed within the families. The mode pursued is to number every sheep and give the same number to all her increase; an exact record is kept in books, and thus Mr. Fischer is enabled to give the pedigree of every sheep he owns, running back to 1811, which is positive proof of their entire purity of blood. The sheep are perhaps not as large as they would be if a little other blood were infused; but Mr. F. claims that entire purity of blood is indispensably necessary to insure uniformity of improvement when crossed on ordinary wool growers' flocks; and such is the general opinion of wool growers in Germany, Poland and Russia, which enables Mr. Fischer to sell at high prices as many bucks and ewes as he can spare, and as he and his father have enjoyed this reputation for so many years, I am fully of opinion that he is right. From these facts you will observe that my sheep are pure Spanish.

"4th. Medium aged ewes shear from 8 to 11 pounds; bucks from 12 to 16 pounds; but in regard to ewes, it must be borne in mind that they drop their lambs from November to February, which lightens the clip somewhat. I do not wash my sheep.

"5th. I have sold my clip from 30 to 45 cents, according to the market.

"6th. We have measured the wool on quite a number of sheep, and find it from one and a half to two inches long, say eight months' growth, but I have no means of knowing what it would be at twelve months' growth.

"7th. Their external color is dark. The wool has oil but no gum whatever, they having been bred so as to make them entirely free from gum—German manufacturers always insisting on large deductions in the price of wool where gum is found.

"8th. As above stated, the Silesians have oil, but no gum like those which are sold for Spanish and French, and the oil is white and free; the wool does not stick together.

"9th. We have weighed five ewes. Three dropped their lambs last month; the other two have not yet come in. Their weights are 115, 140, 130, 115 and 127 pounds; three bucks weighing severally 145, 158, 155 pounds; one yearling buck weighing 130 pounds; but this would be more than an average weight of my flock when young and very old sheep were brought into the average. My sheep are only in fair condition, as I feed no grain. They have beets, which I consider very good for milk, but not so good for flesh as grain.

"10th and 11th. For the first time my shepherd has measured some sheep: ewes from 24 to 28 inches high, fore-leg 11 to 12 inches; bucks, 27 to 28 inches high, fore-leg 12 to 13½ inches.

"12th. We find the Silesians hardy, much more so than a small flock of coarse mutton sheep that I keep and treat quite as well as I do the Silesians.

"13th. They are first-rate breeders and nurses.

"Some of these facts I have given on the statement of my shepherd, Carl Heyne, who was one of Mr. Fischer's shepherds, and came home with the sheep I purchased in 1854, and a man whose honor and integrity I can fully indorse.

"My sheep do not deteriorate in this country, but the wool rather grows finer without any reduction in the weight of fleece."

In a subsequent letter Mr. Chamberlain wrote to me:

"Carl has weighed a few more of our Silesian sheep, and their weights are as follows: Four full aged ewes, respectively, 120, 125, 107, 107 pounds; two ewe lambs, 90, 87 pounds; two two-year old bucks, 124, 122 pounds; one three-fourths blood, 143 pounds.

"I attended to the weighing and selection myself, and am of opinion that our ewes from three to eight years old average fully 115 pounds, say before dropping their lambs. Our younger sheep do not weigh as much. Silesians do not get their full size till four years of age, and after eight or nine years they are not as heavy. * * * Mr. Fischer's sheep are large, say larger than any flock of Vermont Merinos that I have seen. * * * I have the lambs come from November to March, because Carl says it is the best way, and I let him do as he pleases. * * * The ewes do not give quite as much wool, but I think the lambs make stronger sheep, as they get a good start the first summer."

The Silesian ram, a portrait of which is given on page 38, was bred by Mr. Chamberlain, and is now the property of James Geddes, of Fairmount, N. Y. He is regarded as an extraordinarily valuable animal of the family. He is large in size and yields an unusually heavy fleece.

The following cut represents a group of Silesian ewes imported by Mr. Chamberlain.

GROUP OF SILESIAN EWES.

I visited Mr. Chamberlain's flock in February, 1863. Most of the lambs were then dropped and the ewes appeared to be excellent mothers. They were fed beets but no grain. They are housed constantly in cold weather, except when let out to drink—housed nights throughout the year, and from all summer rain storms. From the limited quantity of his available pasturage, Mr. Chamberlain restricts them far more than is usual in that particular *in summer*, but allows them to eat what hay they wish at night. He considers this more profitable than devoting more of his high-priced lands to pasturage, and quite as well if not better for the sheep.

The carcasses of his sheep are round and symmetrical. Some of them are taller in proportion to weight than is desirable—because German breeders pay less attention to this point—but this tendency could be readily changed without going out of the flock for rams. The wool is of admirable quality and uniformity, and opens most brilliantly on a mellow, rose-colored skin. The fleece is very dark externally.

Wherever it is most profitable to grow very fine wool, this variety, or rather this family, ought to stand unrivaled. Whether they have ever been tested under the common rough usage of our country I am not advised. There is nothing in their forms or general appearance to indicate that they would not generally conform to it. They would doubtless lose much of their external color and early maturity, and perhaps something of their ultimate size. But the same would be true of all the summer-housed, high kept and carefully tended Merinos of our country.

CHAPTER V.

BRITISH AND OTHER LONG AND MIDDLE-WOOLED SHEEP IN THE UNITED STATES.

LEICESTERS, COTSWOLDS, LINCOLNS, NEW OXFORDSHIRES, BLACK-FACED SCOTCH, CHEVIOT, FAT-RUMPED, BROAD-TAILED, PERSIAN AND CHINESE SHEEP.

No breed of domestic sheep were indigenous to the United States; nor is it deemed necessary here to attempt to trace the origin or subsequent history of the various breeds and families, imported by our ancestors when they colonized this Continent, and which, being mixed promiscuously together, constituted what it became customary to speak of as the "Native Sheep," when the Merino and the improved British breeds were afterwards introduced. They were generally lank, gaunt, slow-feeding, coarse, short-wooled, hardy, prolific animals—not well adapted to any special purpose of wool or mutton production. A family of them, the Otter Sheep—so termed from their short, crooked, rickety legs, a mere perpetuated monstrosity—and the descendants of some English long-wools, on Smith's Island, imagined by a few persons to be indigenous there—are the only sub-varieties which have ever attracted special notice; and they were wholly unworthy of it.

Not having bred English sheep of late years, and never having bred them extensively, I can entertain little doubt that I shall give more satisfaction to the readers of this volume if I select descriptions of them from British and American sources of recognized authority.

THE LEICESTER SHEEP.*—It is with profound pleasure that I am enabled to trace the first probable importation into the

---

* I leave off the prefix "New," because these sheep have altogether superseded the parent stock, so as to be generally denominated "*the* Leicester." And they are so denominated in the prize lists of the Royal Agricultural Society of England.

United States of improved English Sheep, if not of improved sheep of any kind, to that great man, first in the arts of peace as well as war, GEORGE WASHINGTON. Livingston, writing in 1809, says of the "Arlington Long-Wooled Sheep" that they were "derived from the stock" of General Washington —being bred by his step-son, Mr. Custis, from a Persian ram and *Bakewell* ewes. Gen. Washington died near the close of 1799.*

A Mr. Lax, who resided on Long Island, "smuggled" some Leicesters into the United States not far from 1810; and from these Christopher Dunn, of Albany, New York, obtained the origin of his long celebrated flock.† During the war of 1812 with England, some choice Leicesters, on their way to Canada, were captured by one of our privateers, and sold at auction in New York, and thus became scattered throughout the country. Some sheep of this family were also early introduced by Captain Beanes, of New Jersey.‡

The elaborate descriptions of the Leicesters, by Youatt and Spooner, have been made so familiar to American readers, that I shall use that of Mr. John Wilson, Professor of Agriculture in the University of Edinburgh, in a paper "On the Various Breeds of Sheep in Great Britain," published in the Journal of the Royal Agricultural Society of England, in 1856:

---

* Livingston (see his Essay on Sheep, p. 58,) does not expressly say that Gen. Washington introduced the "Bakewells," but this is to be inferred from his statement that the Arlington Sheep "were derived from his stock," without making an exception of the Bakewells. Mr. Livingston speaks of the Arlington's as an existing family, when he wrote. I have not Mr. Custis's pamphlet before me from which he appears to have derived his facts.

† He commenced crossing it with a Cotswold ram in 1832, and from that period it became a grade flock between the two families. But it was an excellent one. His wethers weighed 35 lbs. per quarter and carried 8 lbs. of wool per head. His first Cotswold ram weighed alive 250 lbs., and yielded at one shearing 15½ lbs. of wool 14 inches long. In 1835 he sold ewes from $12 to $15 a head, and rams from $30 to $50 a head. Several eminent flocks in the vicinity, like those of Mr. Duane and Mr. North, in Schenectady, &c., &c., originated from these. I have obtained most of my facts about Mr. Dunn's sheep from a communication signed B. in the Albany Cultivator, March, 1835. It was undoubtedly written by Caleb N. Bement—entirely reliable authority; but whoever wrote the article, Judge Buell, then editor of the Cultivator, who was perfectly conversant with Mr. Dunn and his flock, would not have published any erroneous statements in regard to either; and had any errors crept into his columns by oversight, he would have promptly corrected them.

Mr. William H. Sotham, in a communication to the Cultivator in 1840, states the following facts of six wethers bred and fed by Mr. Dunn that year. The heaviest weighed 210 lbs., and the fat on the ribs measured 5¼ inches. The thickness of fat on the smallest was 4¼ inches. They were sold to Mr. Kirkpatrick for $22 a head, and the meat sold rapidly in the market for 12½ cents a pound. The fleeces averaged about 10 lbs. each in weight.

‡ Capt. Beanes also introduced Teeswaters and South Downs, but they were not long kept distinct from the surrounding varieties and families. It has been said that some Teeswaters were included among the sheep captured, as above stated, by a privateer in 1812.

LEICESTER RAM.

"It was about the middle of the last century when Mr. Bakewell, of Dishley, in Leicestershire, began his experiments in the improvement of the breed of long-wooled sheep, at that time common to the midland counties. The old Leicesters were then considered as possessing many valuable properties; at the same time they possessed many defects. These defects Bakewell sought by a judicious crossing with other breeds to remedy, while at the same time he retained the good points of the original breed. Up to this period the great object of breeders seems to have been confined to the production of animals of the largest size possible, and carrying the heaviest fleece. The old Leicesters are described as large, heavy, coarse-grained animals, the meat having but little flavor and no delicacy—the carcass was long and thin, flat-sided, with large bones on thick rough legs. The fleece was heavy and long, and of coarse quality. The sheep were slow feeders, and when sent to market at two and three years old, weighed about 100 to 120 lbs. each. Such were the characteristics of the stock upon which Bakewell commenced his improved system of breeding. Recognizing the relation which exists between the form of an animal and its physical tendencies, he sought to cross his sheep with such breeds as he considered would be most likely to insure those points in the animal frame which were defective in the old breed, and thus to introduce an aptitude to lay on the largest possible amount both of flesh and fat in the shortest space of time, and at the least expenditure of food. The fleece too was not forgotten, as that would necessarily share in the general improvement of the animal. * * * * * * *

"In order to obtain a permanent character to his breed, after he had by continued crossing secured all those points he considered desirable, Bakewell carried on his breeding with his own blood, and did not scruple to use animals closely allied to each other. This system, adhered to more or less during a course of years by his successors and by later breeders, while sustaining the purity of the breed, had the effect of lessening its value to the farmer. It gradually exhibited a weakened constitution, became reduced in size and more delicate in form—the ewes were less prolific and less generous to their offspring. These prominent and serious defects soon craved the attention of enlightened breeders, who, by a judicious introduction of new blood, have again restored the original character of the breed, with all the improvements resulting from the advanced system of cultivation and the enlarged area of sheep farming of the present day.

"The New Leicester is now perhaps the most widely extended and most numerous of all our native breeds. The sheep are without horns, with white faces and legs; the head small and clean; the eye bright; neck and shoulders square and deep; back straight, with deep carcass; hind quarters tapering toward the tail and somewhat deficient when compared with the Cotswold sheep; legs clean, with fine bone. The flesh is juicy but of moderate quality, and is remarkable for the proportion of outside fat it carries.

"They are not considered so hardy as the other large breeds, and require shelter and good keep. The ewes are neither very prolific nor good mothers, and the young lambs require great attention. Early maturity and aptitude for fattening are the principal characteristics of the breed; a large proportion of the wethers finding their way to market at twelve or fifteen months old, and weighing from 80 to 100 lbs. each; at two years old they average 120 to 150 lbs. each. The wool is a valuable portion of the flock, the fleeces averaging 7 lbs. each.

"The occasional introduction of a little Cotswold blood into a Leicester flock has the effect of improving both the constitution of the animal and also the hind quarters, in which the Leicester is somewhat defective. Ram-breeding is carried out to a much larger extent with this breed than with any other."

LEICESTER EWE.

The accompanying cuts are from drawings of a pair of Leicesters imported by Mr. Samuel Campbell, of New York Mills, Oneida County, New York, and Mr. James Brodie, of Rural Hill, Jefferson County, New York. They were imported in the spring of 1861. The ram was bred by Mr. Simpson and the ewe by John Thomas Robinson, both of Yorkshire, England. The ram weighs 276 lbs.* Messrs. Campbell and Brodies' ewes weigh from 200 lbs. to 250 lbs. Their "yearlings and wethers yield from 10 lbs. to 15 lbs. of wool and their breeding ewes about 8 lbs."

---

* His weight of fleece was not sent to me, nor was the seperate weight of the fleece of the ewe of which a cut is given. Messrs. C. and B. sold a ram to Sanford Howard, Esq., of Boston, which at 21 months old weighed 273 lbs., and they have a two year old which weighs 300 lbs.

COTSWOLD RAM.

THE COTSWOLD SHEEP.—The Cotswold Sheep were introduced into the United States about thirty-five years ago. Mr. Dunn imported a ram to cross with his New Leicesters in 1832, and I think some other importations of pairs or single ones took place not far from the same period. The first considerable importation of which I have any information was made in 1840, by Hon. Erastus Corning, of Albany, New York, and William H. Sotham, then of Jefferson County, New York, whose sheep, twenty-five in number, were bred by Mr. Hewer, of Northleach, Gloucestershire, England. Like all the improved Cotswolds, they had a dash of New Leicester blood, and they were very superior animals of the family. The same gentlemen purchased later in 1840 fifty ewes in lamb from Mr. Hewer, and twenty from Mr. William Cother, of Middle Aston, England. These were also prime sheep. From Messrs. Corning and Sotham's stock have originated many valuable flocks, now widely scattered throughout the country. Quite a large number of Cotswolds have since been imported from Canada, a considerable portion of them from the flock of Mr. Frederick William Stone, of Moreton Lodge,

Guelph, Canada West. "Pilgrim," the ram, of which a cut is given on preceding page, was bred by Mr. Stone, and is now the property of Mr. Henry G. White, of South Framingham, Massachusetts. Pilgrim, just off his winter feed, weighs 250 lbs. He would weigh considerably more in the fall. He yielded 18 lbs. of wool in 1862.

The ewe, "Lady Gay," a portrait of which is given on next page was also bred by Mr. Stone, and is owned by Mr. White. She weighs 200 lbs., suckling a lamb. She yielded 16 pounds of wool in 1862. Pilgrim, and five ewes belonging to Mr. White, yielded an average of 16 lbs. of wool per head.

The Cotswolds are thus described by Mr. Spooner in his work on Sheep:—"The Cotswold is a large breed of sheep, with a long and abundant fleece, and the ewes are very prolific and good nurses. Formerly they were bred only on the hills, and fatted in the valleys of the Severn and the Thames; but with the inclosure of the Cotswold Hills and the improvement of their cultivation they have been reared and fatted in the same district. They have been extensively crossed with the Leicester sheep, by which their size and fleece have been somewhat diminished, but their carcasses considerably improved, and their maturity rendered earlier. The wethers are now sometimes fattened at 14 months old, when they weigh from 15 lbs. to 24 lbs. per quarter, and at two years old increase to 20 lbs. or 30 lbs. The wool is strong, mellow, and of good color, though rather coarse, 6 to 8 inches in length, and from 7 lbs. to 8 lbs. per fleece. The superior hardihood of the improved Cotswold over the Leicester, and their adaptation to common treatment, together with the prolific nature of the ewes and their abundance of milk, have rendered them in many places rivals of the New Leicester, and have obtained for them, of late years, more attention to their selection and general treatment, under which management still further improvement appears very probable. They have also been used in crossing other breeds, and as before noticed, have been mixed with the Hampshire Downs. It is, indeed, the improved Cotswold that, under the term New or Improved Oxfordshire Sheep, are so frequently the successful candidates for prizes offered for the best long-wooled sheep at some of the principal agricultural meetings or shows in the Kingdom. The quality of the mutton is considered superior to that of the Leicester, the tallow being less abundant, with a larger development of muscle or flesh. We may, therefore, regard this breed as one of established reputation, and extending itself throughout every district of the Kingdom."*

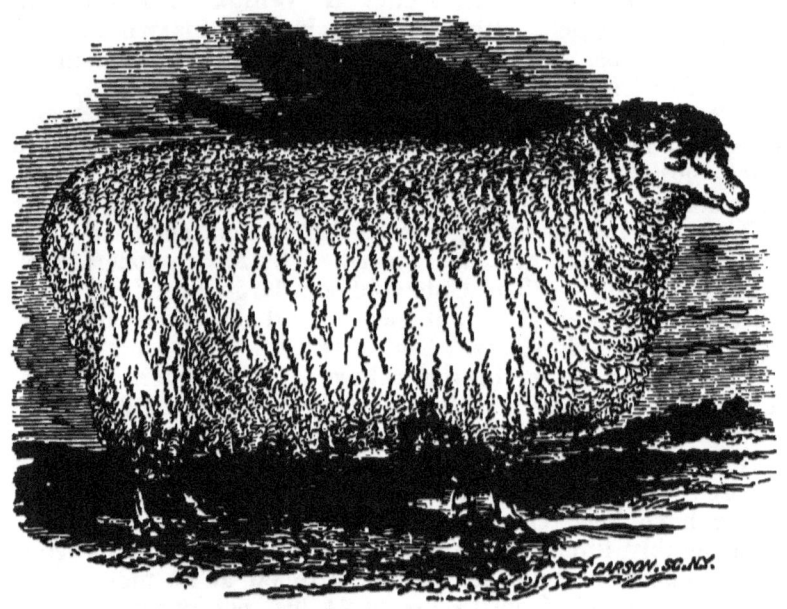

COTSWOLD EWE.

THE LINCOLNS.—The Lincolns are a less improved and larger variety of long-wools than either of the preceding, and those introduced into the United States, having been mostly or entirely merged by cross-breeding with the Leicesters and Cotswolds, they do not demand a separate description. Mr. Leonard D. Clift, of Carmel, Putnam County, New York, imported a ram and ewe of this variety, in 1835, "from the estate of the Earl of Lansdowne, Yorkshire, England." Messrs. George H. Gossip & Brother imported a number in 1836 from Lancashire. From these Mr. Clift obtained sixteen ewes and a ram, and established a flock which was generally regarded as highly valuable. They were hardy, gross feeders, and very prolific. They yielded from 6 lbs. to 10 lbs. of wool per head. Mr. Clift sold a lot of half-blood two year old wethers in February, 1839, which weighed 125 lbs. to the carcass, and he obtained 25 cents a pound for them.

* Spooner on Sheep, p. 99.

THE NEW OXFORDSHIRES, OR IMPROVED COTSWOLDS.—
These were first introduced into this country by Mr. Charles
Reybold, of Delaware, in 1846. They are the result of a cross
between the New Leicesters and Cotswolds, the preponderance being given to the blood of the latter. We have seen
the very high character given of them by Mr. Spooner, in his
description of the Cotswolds, already quoted.

In Mr. James S. Grennell's Report, as Chairman of the
Committee on Sheep Husbandry appointed by the Massachusetts State Board of Agriculture, 1860, is given the following
communication in regard to these Sheep by an American
breeder of them, then of eight years standing—Mr. Lawrence
Smith, of Middlefield:

"I doubt whether they are as hardy as the old-fashioned
Cotswolds or South Downs. I have never had any trouble
with them in regard to cold weather, or changes of climate;
indeed, they prefer an open, cool, airy situation to any other,
and nothing is more destructive to their health than tight, illventilated stables. My present experience warrants me in
saying that one-half the ewes will have twins; they are capital
nurses and milkers; I have not had for the past seven years a
single case of neglect on the part of the dam, nor have I lost
a single lamb from lack of constitution. Yearling ewes will
weigh in store condition from 125 lbs. to 175 lbs.; fat wethers
at three years old, from 175 to 250 lbs. My heavist breeding
ewe last winter weighed 211 lbs. My flock of store sheep
and breeding ewes generally shear from five to seven pounds.
My ram fleeces sometimes weigh ten pounds unwashed, and
will sell in this condition for twenty-five cents per pound. I
never feed any store sheep and lambs with grain, but give
them early cut hay, and occasionally a few roots."

The New Oxfordshires are not to be confounded with the
Oxfordshire Downs, which are cross-breeds between the
Cotswolds and South or Hampshire Downs, and which have
dark faces.

THE BLACK-FACED SCOTCH SHEEP.—These are a small,
active, hardy, but for a mountain family, rather docile sheep,
which have open, hairy fleeces, and black legs and faces.
They can endure great privations, and can even subsist on
heather. Hence they are often called the heath sheep. Their
mutton is of excellent quality. They weigh on an average
from 60 lbs. to 65 lbs. each at three or four years old; and
they yield about 3 lbs. per head of washed wool. They have

been introduced into the United States by Mr. Samuel Campbell, of New York Mills, New York, and by Mr. Sanford Howard, of Boston, Massachusetts, for Mr. Isaac Stickney, of the same State. Mr. Campbell's sheep must be a cross, for he writes me that he should think their weight of fleece would be from 6 lbs. to 8 lbs., and that on the 13th of May, 1863, they weighed alive as follows: old ram, 132 lbs.; old ewe, 103 lbs.; yearling ram, 102 lbs.; two yearling ewes, 99 lbs. and 100 lbs. They have often been crossed successfully in Scotland and the North of England, with larger families. On the bleak, sterile mountain ranges of North-Eastern New York, and portions of New England, they probably would prove a profitable acquisition.

THE CHEVIOT SHEEP.—Some of these (middle-wooled) sheep were introduced into the State of New York a number of years since, and were thus mentioned by me in Sheep Husbandry in the South (1848):

"Sheep of this kind have been imported into my immediate neighborhood and were subject to my frequent inspection for two or three years. They had the appearance of small Leicesters, but were considerably inferior in correctness of proportions to high-bred animals of that variety. They perhaps more resemble a cross between the Leicester and the old Native or common breed of the United States. Their fleeces were too coarse to furnish a good carding wool—too short for a good combing one. Mixed with a small lot of better wool, their this year's clip sold for 29 cents per pound, while my *heavier* Merino fleeces sold for 42 cents per pound. They attracted no notice, and might at any time have been bought of their owner for the price of common sheep of the same weight. I believe the flock was broken up and sold to butchers and others this spring, after shearing. They were certainly inferior to the description of the breed by Sir John Sinclair, even in 1792, quoted by Mr. Youatt,[*] and had all the defects attributed to the original stock by Cully.[†] They might not, however, have been favorable specimens of the breed."

Mr. Spooner thus describes the improved family:—"This breed has greatly extended itself throughout the mountains of Scotland, and in many instances supplanted the black-faced breed; but the change, though in many cases advanta-

---

[*] On Sheep, pp. 285–6.   [†] Cully on Live Stock, p 150.

geous, has in some instances been otherwise, the latter being somewhat hardier, and more capable of subsisting on healthy pasturage. They are, however, a hardy race, well suited for their native pastures, bearing with comparative impunity the storms of winter, and thriving well on poor keep. Though less hardy than the black-faced sheep of Scotland, they are more profitable as respects their feeding, making more flesh on an equal quantity of food, and making it quicker. They have white faces and legs, open countenances, lively eyes, without horns. The ears are large, and somewhat singular, and there is much space between the ears and eyes. The carcass is long; the back straight; the shoulders rather light; the ribs circular; and the quarters good. The legs are small in the bone and covered with wool, as well as the body, with the exception of the face. The Cheviot wether is fit for the butcher at three years old, and averages from 12 lbs. to 18 lbs. per quarter—the mutton being of a good quality, though inferior to the South Down, and of less flavor than the black-faced. * * * The Cheviot, though a mountain breed, is quiet and docile, and easily managed. The wool is fine,* closely covers the body, assisting much in preserving it from the effects of wet and cold; the fleece averaging about $3\frac{1}{2}$ lbs. Formerly the wool was extensively employed for making cloths, but having given place to the finer Saxony wool, it has sunk in price, and been confined to combing purposes. It has thus become altogether a secondary consideration."

FAT-RUMPED, BROAD-TAILED, PERSIAN AND CHINESE SHEEP.—All of these breeds of sheep have been introduced into the United States from Asia and Africa, but as a general thing perhaps rather for the indulgence of curiosity than from any expectation of establishing valuable flocks from them. A variety of the Broad-Tailed sheep, however, sent home by Commodore Porter from Smyrna, was bred for a considerable period in the United States, and kept pure in South Carolina.† A family of them, termed the "Tunisian Mountain Sheep," were received "in a national ship" by Col. Pickering, who caused them to be distributed in Pennsylvania; they were bred there for some time, and were very highly commended by Mr. John Hare Powell.‡ A

---

* Mr. Spooner undoubtedly employs this term relatively, meaning fine for a middle-wooled sheep.

† I received this information from Hon. R. F. W. Allston, late Governor of that State.

‡ See his Letter on Various Breeds of Sheep, 1826, in Memoirs of N. Y. Board of

Persian ram, "very large and well formed, carrying wool of great length, but of a coarse staple," crossed with New Leicester ewes, formed, as we have already seen when speaking of the New Leicesters, the "Arlington long-wooled sheep" of Mount Vernon, a sub-variety which attracted considerable notice in its day.

The Chinese, or Nankin sheep, have recently been brought into this country and England, and have attracted some notice from the fact that they frequently give birth to three or four lambs at a time and breed twice a year—facts which have led to the expectation that they may prove profitable for lamb raising in the vicinity of cities. I have seen no description of their qualities in any other particulars. None of these breeds have proved, or probably will prove, of much value as mutton sheep, compared with the improved English families, and as wool-producing sheep they are all worthless compared with the Merino. I have therefore thought that particular descriptions of them would not be worth the space they would occupy.

Agriculture, vol. 3, p. 377. Mr. Peters, of Pennsylvania, also imported Tunis sheep, and thought well of them.

CHAPTER VI.

BRITISH SHORT-WOOLED SHEEP, ETC., IN THE UNITED STATES.

THE SOUTH DOWNS, HAMPSHIRE DOWNS, SHROPSHIRE DOWNS, AND OXFORDSHIRE DOWNS.

The principal Short-Wooled British families of Sheep which have been introduced in any considerable numbers into the United States since the period of the early settlement of the country, are the South Downs, the Hampshire Downs, the Shropshire Downs and the Oxfordshire Downs. I include the last under this designation only because they are classed among the Downs,—for those introduced into the United States are really a middle if not almost a long-wooled sheep.

THE SOUTH DOWNS.—Professor Wilson, in his paper already cited, thus describes the South Downs:

"The South Downs of the present day present probably as marked an improvement upon the original breed as that exhibited by the Leicesters or any other breed. To the late Mr. Ellman, of Glynde, they are indebted for the high estimation in which they are now generally held. When he commenced his experiments in breeding he found the sheep of small size and far from possessing good points; being long and thin in the neck; narrow in the fore quarters; high on the shoulders; low behind, yet high on the loins; sharp on the back; the ribs flat, drooping behind, with the tail set very low; good in the leg, though somewhat coarse in the bone. By a careful and unremitting attention during a series of years to the defective points in the animal, and a judicious selection of his breeding flock, his progressive improvements were at length acknowledged far and wide; and he closed an useful and honorable career of some fifty years with the satisfactory conviction that he had obtained for his favorite breed a reputation and character which would secure them a place as the first of our short-wooled sheep.

"The South-Down sheep of the present day are without horns, and with dark brown faces and legs; the size and weight have been increased; the fore quarters improved in width and depth; the back and loins have become broader and the ribs more curved, so as to form a straight and level back; the hind quarters are square and full, the tail well set on, and the limbs shorter and finer in the bone. These results are due to the great and constant care which has been bestowed on the breed by Ellman and his contemporaries, as well as by his successors, whose flocks fully sustain the character of the improved breed.

SOUTH DOWN RAM.

"The sheep, though fine in form and symmetrical in appearance, are very hardy, keeping up their condition on moderate pastures and readily adapting themselves to the different districts and systems of farming in which they are now met with. They are very docile, and thrive well, even when folded on the artificial pastures of an arable farm. Their disposition to fatten enables them to be brought into the market at twelve and fifteen months old, when they average

80 lbs. weight each. At two years old they will weigh from 100 to 120 lbs. each. The meat is of fine quality and always commands the highest price in the market. The ewes are very prolific, and are excellent mothers, commonly rearing 120 to 130 lambs to the 100 ewes. The fleece, which closely covers the body, produces the most valuable of our native wools. It is short in the staple, fine and curling, with spiral ends, and is used for carding purposes generally."*

* Mr. Jonas Webb, of Babraham, Cambridgeshire, was the most successful follower of Ellman, and carried the breed to that perfection which is now seen in its best specimens. The average weight of his sheep, at from 13 to 15 months old, was about 126 lbs., and the average yield of wool per head, about 6 lbs.

SOUTH DOWN EWES.

Choice specimens of Mr. Ellman's sheep were imported into the United States some years since by Mr. John Hare Powell, of Pennsylvania, Francis Rotch, Esq., of New York, and various other breeders. Mr. Webb's have also been exten-

*Journal of the Royal Agricultural Society, vol. 16, p. 233.

sively imported by Mr. Thorne of New York, Mr. Alexander of Kentucky, Mr. Taylor of New Jersey, and others. It is understood that the leading American importers left no sheep in England superior to those purchased by them.

Mr. Thorne furnished me the following facts in regard to his flock, in answer to inquiries which embraced all the subjects touched upon by him:

"My flock of South Downs consists of something over 200 head, exclusive of lambs. They are descended from fourteen different importations, principally from the flock of the late Jonas Webb. Those not of his breeding were prize pens at the Show of the Royal Agricultural Society of England, and bred by Henry Lugar, of Hengrave, near Bury St. Edmunds. The rams used have all been selected with the greatest care from the celebrated Babraham flock. 'Archbishop' is the one which is now being principally used. He was the first prize yearling at the Royal Show at Canterbury in 1860, and was chosen by myself from Mr. Webb's folds as the best ram he then had. His price there was $1,250. He was imported in December, 1860.

"The breeding ewes average from 80 to 100 in number. They usually lamb in March. The rate of increase for the past six years has been 142 per cent. This year (1863) it has been 158. As soon as the lambs straighten up, they are docked, and the males that are not to be kept for service are castrated. They are weaned at about four months old. The ewe and wether lambs are given good, short pastures,* and the ram lambs are folded on rape and kept there until all stock is housed. Frost (unless perhaps a very severe one) does not appear to injure the plant, and hence they can be kept upon it longer than on grass. They are confined to this feed, unless a few small ones may require grain, which sometimes is given to the lot. When put in winter quarters the wethers have hay and roots: the others have in addition a little grain. The breeding ewes are kept on hay until two months before lambing, when they are given a small feed of corn which is soon increased to half a pint each per day. When they lamb they are given turnips instead of grain. The wethers [yearlings] are given good pasturage the next season and feed is commenced as soon as the slightest frost makes its appearance, half a pint of corn to each. When put in the

---

* In another letter, Mr. Thorne says: "My own experience has convinced me that it is not advisable to put lambs upon new seeds, or after growth from new meadows, where the growth has been very rank."

sheds they are given turnips and the corn is increased to a pint each. They are marketed generally at Christmas. They usually dress from 75 to 100 lbs. This year 75 that were sold to Bryan Lawrence of New York averaged in weight 87½ lbs.

"With regard to the wool-producing qualities of the South Down, the one year that I kept an accurate account, the ewe flock, including among the number sheep eight and nine years old, all having suckled lambs, gave 6 lbs. 5½ oz.; the yearling ewes 8 lbs. 12 oz.; the yearling rams from 8 to 12 lbs. This was unwashed wool, though as you are aware, their wool is not of a greasy character, and should not be shrunk at the most over one-fourth, by the buyer.

"You may remember to have seen some notices of the sales of Jonas Webb's South Downs. The first sale, in 1861, included all the flock except lambs, and numbered 200 rams and 770 ewes. They brought £10,926. The balance were sold in 1862, and numbered 148 rams and 289 ewes. Amount of sale, £5,720. Total two years sales, more than $80,000."*

Mr. Thorne further writes me:—"Breeding ewes require exercise; I have always considered it more to the advantage of meadows than of sheep that they should be yarded." His sheep have been extremely healthy. The only prevalent disease among them has been puerperal or parturient fever, at lambing. Prior to 1859 he had but one or two cases a year, but that year twenty, and four ewes died. This was his worst year, and under a new mode of treatment the disease is apparently entirely disappearing from his flock. It never, however, was confined to his flock or family of sheep, he informs me, but has been a prevalent disease among sheep of all kinds in the neighborhood, though often called by other names.

The ram, a cut of which is given on page 56, is "Archbishop," already mentioned, bred by Mr. Jonas Webb, and owned by Mr. Thorne. The ewes, cuts of which are given on page 57, are a pair of two-year olds bred by Mr. Thorne from his imported stock.

HAMPSHIRE DOWNS.—Professor Wilson thus describes the Hampshire Downs:

"This rapidly increasing breed of sheep appears to be the result of a recent cross between the pure South Down and the old horned white-face sheep of Hampshire and Wiltshire, by which the hard-working, though fine quality, of the former is

---

* This letter is dated Thorndale, Washington Hollow, N. Y., April 3, 1863.

combined with the superior size and constitution of the latter. The breed was commenced at the early part of the present century; and by a system of judicious crossing now possesses the leading characteristics of the two parent breeds. In some of the best farmed districts of Wiltshire, Hampshire and Berkshire, they have gradually displaced the South Downs, and have in themselves afforded another distinct breed for crossing with the long-wooled sheep. Their leading characteristics are, as compared with the South Down, an increased size, equal maturity, and a hardier constitution. The face and head are larger and coarser in their character; the frame is heavier throughout; the carcass is long, roomy, though less symmetrical than the South Down, and the wool of a coarser though longer staple. Their fattening propensity is scarcely equal to that of the South Down. These points have all received great attention lately from the breeders; and the *improved* Hampshire Down now possesses, both in shape, quality of wool, aptitude to fatten and early maturity, all the qualities for which the pure South Down has been so long and so justly celebrated. The lambs are usually dropped early and fed for the markets as lambs, or kept until the following spring, when, if well fed, they weigh from 80 to 100 lbs., and command a good market.

"The Hampshire Downs are used like the South Downs for the purpose of crossing with other breeds; being hardier in constitution they are perhaps better calculated for the Northern districts, where the climate is sometimes very severe."

Mr. Spooner, in a paper "On Cross Breeding," published in the Journal of the Royal Agricultural Society of England, 1859, expresses opinions of this variety of sheep very similar to those above given by Professor Wilson, and he makes the following remarks in relation to their origin and blood:

"We have no reason to suppose that after a few generations the Hampshire breeders continued to use the South Down* rams; as soon as the horns were gone, to which perhaps the Berkshire Notts contributed, and the face had become black, they employed their own cross-bred rams with the cross-bred ewes. If then we were asked what original blood predominated in the Hampshire sheep, we should unquestionably say the South Down; but if the further question were put, is the present breed derived from the South Down and the original Hampshire alone, we should express a doubt as to such a

---

* Mr. Spooner in several instances terms them "Sussex" in the remarks I quote, meaning thereby South Down; and to prevent confusion among those not used to the former name, I have changed it in every instance to South Down.

conclusion, as there is good reason to consider that some improved Cotswold blood has been infused."

After giving some facts to prove that this last cross was taken, Mr. Spooner continues:

"Although after dipping once or twice into this breed, they then ceased to do so, yet they have continued breeding from the descendants of the cross, and thus in very many of the Hampshire and Wiltshire flocks, there is still some improved Cotswold, and consequently Leicester blood.* Probably an increase of wool has thus been obtained. Some say that on the borders of Berkshire the Berkshire Nott was also used, and others contend, although without proof, that a dip of the Leicester has been infused. Be this as it may there is no doubt that, although for some years past the Hampshire sheep have, for the most part, been kept pure, yet they have been very extensively crossed with other breeds before this period."†

A ram and five ewes of this family, bred by Francis Budd, Esq., of Hampshire, England, and which had been successful competitors at the Exhibition of the Royal Agricultural Society, were imported in 1855 by Mr. Thomas Messenger, of Clarence Hall, Great Neck, Long Island. They have received first prizes from the State Agricultural Society, from the American Institute, and from various other Societies; and they found a rapid sale in the South prior to the present war. Mr. Messenger writes me that he finds them better suited to the climate where he resides, and more hardy, than the South Downs. He breeds them pure, and also crosses them with Cotswolds and Leicesters, with great advantage, in his opinion, to both the latter families of sheep.

THE SHROPSHIRE DOWNS,—Shropshire or Shrops, as they are variously called, are thus described by Professor Wilson:

"In our early records of sheep farming, Shropshire is described as possessing a peculiar and distinct variety of sheep, to which the name of 'Morfe Common' sheep was given, from the locality to which the breed was principally confined. * * In 1792, when the Bristol Wool Society procured as much information as possible regarding sheep in England, they reported as follows in reference to the Morfe Common breed:—'On Morfe Common, near Bridgenorth, which contains about 600,000 acres, there are about 10,000 sheep kept

---

* In a note Mr. Spooner here states that it is "generally acknowledged that the Cotswold sheep have been improved by crosses from the Leicester ram."

† Journal of Royal Agricultural Society, Vol. 20, page 302.

during the summer months, which produce wool of superior quality. They are considered a native breed — are black-faced or brown, or a spotted faced, horned sheep, little subject to either rot or scab — weighing, the wethers from 11 to 14 lbs.,

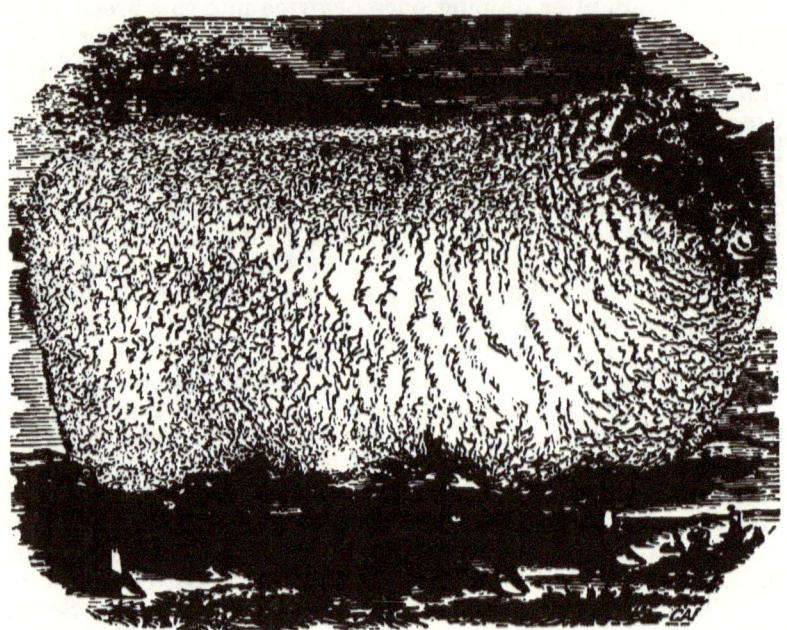

SHROPSHIRE RAM.

and the ewes from 9 to 11 lbs. per quarter, after being fed with clover and turnips; and clipping nearly 2 lbs. per fleece, exclusive of the breeching, which may be taken at one-seventh or one-eighth part of the whole.' * * This appears to have been the original stock from which the present breed of Shropshire Downs has sprung. As the county advanced, and the breeds became valuable for their carcasses as well as for their wool, the Morfe Common sheep were crossed with other breeds, but more particularly with the long-wooled Leicesters and Cotswolds, or the short-wooled South Downs. The admixture of such different blood has produced a corresponding variation in the characters of the present breed of Shropshire Downs, and has tended materially to sustain the hesitation which still exists to allow them a place as a distinct breed.*

---

* This was written in 1856.

Where, however, the original cross was with the South Down, and the breed has been continued unmixed with the long-wooled sheep, they present the characteristics of a short-wooled breed, and as such are already recognized in the Yorkshire and other markets. * * * These sheep are without horns, with faces and legs of a gray or spotted gray color; the neck is thick with excellent scrag; the head well shaped, rather small than large, with ears well set on; breast broad and deep; back straight, with good carcass; hind quarters hardly so wide as the South Down, and the legs clean with stronger bone. They are very hardy, thrive well on moderate keep, and are rapidly prepared for market as tegs, [between weaning and shearing,] weighing on the average 80 lbs. to 100 lbs. each. The meat is of excellent quality, and commands the best prices. The ewes are prolific and good mothers. The fleece, which is heavier than the South Down, is longer and more glossy in the staple than the other short wools, and weighs on the average 7 lbs."

Mr. Spooner says of them that they were first brought into national repute at the Shrewsbury Meeting, in 1845. He remarks:—"At the Chester Meeting they beat the Hampshire Downs as old sheep, but in their turn were conquered by the latter in the younger classes. They present themselves to our notice in a more compact form; though shorter they are wider, broader on the heart and deeper through the heart." Mr. Spooner quotes Mr. J. Meire, as having stated at a meeting of the Farmers' Club in Shropshire, [in 1858 or 1859,] that the sheep produced by the cross between the original sheep and South Down "was well adapted for the downs, but for the inclosures of Shropshire something more docile was required, consequently recourse was had to the Leicester." And Mr. Spooner adds:—"This crossing and recrossing at length gave place to the practice of careful selection, and thus uniformity was sought for and attained, and the present superior breed was established. It is now held that no further cross is required."

Mr. Charles Howard of Biddenham, Bedfordshire, in an address delivered before the London or Central Farmers' Club, in 1860, said:

"This breed has been established by a prudent selection of the breeding animals, and I learn from a gentleman who kindly favored me with information upon the point, that the late Mr. Meire was the first to improve upon the original type. This he did in the first place by the use of the Leicester;

as their faces became white he would then have recourse to a South Down or other dark-faced sheep. It was, however, left to the son to carry out and to bring to a successful issue what the father had commenced, and Mr. Samuel Meire no doubt may be looked upon as the founder of the improved Shropshire Downs. We gather from his address to the Wenlock Farmers' Club that he accomplished this, not by resorting to any of the established breeds, but by using the best animals from his own large flock. * * Lately a very great change has come over the breeders of Shropshire; they have availed themselves of larger sheep of heavier fleece and earlier maturity, so that the only affinity they bear to the original Shrop are dark faces and legs; they now pride themselves in exhibiting some well fatted shearlings [yearlings past,] weighing upon times 22 lbs. to 24 lbs. per quarter, but this is not general.

SHROPSHIRE DOWN EWE.

Very fine specimens of this variety have been imported into the United States and Canada. The two animals represented in the foregoing cuts are owned by Hon. N. L. Chaffee, of Jefferson, Ashtabula County, Ohio. The ram, "Lion," now three years old, was bred by Lord Berwick, of Shrewsbury, England, and imported in 1861. His live weight is 334 lbs., and he yielded on the 16th of May, 1863, 17 lbs. 5 oz. of washed wool of 11½ months growth. The ewe, "Nancy," was bred by Lord Berwick, and imported at the same time. She is three years old, and her live weight is 241 lbs. On the 16th of May, 1863, she yielded 9 lbs. 3 oz. of washed wool of 11½ months growth. Six ewes at the same time, and under the same circumstances, yielded 42 lbs. 5 oz. of wool. They were sheared the fifth day after washing in clear brook water.

In answer to my inquiries on the subject, Judge Chaffee writes me that these sheep were imported by Mr. George Miller, of Markham, Canada West; that they are very hardy, healthy and easily kept; and that they excel in these particulars all his other sheep, of which he has four kinds. He says:

"They are nearly as large as the long-wooled breed, say Cotswolds or Leicesters, and yielding just about the same quantity of wool, are in my judgment much more hardy and healthy. They have the dark colored legs and face of the South Down; much longer, thicker and more compact fleeces than the South Downs, and much thicker and more compact ones than the long-wooled breeds. They have all the nice, round, compact frame, and even, uniform symmetry of appearance of the South Down, and are about 33 per cent. heavier. I have never slaughtered any of this breed, and cannot speak from personal knowledge as to the quality of their mutton, but it is said, by those who do know, to be very superior and hardly to be excelled by the South Down."

The Oxfordshire Downs.—This is a new family of sheep, and I take the following account of its origin from the already quoted address of Mr. Charles Howard, delivered before the London Farmers' Club. Mr. Howard is a well known breeder of them. He says:

"The 'Oxfordshire Downs' are what are commonly styled cross-bred sheep; but their patrons, in 1857, determined upon giving them a definite name. Hence their new title, the propriety of which is demurred to by some; for its only similarity to a Down is its color, while its size and fleece partake more of the long-wool—important qualities, which have been long and carefully cultivated by the promoters of this breed. They were originally produced by crossing the

Hampshire and in some instances South Down ewe with a Cotswold ram—most commonly the former, for it gave increased size—and the putting the crosses together: by constant attention and weeding, a most successful result has been accomplished, producing a kind of sheep that possess, with uniformity of character and hardiness of constitution, large frames, good fleeces, aptitude to fatten, and mutton of superior quality."

Mr. Howard quotes the Messrs. Druce, father and son, who were among the leading originators and most successful exhibitors of the variety, as publishing the fact that their flock originated from a cross between the South Down and Cotswold. The younger Druce says:—"The flocks generally drop their lambs in the month of February, and at 13 or 14 months old they are ready for market, weighing upon an average 10 stones [140 lbs.] each, with a fleece varying from 7 to 10 lbs. The ewes are good mothers and produce a great proportion of twins." Mr. Druce, senior, commenced this cross in 1833. Mr. Hitchman, an extremely successful breeder and exhibitor of them, started five years earlier, crossing the Hampshire Down and Cotswold. His tegs [weaned lambs] when shorn would average, in 1860, eleven stone [154 lbs.,] and his entire clip of wool 7 lbs. per fleece.

These sheep were first introduced into the United States by Richard S. Fay, Esq., of Lynn, Massachusetts, and the Hon. William C. Rives, of Virginia, who selected and imported their sheep together. Mr. Fay had a considerable extent of rough pasturage better adapted to sheep than other animals, and he first stocked it with fine-wooled sheep and subsequently with crosses between them and South Downs. Neither experiment resulted satisfactorily. A residence of several years in England induced him to turn his attention to the English breeds, and he came to the conclusion that they would better answer his purposes. Living two years among the Shropshires he was highly pleased with them, but on going to see Mr. Gillet's and Mr. Druce's Oxfordshire Downs he gave them the preference, and purchased and sent home a ram and ten ewes of this family. He subsequently imported several other lots for David Sears, Jr., of Boston, and for himself. Mr. Fay, in answer to my inquiries, informs me that these sheep fully meet his expectations—that they are of good constitution, and "take to his briars and rough pastures as if 'to the manor born.'" He has no difficulty in raising all their lambs, dropped in March, and the ewes are many of them then fit for the butcher. The mutton, killed from his rocky, rough pastures, in November, is of very high quality. His ewes, in 1862, averaged 8¼ lbs. to the fleece, unwashed — the average weight of the shorn ewes being 135 lbs. and rams 220 lbs. The yield of lambs was 160 per cent. on the number of breeding ewes. In 1863 the yield of wool fell to a small fraction under 8 lbs., and the increase of lambs rose to 175 per cent.* His wethers yield on the average fully 10 lbs. of wool. At my request, Mr. Fay forwarded me specimens of their wool. The first was taken from a ram two years old, weighing 220 lbs., and his fleece this year weighed 12 lbs. 10 oz. The wool is about 8 inches long. The ewe, three years old, with two ram lambs at her side nearly two months old, weighed 136 lbs., and her fleece 8 lbs. The wool is over 7 inches long; the quality in both instances is rather fine for wool of such length; it has a good luster; is neither hairy nor harsh; and it has a very desirable quality for certain fabrics, and will always command a ready sale.†

These sheep have gray faces and legs, lighter colored than those of the South Downs. They partake of the admirable forms of their parent stocks; are gentle and disinclined to rove; but they are willing to work hard for their feed, and are very promiscuous feeders. They make excellent returns for their feed and mature very early.

---

* Every practical sheep farmer understands of course that a nursing ewe yields considerably less wool than a dry one, and that the fleece is still more diminished by a ewe's nursing two lambs.

† I made special inquiries in regard to this wool, and detail the result, when I have not done so in regard to the other English families, because the Oxfordshire Downs are of more recent origin, and far less is generally known of them in our country, in this particular.

CHAPTER VII.

THE POINTS TO BE REGARDED IN FINE-WOOLED SHEEP.

CARCASS—SKIN—FOLDS OR WRINKLES—FLEECE—FINENESS—EVENNESS — TRUENESS AND SOUNDNESS — PLIANCY AND SOFTNESS—STYLE—AND LENGTH OF WOOL.

Whether in purchasing sheep for the establishment of flocks, or in carrying on the breeding of existing flocks, it is necessary to have a clear knowledge of those points which constitute the peculiar excellencies of the chosen variety. With respect to the English mutton breeds, this information was placed before the world with all the precision and accuracy of combined scientific and practical knowledge, by the late Mr. Youatt—by far the most comprehensive and able investigator in this department of knowledge, and also in the veterinary art, the world has yet known. The new discoveries, advances, or changes in public taste, which have taken place in breeding the English sheep since his day, have been carefully described by Mr. Spooner, Professor Wilson, and various other writers, in the English Agricultural periodicals, particularly by the authors of the prize essays published in the Journal of the Royal Agricultural Society. In one form or another, all these publications have become widely known to the American public. They are to be found in every considerable library. Our American works on sheep have been—at least so far as English breeds are concerned—but reprints of them. Our universally disseminated Agricultural Journals have spread all their most important contents broadcast throughout our country.

The fine-wooled or Merino sheep has been made the subject of comparatively little accurate and detailed investigation and description. Spain, the native land of this breed, has no literature which pertains to sheep.* In Great

---

* Though much that pertains to shepherds and shepherdesses! Cervantes several times makes himself merry over the pastoral literature of Spain. Speaking of his own

Britain the Merino was soon found not to meet the requirements of the market and prevailing systems of agriculture; and its breeding has been but little pursued there. In France and Germany considerable has been written concerning it, but most of it is inapplicable here, because the standards of excellence adopted in each of those countries differ essentially from those accepted in our own. Indeed, our own standards have materially changed within a few years, owing to circumstances already mentioned. It is for this last reason that the valuable works on Sheep Husbandry which have appeared in the United States do not furnish full information in regard to those points of the Merino sheep which now best meet the requirements of the market and the interests of the grower. This information is the more needful at a moment when multitudes of comparatively inexperienced persons, under the stimulus of an extraordinary demand for wool, are engaging in its production.*

CARCASS.— Carcass is undoubtedly the first point to be regarded, even in the fine-wooled sheep, for on its form and constitution depends the health of the animal. Good medium size, *for the family*, is the most desirable one under ordinary circumstances, for with that size generally go the best development of the parts and the greatest degree of vigor. The body should be round and deep, not over long, and both the head and neck short and thick. The back should be straight and broad; the bosom and buttock full; the legs short, well apart, straight and strong, with heavy forearm and fulness in the twist. I decidedly incline to the opinion that it is not advisable to attempt to bring all our American Merinos to the same standard of size. There are now two well marked families—the Infantado, which have been bred large, and the Paulars, which have been kept a size or two smaller and shorter. The former are for the rich lands, the

---

"Galatea," he says many of its shepherds and shepherdesses are only such in their costume; and this describes all the pastoral romance and poetry of Spain from Montemayor's "Diana Enamorada" down to Lope de Vega's "Arcadia." If there is a book in the Spanish tongue on the practical topics of Sheep Husbandry I have never heard of it!

* The prices of pure Merino sheep were nearly as high, and in some cases higher, during the fall and winter of 1862–63 than they were between 1808 and 1815. Considerable flocks of ewes were sold at $100 a head, and small numbers at every intermediate price between this and $300, $400, or even $500, a head. One breeder sold some ewes at $600 and declined much higher offers for favorite individuals. He declined an offer of $20,000 for 50 ewes. Had they been sold, the purchaser was to receive $15,000 for half of them from other parties. I state this on the authority of the person making the offer, Mr. A. M. Clark, of St. Albans, Vermont. Choice rams sold for $500 to $600, and for one or two very celebrated ones $2,500 a piece could have been taken.

latter for the more elevated and sterile ones. They bear the same relation to each other in this particular that is reciprocally borne by the Short-Horn and Devon cattle. Of the crosses between them, I shall have occasion to speak hereafter.

The Skin.—The skin should be of a deep, rosy color. The Spaniards justly regarded this a point of much importance, as indicative of the easy-keeping and fattening properties of the animal, and of a healthy condition of the system. The skin should be thinnish, mellow, elastic, and particularly loose on the carcass. A white skin, when the animal is in health, or a tawny one, is rarely found on a high bred Merino. A thick, stiff, inelastic skin, like that found on many badly bred French sheep, is highly objectionable.

Folds or "Wrinkles."—The Spanish, French and German breeders approved of folds in the skin, considering them indications of a heavy fleece. The French have bred them over the entire bodies of many of their sheep. To this extent, and especially when prominent, firm to the feel, and incapable of being drawn smooth under the shears, they are an unmitigated nuisance, both in appearance and reality. If they bear additional wool, this is counter-balanced by its defective quality on the upper edges of the folds and the great unevenness they thereby give the fleece; and were this otherwise, the additional amount would not half compensate for the loss of time in shearing, in the "catching" weather of the spring, when good shearers are so difficult to obtain. It would be vastly more economical to keep one or two per cent. *more sheep*, to obtain the same amount of wool. But I must confess that among the thousands of these disfigured animals which I have examined, I never yet saw one which presented the maximum of both length and density of wool, or yielded the maximum in weight of fleece. For reasons which I cannot explain, the wool, though often very thick between the folds, is never very long; and it is usually comparatively loose, dryish and light as well as coarse, on the outer edges of the folds.

A wide dew-lap, plaited or smooth, single or branching into two parts under the jaws, with "the cross" on the brisket, were all that the older breeders of Merinos desired in this way, on ewes. To these might be added moderate corrugations on the neck of the ram. Now, fashion calls for heavy folds on the neck of the ram and more moderate sized ones on the neck of the ewe—but few besides a class of extremists desire these to extend in great, prominent rolls over the upper side of the neck. The cross extended into a pendulous "apron"—a short fold or two on and immediately back of each elbow—some small curling ones on and uniting with the edges of the tail, (so as to give it a corrugated appearance, and twice its natural breadth,) some smallish ones uniting on the breech under the tail, and running in the direction of lines drawn from the tail to the stifle, or perpendicular ones up and down the back edges of the thighs, which, when the wool is grown, close over the twist—a wide plaited fold of loose corrugated skin running up the front edge of the thigh and across the lower edge of the flank, so as to give both the appearance of extraordinary breadth—and finally a general looseness of the skin, which disposes it to lie in small, rounded, very slightly elevated and *perfectly soft* ridges over the body, giving it a *crinkled* appearance, but offering no obstruction whatever to the shears, and not showing on the surface of the fleece—are now the points, in these regards, which constitute the ideal of the Merino breeder.

FLEECE.—The greatest attainable combination of length and thickness of wool, of the given quality, is the first point to be regarded in a market where all lengths are in equal demand. And the more evenly this length and thickness extend over every covered part, unless below the knees and hocks, the higher the excellence of the animal. It is in this point especially that the modern breeder has improved on his predecessors; and it is this, in a very considerable degree, which gives the improved American Merino its vast superiority in weight of fleece over all other fine sheep, of the same size, in the world.

Wool of full length below the knees and hocks would hardly be desirable on account of its liability to become filthy,—but a thick, shortish coat, particularly on the hind legs, making them appear "as large as a man's arm," is regarded by most as a fine, showy point—though it does not add much to the value of the fleece. The wool should extend in an unbroken and undivided mass from the back of the neck over the top of the head and down the face for an inch or two below the eyes, and there abruptly terminate in a square or rounded shape; it should cover the lower side of the jaws nearly to the mouth, and rise on the cheeks so as to leave only the front face bare, terminating abruptly like the forehead wool. The cheek and forehead wool should meet unbroken immediately over the eye and between it and the horn and ear.* But it must by no means *unite* under the eye—though its outside ends may touch there for a little way. The eye should have just naked space enough about it to leave the sight unimpeded, without any resort to the scissors. The nose should be covered with short, soft, thick, perfectly white hair. Pale, tan-colored spots or "freckles" about the mouth, and the same color on the outer half of the ear,† are not objected to by the breeders of the Paulars—but Infantado breeders usually prefer pure white. Wool on the lower part of the front face, as is often seen in the French Merinos, whether short or long, is regarded as decidedly objectionable, and any wool which obstructs the sight in any degree, is a fault.

The cavities of the fleece at the arm-pits, at the base of the scrotum, and inside of the arms and thighs, should be as small as the proper freedom of movement admits. The scrotum should be densely covered with wool to its lower extremity, and the wool on the front of it should extend up so as to unite with the belly wool.

The wool should stand at right angles to the surface, except on the inside of the legs and on the scrotum (and the nearer it approaches doing so on the scrotum the better); it should present a dense, smooth, even surface externally, dropping apart nowhere; and the masses of wool between those natural cracks or divisions which are always seen on the surface, should be of medium diameter. If they are too small, they indicate a fineness of fleece which is incompatible with its proper weight; if too large, they indicate coarse, harsh wool.

FINENESS.—Without having regard to the present anomalous state of affairs, which has temporarily so changed the

---

* If it unites in a thick, solid mass of full length, it is a beautiful and now rather rare point.

† These spots were highly characteristic of several of the families of Merinos originally imported from Spain; and the lambs of some of them were occasionally covered over the carcass at birth with larger spots of the same color, or of a deeper tawny red. Sometimes the whole body was thus colored. But all these tints disappeared on the body when the wool grew out, and were seen no more. Small black spots were frequently seen about the mouths of Spanish sheep and larger ones on different parts of the body, and coal-black lambs were sometimes yeaned. This color often fades but never disappears. Black lambs are now exceedingly rare in pure American Merino flocks, yet they continue to appear. They are always excluded from the flock to prevent their increase, as they are regarded as unsightly and their wool is less valuable. All the different colors above mentioned are inherited by the Spanish sheep from their original stocks,—from the black, red, and tawny sheep which Pliny, Columella, and other contemporaneous writers describe as existing in Spain about the opening of the first century.

relative value of our fine and coarse wools, it is known to all conversant with the subject, that uniformly and under all circumstances, there has been a much greater demand for medium than for very fine wools in the American Wool Market; and the table of prices presently to be given will show that the former have always borne a more remunerating price than the latter to the producer. This was true even before our broad-cloth manufactories sunk under the horizontal tariff of 1846. Before that time, by far the greater portion of our home manufactured woolens did not require staples above medium in quality. And of late years fashion has lent its aid still further to reduce the demand for the finer staples. There has been a steadily increasing tendency among our best dressed and most fashionable population to substitute for the broadcloths and fine black cassimeres formerly worn for dress, comparatively coarse cassimeres of various, and among the young, of "fancy" colors.

All these causes combined have turned the domestic demand for wools above the grade of coarse, principally into a channel where the requirements of the market are met, and most profitably met for the producer, by the heavy-fleeced American Merino. Should our manufactories of broadcloths and other fine textures revive, as it is to be hoped they may, so far as to supply the domestic demands for such fabrics, there will be an additional call for finer wool, and this will necessarily increase the demand for finer sheep.

EVENNESS.—Evenness of quality throughout the fleece, so far as it is attainable, is one of the best results as well as proofs of good breeding. Those usually short, detached, not very coarse, glistening particles of hair found in the fleece, termed "jar," are very objectionable—though they mostly drop out in the different processes to which wool is subjected in manufacturing. They are not so objectionable, however, as that long, strong, rooted hair which crops out through the wool on the thighs and on the edges of the folds—particularly where the latter run *over* the neck and shoulders in very large prominent rolls. I would not reject an otherwise valuable ewe, of known purity of blood, because half a dozen hairs barely showed themselves on the back edge of and half way down the thigh—though I would much prefer not to see them there, and I would breed such a ewe to a ram which would be sure to leave no such bad mark on the common progeny. But I would much dislike to breed from a ram exhibiting

4

that defect to the least degree. Rams which have very *large* folds on the *upper side* of the neck, are very apt to exhibit more or less hairs on them, and I have occasionally seen this in animals of good blood and good reputation as sire rams. It must be regarded, however, as a serious defect — though not as inexcusable as the cropping out of hairs on other parts of the body, either singly or in masses. This indicates bad blood or breeding.

TRUENESS AND SOUNDNESS.—Wool should be of the same diameter or fineness from root to point. This is termed "trueness." On a poor sheep it grows finer, on a fat one coarser. Consequently a change of condition in either direction correspondingly changes the diameter of the same fiber during different stages of its growth. The difference is sometimes visible to the naked eye. When the change of condition has been great — especially when it takes place from a low and unhealthy state to a healthy and fleshy one — it generally occasions "a joint" in the wool,— i. e., the place in the fibers where the change began, is so weak that a slight pull will detach the two parts. Indeed, they often separate on the back of the animal and the whole outer part is shed off. Untrue or jointed wool is not so valuable for various manufactures, and the different parts of it do not receive certain dyes equally. The entire fiber of the wool produced on a diseased sheep, whether it is true or not, usually lacks the proper strength. The same is the case with the wool of very old and very lean sheep. Wool to be "sound" must be strong, firm and elastic.

PLIANCY AND SOFTNESS. — Among full-blood, healthy animals, in fair condition, the pliancy and softness of wool usually correspond in degree with its fineness. Where they do not, I should always seriously distrust pretentions to purity of blood. Some allowances, however, are to be made for modes of keeping. Sheep sheltered from storms and violent atmospheric changes, have softer wool than those habitually exposed to them. Disease, old age and excessive leanness give a drier and "wirier" feeling to wool. But whether this feeling arises from natural or artificial causes, it indicates inferiority of quality. Fabrics made of such materials have less softness and elasticity, fret or fray more readily, and break sooner at corners and on the edges of folds. They admit of less finish, and take less rich, lustrous colors. They are therefore neither so beautiful, nor so good for actual wear. Pliancy and softness are so inseparably connected with the other best properties of wool, that a thoroughly practiced person can readily determine its general quality by handling it in the dark. Indeed, where the quality is very high, it can be detected by the first touch of the hand. It has an exquisite downiness of feel which is unmistakable.

STYLE.—Style means that combination of appearances which indicates choice wool—viz., fineness, clearness of color, luster, regularity and distinctness of "crimp"— that curved and graceful form and arrangement of the locks and fibers in the sheared fleece which indicate extreme pliancy (stiff, harsh wool is straighter,) and that life-like movement on handling and peculiar re-adjustment of the fibers after handling which is occasioned by their spiral form and exquisite elasticity. Style cannot be satisfactorily described in words, but it is as palpable to experienced organs, and is as indicative of actual quality, as the most gross properties of wool — such as length, fineness, or coarseness, etc.

I should remark that the highest style, like the highest fineness, softness, etc., belongs only to the smaller and more delicate families of the Merino, like the Electoral Saxon. Prime American Merino wool only approximates to these qualities. And another remark may not be out of place, in passing. The qualities of wool, even including fineness, can be more accurately determined by the natural eye than by the aid of powerful magnifying glasses.

LENGTH.—It has already been incidentally mentioned that fine wools of all lengths find an equally ready sale in our markets. Those which would have been regarded as too long for broadcloths when they were manufactured in this country, are more desirable for delaines, shawls, etc., than shorter wools. The American Merino wool, generally, I think, exceeds all other Merino wools in length.

Mr. George Campbell, of West Westminister, Vermont, who recently, (June, 1863,) started with some sheep to exhibit at the World's Fair, at Hamburgh, some time before his departure inclosed me specimens of the wool of the ewes taken out by him. It was of about a year's growth. The longest sample, lying naturally on paper without a particle of stretching, measures $3\frac{1}{2}$ inches in length; another measures $3\frac{1}{4}$; another $3\frac{1}{8}$; two of them 3; the shortest $2\frac{5}{8}$. Mr.

Campbell wrote to me:—"The sheep are nearly all of my own stock, which have been bred from the Jarvis and Humphreys importation, and recently from Mr. Hammond's flock."

Mr. Prosper Elithorp, of Bridport, Vermont, recently sent me a number of samples of his own wool and that of Mr. O. B. Cook, of Charlotte, Vermont. Mr. Elithorp's, from ewes over one year old, and all having lambs, range from $2\frac{1}{4}$ to $2\frac{3}{4}$ inches long, and that of a ram is $3\frac{1}{8}$ inches long, though all lack 45 days of a year's growth. A part of these ewes are Paulars and a part Infantados. Two of Mr. Cook's (one from a yearling and the other from a two year old ewe,) measure $3\frac{1}{4}$ inches long, and the rest (from yearlings,) from $2\frac{5}{8}$ to $2\frac{3}{4}$ inches. The sheep are pure Infantados.

Mr. A. J. Stow, of West Cornwall Vermont, has forwarded me numerous specimens. The longest is $3\frac{3}{8}$ inches long, two of them are 3, and most of the remainder are about $2\frac{3}{4}$ inches long. They are all from ewes over one year old, and the wool lacks three or four days of a year's growth. Mr. Stow says "they are all from his Hammond sheep."

I have an old specimen of wool from a Paular ram, bred by one of the Robinson's, of Shoreham, Vermont, (and owned by Myrtle & Ackerson, of Steuben County, New York,) which measures $3\frac{1}{4}$ inches long.

The recent Vermont specimens above given are fairer tests of the length of the longer stapled American Merino wool, from the fact that they were not sent in any case as specimens of mere length, but of fleeces of extraordinary weight. And I think great length is not now usually particularly valued in any other connection. The sheep which yield the most extraordinary weights of fleece, indeed, rarely have *extremely* long wool, because such length is rarely accompanied by sufficient thickness. Mr. Hammond's "Sweepstakes," whose weight of fleece has probably never been excelled, yields wool not exceeding $2\frac{1}{4}$ inches long, and "21 per cent.," several times named in this volume, probably never excelled in the proportion of wool to meat, yields wool $2\frac{5}{8}$ inches long.

CHAPTER VIII.

THE SAME SUBJECT CONTINUED.

YOLK—CHEMICAL ANALYSIS OF YOLK—ITS USES—PROPER AMOUNT AND CONSISTENCY OF IT—ITS COLOR—COLORING SHEEP ARTIFICIALLY—ARTIFICIAL PROPAGATION AND PRESERVATION OF YOLK.

YOLK.—This is that oily feeling fluid, or that sticky, pasty or half-hardened substance, within the wool, or that hard substance on the outer ends of the wool, which commonly receives the name of oil, grease, or gum. These appellations are obvious misnomers when we take its chemical constituents into consideration.

CHEMICAL ANALYSIS OF YOLK.—Vauquelin, a celebrated French chemist, found that various specimens of yolk contained about the same constituents:—1. A soapy matter with a basis of potash, which formed a greater part of it. 2. A small quantity of carbonate of potash. 3. A perceptible quantity of acetate of potash. 4. Lime, whose state of combination he was unacquainted with. 5. An atom of muriate of potash. 6. An animal oil, to which he attributed the peculiar odor of yolk. He found the yolk of French and Spanish Merinos essentially the same. He assumed that the yolk in sheared wool injures it after a few months, if not scoured out.

USES OF YOLK.—Yolk has been believed in all countries and times to promote the growth of wool and render it soft, pliant and healthy. It seems to me to have other and obvious uses.* The small, irregular-shaped masses of wool which adhere together in the unshorn fleece of the Merino sheep, and which are bounded externally by visible, permanent cracks,

---

* I suggested these uses in my Report on Fine-Wool Husbandry, made in February, 1862.

slide on each other with every movement of the animal; so that, in effect, the cracks are the joints of the fleece. If dry and unlubricated by the yolk, the friction of these sliding masses would, on the sides subjected to abrasion, wear or break off the tooth-like processes on the wool on which the felting property depends; and this same effect would follow, whether to a greater or lesser degree, I am unable to say, on those coarse open fleeces in which, as in the covering of hairy animals, there is no such massing of the fibers and each slides separately on the surrounding ones. Again: if the wool was unlubricated, heavy rains, and the contact of the sheep with each other, with the ground and other substances, would cause felting on the back — a result now sometimes witnessed to a limited extent, and termed "cotting."

PROPER AMOUNT AND CONSISTENCY OF YOLK.—Different opinions are entertained of the amount of yolk it is profitable to propagate in wool. If the fleece is sold unwashed, and according to the present general mode, at a fixed rate of shrinkage on that account, it is obviously the interest of the wool grower to produce as much yolk as is consistent with the greatest united production of wool and yolk. And even if wool is sold nominally "washed," it is evident that the same amount of washing will leave very yolky fleeces heavier than unyolky ones. Farmers have learned that if they can only *say* their wool is washed — no matter *how* washed — ten or fifteen per cent. more yolk than would be left by thorough washing, will not cause any corresponding deduction in the price. There are a class of experienced buyers, certainly, who do not purchase in this indiscriminate way, but as the wool business has constantly expanded and opened new opportunities for the profitable investment of money, every year brings its fresh horde of raw, eager buyers — the agents of manufacturers or speculators, or persons speculating on their own account — and some of these always take the heavy, dirty wools at about the price of the clean ones. I shall allude to this topic again under subsequent heads.

I esteem it particularly fortunate for the preservation of the intrinsic value of our Merino sheep, and fortunate for the public interest, that it is already incontestibly ascertained that the greatest amount of yolk is not consistent either with the greatest amount of wool, or with the greatest aggregate amount of both yolk and wool. The black, miserably "oily," "gummy" sheep, looking as if their wool had been soaked to saturation in half inspissated oil, and then daubed over externally with a coating of tar and lamp-black, never exhibit that maximum of both length and density of wool which, with a proper degree of yolk, produces the greatest aggregate weight. Yolk has been generally thought to be the pabulum of wool and if so, its excessive secretions, as a separate substance, may diminish its secretions in the form of wool. Be this as it may, the fact I have stated stands without an exception. And animals exhibiting this marked excess of yolk, are invariably feebler in constitution, less easily kept, and especially less capable of withstanding severe cold. Such excessive secretions appear, then, to cause, or else to be the results of an abnormal or defective organization. For these reasons, these comparatively worthless animals, once so eagerly sought, have already gone out of use among the best informed breeders; and where they linger, it is, like antiquated fashions, in regions where the current ideas of the day penetrate slowly!

There should be enough fluid yolk within the wool on the upper surfaces of the body, to cover every fiber like a brilliant, and, in warm weather, like an undried coat of varnish — but not enough to fill the interstices between them, so that the fleece shall appear, as it sometimes does, to be growing up through a bed of oil. And if there is a sufficiency of yolk above, it must be expected that underneath where the fleece is less exposed to evaporation and the washing of rains, and to which part gravitation would naturally determine a fluid substance, a considerably greater quantity of it will be found. But hardened or pasty masses of it within the wool are to be avoided, on all parts of the body. A portion of the fluid yolk will necessarily inspissate or harden on the outer ends of the wool. It is proper that it should sensibly thicken those ends, and clot them together in small masses on the upper parts of the body — forming a coat considerably thicker, firmer and harder to the hand than would the naked wool, and quite rigid when exposed to cold; but it should not cover the wool in rounded knobs, or in thick, firmly adhering patches, bounded by the fleece cracks — sticking to the hand in hot weather like a compound of grease and tar, and in cold having a "board-like" stiffness. Underneath, for the same reasons given in reference to inside yolk, a greater quantity of it must be tolerated. It should stick the masses of wool together in front of the brisket and scrotum, and large rounded knobs of it inside the legs and thighs and on the back side of the scrotum, are considered desirable.

COLOR OF YOLK.—The external yolk is occasionally somewhat yellowish — of the tinge of dirty bees-wax — but more generally of some dark shade of brown, or what would more commonly be termed black. The darker color is preferred. All American Merino sheep having what is esteemed a sufficient amount of yolk, become very dark colored each year before the winter is far advanced, if they are housed from summer and winter storms after shearing. Rains wash away the yolk and with it the color. But the yolk is soluble in different degrees in different families, and even on different animals of the same flock. The Paular (Rich) sheep hold their color uncommonly well; the French rapidly bleach. It has been supposed that the black color is communicated to external yolk by dust, the pollen of hay, etc. These may contribute to the result, but I have recently learned from entirely reliable persons, who house their sheep in summer, that if kept entirely dry, they never assume their darkest color — that to obtain this, they must be exposed to dews, light sprinkles of rain, or the contents of the watering pot. The change in color, accordingly, is partly chemical.

Internal yolk varies in color from a pure white to a deep yellow. It has been rather the fashion, in this country, since the days of the Saxon sheep, to breed for the former, and this is the prevailing color in the American Paulars. The breeders of the American Infantados, and of the Silesians, generally follow the old Spanish custom of giving preference to shades of yellow. A brilliant "golden tinge," faint or imperceptible near the roots of the wool, but deepening towards its outer extremities, is the one sought after. The founder of the improved Infantado family has, as already stated, bred steadily for that color; and he has done so not merely as a matter of taste, but under the impression that it betokens a vigorous growth of wool and general vigor of constitution — and particularly vigor of that kind, which exhibits itself in the forcible transmission of individual properties to progeny. But this "golden tinge" is not to be mistaken for the deep saffron yellow which attends cotting — or for a dull, dead yellow — or for a tawny bees-wax hue — or for the hue of "nankeen" cloth, sometimes seen in imperfectly bred animals. The favorite color among the French breeders is a creamy one. In answer to inquiries made by me, in 1862, several experienced manufacturers — all I consulted — concurred in the statement that the color of the yolk is not, in itself, a matter of any consequence, in reference to any of the objects of manufacturing; and that its quantity and consistency are only important in so far as they affect its weight and cause a loss in scouring.

I have been speaking of the natural color of yolk. In many regions where sheep are not pastured on thoroughly sodded ground, the whole interior of the fleece becomes stained by dust to the prevailing color of the ground. This often occurs on our Western prairies.

COLORING SHEEP ARTIFICIALLY.—To give Merinos destitute of it, a dark external color, they are sometimes painted. A coating of linseed oil and burnt umber, slightly darkened with lamp-black, neatly applied within a few weeks after shearing, can be distinguished from the natural dark coat of a *housed* sheep with some difficulty, by inexperienced eyes. But generally the sheep jockey overdoes the thing and excels nature! He lays on the coat more evenly and more uniformly dark. It is said there are other preparations, with or without coloring matter, intended to give the fleece a thick, firm feeling, but I have not learned their composition. It is not necessary to remark that all such practices are rank frauds.

ARTIFICIAL PROPAGATION AND PRESERVATION OF YOLK.—Yolk is greatly increased in the fleece by high keep; and careful housing in summer, as well as winter, as I have repeatedly remarked, preserves it there. The objects and effects of these practices will be alluded to hereafter.

CHAPTER IX.

ADAPTATION OF BREEDS TO DIFFERENT SITUATIONS.

MARKETS — CLIMATE — VEGETATION — SOILS — NUMBER OF SHEEP TO BE KEPT — ASSOCIATED BRANCHES OF HUSBANDRY.

Persons desirous of engaging in Sheep Husbandry are frequently at a loss to decide what breed of sheep is best adapted to their particular wants and circumstances. The first and leading point to determine is whether it would be most profitable to make mutton the prime consideration and wool the accessory—or wool the prime consideration and mutton the accessory. If the first conclusion is adopted, some of the improved English mutton varieties are undoubtedly to be preferred; if the last, the Merino has no competitor.

Markets.—Where other circumstances equally admit of either husbandry, it is the market that determines which product is most profitable to the producer. Wool has a vastly greater and more universal consumption than mutton, because it is a prime necessary of life to every man outside the tropical zone. As such a necessary, it can never find any practical substitutes. Mutton is not a necessary of life, although it is made to contribute largely towards one—human food. It readily admits of substitutes. It is scarcely used by large classes of men and even by whole nations. Yet it is demonstrable that it can be produced more cheaply than any other meat. No meat, not even the choicest of beef, is more palatable to those accustomed to its use; and none is more nutritious and healthful. The prize-fighter, whose success depends upon the perfect integrity of all his physical tissues and functions, is as often trained on mutton as on beef; the physician as often recommends it to the invalid. And finally, it wastes less than beef in being converted into food.* Every-

---

* The Report on Sheep Husbandry made to the Mass. Board of Agriculture in 1860, by Mr. James S. Grennell, thus condenses the results of various experiments

thing therefore marks it as one of the most valuable articles of human consumption; and where its use is once established, there is no one which finds a steadier demand or more uniformly remunerating prices.

In England mutton is the favorite animal food from the peer to the peasant—the former preferring the choicer qualities as a matter of taste, the latter the cheaper and fatter ones as a matter of economy. A pound of Leicester mutton which has an external coating of fat as thick as that on well fattened pork, will go as far to support life as a pound of pork, eaten simply in the condition of cooked meat; and eaten partly as meat and used partly to convert vegetables into soups having the flavor and to some extent the nutritive qualities of meat, it will not only produce more palatable nutriment than the pork, but nutriment capable of being distributed so as to supply more wants.

Thirty or forty years ago but very little mutton was consumed in the United States. Our people had not learned to eat it. Colonizing a new country covered with forests containing animals that prey on sheep, and in which the necessary labor for guarding them was scarce and high, our forefathers kept only enough to meet pressing wants for wool for household uses. Few were used for food, and the early sheep of our country did not constitute very palatable food. Beef and pork were more easily grown and better relished. This state of things continued until mutton became a stranger to American tables. When at length the country became better adapted to the production of sheep, there was no call for mutton. I can myself remember when it was rarely seen and never habitually used on the table, except perhaps in cheap school boarding-houses of the "Dotheboy's Hall" order. This prejudice continued until the comparatively recent general introduction of the improved English mutton sheep—and until *fashion* in cities, for once, inaugurated a great and useful change in the public taste. Some of the earlier prejudices yet linger among our rural population; yet the same change is making its way, not slowly, into the country. The first quality of mutton now commands a higher price in our

---

on this subject: "English chemists and philosophers, by a series of careful experiments, find that 100 lbs. of beef, in boiling, lose 26½ lbs., in roasting, 32 lbs., and in baking 30 lbs. by evaporation and loss of soluble matter, juices, water and fat. Mutton lost by boiling 21 lbs., and by roasting 24 lbs.; or in another form of statement, a leg of mutton costing raw, 15 cents, would cost boiled and prepared for the table, 18¼ cents a pound; boiled fresh beef would, at the same price, cost 19¼ cents per pound, sirloin of beef raw, at 16¼ cents, costs roasted 24 cents, while a leg of mutton at 15 cents, would cost roasted only 22 cents."

markets than the first quality of beef. The extent and rapidity of the change in our cities receives a striking illustration from the following facts stated in Mr. Grennell's Report to the Massachusetts Board of Agriculture, 1860:

"At Brighton (near Boston,) on the market day previous to Christmas, 1839, two Franklin county men held 400 sheep, every one in the market, and yet so ample was that supply, and so inactive the demand, that they could not raise the market half a cent a pound, and finally sold with difficulty;" and "just twenty years after that, at the same place, on the market day previous to Christmas, 1859, five thousand four hundred sheep changed from the drover to the butcher."

The history of Boston in this respect is but the history of all our larger cities, towns and villages. When this taste fully extends to our rural population — when our laboring farmers learn, as they ought to learn and will learn, that eating fat pork all the year round is not most conducive to health and to an enlarged general economy — when they acquire the habit, as they so conveniently could, of killing mutton habitually for household and neighborhood consumption in its fresh state[*] — our people, now the greatest consumers of animal food among the civilized nations of the world, will become by far the greatest consumers of mutton in the world. I doubt whether the enormous amount which will be annually grown and consumed in this country, within fifty years, has yet occurred to our most sanguine advocates of mutton sheep.

It is a fixed fact, thoroughly settled by the experience of England, and beginning to be well understood in extensive regions of our country, that where the market for mutton is large and near by, and the local circumstances are favorable to its culture, its production, if well understood and conducted, is more profitable as a leading object, than the production of wool. The Merino was introduced into England under the most favorable auspices, and its propagation fostered by kingly example and encouragement. But neither as a wool sheep proper, nor when bred into what may be termed a half mutton sheep, has it been able to compete at all successfully with the pure mutton breeds. Where the soils and surroundings are suitable, it is already becoming more profitable (in

---

[*] The frequent killing of beeves on farms, to be eaten fresh, is not convenient on account of their size. In warm weather, the meat could not usually be disposed of without salting down, unless the farmer should change his occupation to that of a traveling meat peddler. It is not so with the sheep. Three or four farmers could join together to buy all the meat, or to kill alternately and divide the carcass.

ordinary times, when the natural conditions of the market are not unsettled by war,) to grow first-class mutton sheep throughout most of New England, excepting Vermont and the northern halves of New Hampshire and Maine — throughout the eastern portions of New York and Pennsylvania — and throughout a belt of country round every city and village, wider or narrower according to its population — than it is to grow the wool sheep proper. And this area of mutton production must steadily increase, pushing back wool production further from the sea-board and from all dense aggregations of population.

While the preceding facts, in my opinion, admit of no reasonable question, it is nevertheless equally true that the demand for wool in the United States is, as I shall presently show, far less adequately supplied already with the domestic product — and that this demand must of absolute necessity go on increasing forever in the same ratio with the increase of our *entire* population — so that, in the aggregate, the amount of land and other capital, which can be profitably invested in its production will always exceed that which can be profitably invested in mutton production, in the proportion of almost hundreds to one. Our vast interior regions, with the exceptions already indicated in the vicinity of cities, and with certain others which it is not necessary to specify here — in other words, all regions remote from meat markets or from which the transportation to such markets is distant or expensive — can be more profitably devoted to the production of wool as a leading object than mutton.

It will be seen from all the foregoing that there is, properly speaking, no competition whatever between the mutton growing and the wool growing sheep — that their respective profitableness is purely a question of place and some other circumstances which I am about to name — and that to raise that question abstractly, and independently of these local and other considerations, as is often done, is almost as irrelevant and unmeaning as it would be to ask which is the most profitable mode of transportation, ships or locomotives, without having reference to the fact whether such transportation must be made by land or water. I will now proceed to examine the other qualifying local circumstances, besides those of market.

CLIMATE. — The English improved mutton sheep in its present perfect development of all the points which constitute a matchless meat-producing animal, is in some part a product of the temperate, uniform and moist climate of England. It has withstood the effects of acclimation in the United States successfully, but it requires more care and shelter and is not so well adapted to our habitual extremes of heat and cold as the hardier Merino.* Exposed without good, adequate shelter to rapid and excessive variations of temperature, it is subject to colds which tend to various diseases, both of inflammatory and typhoid types: and, at best, it wilts and withers away. It is not adapted to very cold or very warm climates for another reason — on account of the influence they exert on vegetation. But its sustentation will be considered under another head.

The Merino endures vicissitudes and extremes of weather better than any other sheep which approximates to it in value. Its range of habitation extends throughout the temperate zone. It will flourish wherever the ox or the horse will flourish; but, like those animals, thrives better for some degree of winter shelter anywhere, and demands it in regions of severe cold, and especially in those where humidity and cold are liable to follow each other rapidly.

VEGETATION.— The English breeds of sheep require abundant and steady supplies of food properly or profitably to develop their peculiar value as mutton sheep — viz., their fattening properties and early maturity. They are therefore unadapted to regions where the summer is hot enough to dry up the vegetation, as on the plains of Texas and Southern Spain—or regions subject to periodical drouths, like Australia and the Cape of Good Hope — or those where vegetation is locked up by long and rigorous winters, as in various northern inhabited regions of both hemispheres. For the scarcity of succulent food produced by summer drouth, there can be no adequate reparation to these hearty and gross feeding animals. For the long and severe winter, there may be sufficient extra provision made in grain and roots: and where land is comparatively cheap, and mutton in good demand, that extra provision can be profitably made. These are the conditions of New York and New England as mutton producing countries. England presents far more favorable natural, and, in many respects, artificial conditions, for its

---

* I do not of course here include among the improved English mutton sheep, the black-faced Scotch or Heath Sheep, or the Cheviots, though I enumerated them among the English sheep which are residents of the United States.

production, but still the greatly higher cost of land there, more than counterbalances those advantages on the score of actual and direct profit to the grower. While all the mutton sheep are abundant consumers, there is a difference in them in this particular, and in the quality of the food they require. Speaking generally, the long-wools require the richest and most abundant pasturage, and they will consume ranker herbage than would be adapted to upland breeds, or to the Merino. They are much less inclined to travel or work for their food. They are therefore, properly, low-land sheep. Their place is rather the rich, moist plain, than the dry hill-side. The Leicester is the tenderest and the least disposed to work of all. The Cotswold is perhaps the hardiest and best worker of the long-wools which I have described, and thrives on low, moist hills, like those from which it derives its name.* Judging from its blood, the New Oxfordshire should occupy an intermediate place between the two preceding families. All the Down families are hardy and possess good working qualities. In England they are regarded as an upland sheep, adapted to dry and comparatively scanty pasturage when necessary. But this is to be understood with qualifications, in the United States. The words "upland" and "dry," as applied to pasturage, have very different significations from their English ones, in our land of lofty hills and mountains, and of dry, scorching summers.

As a hard working sheep—as a sheep adapted to very scanty, or dried up, or poor pasturage,—none of the heavy English mutton breeds can compare with the Merino. The latter, indeed, work for their food of preference. Where they have an opportunity to choose, they will invariably desert the rich valley a considerable portion of each day to climb the lofty hill-side, and they love to clamber about its steep declivities and among rocks, to crop the scattered tufts of grass, and browse on those bushes and weeds which they are fond of mingling with their food. They have not, in these particulars, been bred away as far from the natural habits of the species as the English sheep. Their annual sojourn among the mountains of Spain, until a comparatively recent period, preserved these habits.

From an observation of these facts, it has been inferred that the Merino requires short verdure, and a considerable variety of it. It is probable, on chemical considerations, that, other things being equal, several kinds of food will furnish

---

* The Cotswold Hills are in Gloucestershire, England.

more of the constituents of wool than will a single kind — and consequently that a variety in it, tends to the development of a heavier fleece. But abundance and richness of food, when the Merino is compelled to accept them, affect its tissues as they do those of all other sheep, and more than compensate for the want of variety. Removed from the pastures of New England, or of North-eastern, Eastern and Southern New York — grazing lands proper — to the rich clover fields of Western New York, Ohio, etc., the Merino increases considerably, both in size and weight of wool — and it continues equally healthy.

SOILS.—The fertility of the soil is a consideration of weight in selecting a breed of sheep to stock it, because on that fertility depends the luxuriance, and, to some extent, the quality of its vegetation. Its nature and condition in other respects are also important. Habitual wetness of the ground, from whatever cause it arises, is highly injurious to most kinds of sheep, and particularly to upland ones. The Merino cannot endure it; and wool growing can never be profitably pursued on such lands. That mutton growing can, is abundantly proved by the example of the English farmers in Lincolnshire, Kent, etc. In such situations, the long-wooled sheep are decidedly preferable.

It is thought, in England, that an occasional or even single visit to some fen or stagnant pool sometimes communicates the fatal rot* to flocks of sheep. I never have heard of an instance of this in the United States. In our Northern and Eastern States I never have known the most free access to swamps, pools, etc., to prove injurious to sheep, provided they had abundant pasturage and pure water without, and only entered the marshy lands voluntarily, as all sheep will occasionally do in quest of a change of food. Constant access to salt-marshes is considered actually promotive of their health and thrift. I have received various accounts of fatal disorders attacking sheep in Texas, in consequence of being kept on what are termed hog-wallow prairies — low, flat, moist, very rich lands. I should expect such results in large flocks restricted to such lands, in all our warm climates; and such pasturages would be decidedly uncongenial to all the short-wooled varieties of sheep, in any climate.

---

* I speak of liver rot, not hoof rot. The names are sometimes confounded in our Northern States where the former disease is mostly unknown.

A very light, sandy or other soil which rises readily in clouds of dust, when not well sodded over, is unfavorable to the cleanliness and beauty of wool — yet some healthy and profitable sheep ranges have this fault. A gravelly loam, or other soil of about equal consistency, readily permeable to surface water, thoroughly drained, abounding in clear, rapid-flowing brooks, elevated and free from malarious influences, dotted with groves or clusters of shade trees, and of about medium fertility, combines the conditions preferred by the Merino. The same conditions would as well meet the wants of the Downs; and greater fertility would not be objectionable to them. Lower and moister soils of the richest quality are congenial to the long-wools.

THE NUMBER OF SHEEP TO BE KEPT.—Mutton sheep consume more, demand a greater variety of artificial feed, and greatly more care than Merinos, and therefore are better adapted to small, high-priced farms, where it is desirable to invest as much capital in sheep as can be rendered remunerative. But the long-wooled families would be wholly unadapted to large farms, where surplus capital is wanting, even were there not a difficulty of another kind. They do not herd well — that is, thrive well when kept together in large numbers. The Down families herd much better, but still do not compare with the Merinos in this respect. In Australia and Texas, a thousand or more Merinos often run in the same flock, summer and winter, throughout the year, occupying the same pastures by day and the same folds by night. And my friend, George Wilkins Kendall, of Texas, used playfully to insist to me that in his Merino flocks of that number, he could not find one poor enough to make palatable mutton! His flocks passed through the terrible winter of 1860 without artificial feed or shelter — when the cold was severer than ever before known in that climate, and when it so arrested the growth of grass that his sheep daily traveled four or five miles from their folds to obtain food;— and he did not lose scarcely one per cent. of their number! A large number of mutton sheep may be kept on the same farm with a sufficient division of the fields and winter shelters; but they cannot profitably or safely be kept together in large flocks.

ASSOCIATED BRANCHES OF HUSBANDRY.—Economy demands that for the most profitable production of mutton there should be associated with it a proportionable amount of convertible husbandry. Mutton sheep demand grain, roots, etc., in large quantities, and in return they supply all the necessary fertilizing materials for those crops. These fertilizers are comparatively wasted if not devoted to those crops. Each husbandry, then, is necessary to the highest profitableness of the other. Without such union, neither the present admirable system of British agriculture, nor the present maximum of population which derives its sustentation from that agriculture, could be kept up. The adaptation of the soil and other circumstances to convertible husbandry, the tastes or wishes of the flock-master in regard to embarking in it in connection with mutton growing, and the local market for its products, all become, therefore auxiliary considerations of weight in choosing between mutton and wool growing.

I have aimed to present, with impartiality, the principal circumstances which determine the adaptability of different kinds of sheep to different situations. There are, however, generally more or less minor ones in every man's case, known only to himself, which somewhat qualify the influence of the major ones; and of these he must be his own sole judge. In closing this branch of my subject, I will only further add that while, in selecting a breed of sheep, every one should keep his eyes firmly fixed on the primary object of production, he never should altogether lose sight of the accessory one. The mutton sheep would probably be nowhere profitable without its wool, and the wool sheep would be much less profitable without its mutton.

CHAPTER X.

PROSPECTS AND PROFITS OF WOOL AND MUTTON PRODUCTION IN THE UNITED STATES.

The subjoined table of the Prices of Wool, in one of the principal Wool Markets of the United States, extending through thirty-eight years — through the most disastrous revulsions in the money market and in the prices of all kinds of property — under tariffs which have at one period given excessive protection to our woolen manufactures, and at others abandoned them unaided to the competition of Europe — presents the best proof I possess, nay, the most unanswerable proof possible, of the steady remunerativeness of wool production. It was prepared for me in 1862, from his own books and those of his predecessors in the same firm, by George Livermore, Esq., of Boston, one of the most eminent wool commission merchants ever in the United States — and his name is an ample guaranty of its accuracy. It has now been published a year, and has circulated throughout the trade without one of its figures being questioned.* I have added a column to it indicating the tariff laws in force at the different periods, but there is not space here to give even a synopsis of those tariffs.†

The average and not the extreme prices for each quarter are given, and it will be observed that these are not given strictly by quarters anterior to 1827.

I have learned, from various reliable sources, that from 1800 to 1807, wool bore low prices in our country; that in 1807 and 1808 full-blood Merino wool sold for $1 a pound; that in 1809, it rose to about $2 a pound, and so continued through the war against England, commenced in 1812 — some choice lots fetching $2.50 a pound; that when our infant

---

* It was published in my Report on Fine-Wool Husbandry, in 1862; and in the Boston trade publications which would place it in the hands of all the leading wool merchants and manufacturers.
† A complete synopsis of them is given in my Report on Fine-Wool Husbandry, 1862.

manufactories were overthrown at the close of that war, in 1815, it again sunk to a low price, and so remained until the Tariff of 1824 was enacted.

PRICES CURRENT OF WOOL IN BOSTON.

Tariff and time of taking effect.	Year.	Quarter ending	Fine.	Medium.	Coarse.
	1824.	January,	..	..	..
		March,	70	45	33
June 30.		July,	..	..	..
Tariff of 1824.		October,	50	40	30
	1825.	January,	60	45	33
		April,	..	..	..
		July,	..	..	..
		October,	..	..	..
	1826.	January,	..	..	..
		April,	52	45	40
		June,	37	30	27
		October,	44	38	33
	1827.	January,	37	33	28
		April,	44	36	30
		July,	36	31	26
		October,	42	32	25
	1828.	January,	40	30	25
		April,	44	36	28
		July,	48	40	33
Sept. 1.		October,	47	40	31
Tariff of 1828.	1829.	January,	55	45	35
		April,	43	35	30
		July,	45	35	30
		October,	38	31	27
	1830.	January,	40	35	30
		April,	48	38	32
		July,	62	50	40
		October,	70	60	47
	1831.	January,	70	60	47
		April,	70	60	50
		July,	75	63	50
		October,	70	60	50
	1832.	January,	65	55	45
March 3.		April,	60	50	40
Tariff of '32.		July,	50	40	30
		October,	50	40	30
	1833.	January,	..	..	..
		April,	..	..	..
		July,	62	55	42
		October,	65	55	45
Dec, 31.	1834.	January,	70	60	47
		April,	65	55	42
		July,	60	50	40
		October,	60	50	40
	1835.	January,	60	50	40
		April,	65	58	45
		July,	65	58	45
		October,	65	58	45
	1836.	January,	65	58	45
		April,	65	58	45
		July,	70	60	50
		October,	70	60	50
Tariff of 1832.	1837.	January,	70	60	50
		April,	70	60	50
		July,	..	..	..
		October,	50	40	33
	1838.	January,	50	42	35
		April,	50	42	35
		July,	45	37	32

TABLE OF WOOL PRICES.

Tariff and time of taking effect.	Year.	Quarter ending	Fine.	Medium.	Coarse.
		October,	55	48	37
	1839.	January,	55	48	38
		April,	55	48	38
		July,	58	50	40
		October,	60	52	46
	1840.	January,	50	45	38
		April,	48	41	36
		July,	46	38	33
		October,	46	38	33
	1841.	January,	52	45	37
		April,	52	45	37
		July,	50	44	35
October 1. Tariff of 1841. August 30.		October,	48	41	33
	1842.	January,	48	43	35
		April,	46	42	33
		July,	43	38	31
		October,	37	31	26
	1843.	January,	35	30	25
		April,	34	29	25
		July,	35	30	26
		October,	36	32	26
Tariff of 1842.	1844.	January,	37	31	26
		April,	45	37	30
		July,	45	37	31
		October,	50	42	33
	1845.	January,	45	38	31
		April,	45	38	33
		July,	40	35	30
		October,	38	34	28
	1846.	January,	40	35	30
		April,	38	33	28
		July,	38	33	28
Doc. 1.		October,	36	30	22
	1847.	January,	47	38	30
		April,	47	40	31
		July,	47	40	31
		October,	47	40	30
	1848.	January,	45	38	30
		April,	43	37	30
		July,	38	33	28
		October,	33	30	22
	1849.	January,	33	30	23
		April,	42	36	30
		July,	40	35	28
		October,	42	36	30
	1850.	January,	47	40	33
		April,	45	38	31
		July,	45	38	32
		October,	45	38	35
	1851.	January,	45	37	32
		April,	50	44	40
		July,	47	42	37
Tariff of 1846.		October,	45	40	33
	1852.	January,	42	37	32
		April,	42	36	31
		July,	45	38	32
		October,	50	42	37
	1853.	January,	58	55	50
		April,	62	55	50
		July,	60	53	48
		October,	55	50	48
	1854.	January,	53	47	42
		April,	57	52	44
		July,	45	37	30
		October,	41	36	32
	1855.	January,	40	35	32
		April,	43	35	32

TABLE OF WOOL PRICES.

Tariff and time of taking effect.	Year.	Quarter ending	Fine.	Medium.	Coarse.
		July,	50	40	33
		October,	52	41	36
	1856.	January,	50	38	35
		April,	57	43	37
		July,	55	43	38
		October,	60	55	45
	1857.	January,	58	50	43
		April,	60	56	43
July 1.		July,	56	48	40
		October,	38	30	26
	1858.	January,	40	33	28
		April,	42	35	30
Tariff of 1857.		July,	42	37	30
		October,	55	42	36
	1859.	January,	60	52	45
		April,	60	46	37
		July,	55	40	36
		October,	60	49	42
	1860.	January,	60	50	40
		April,	52	45	40
		July,	55	50	40
		October,	50	45	40
	1861.	January,	45	40	37
April 1.		April,	45	37	32
Tar. of '61.		July,	40	35	32
		October,	47	47	52

From the beginning of 1827, from which the above prices present the averages of each quarter, to the close of 1861, a period of 35 years, the average price of fine wool was 50 3-10 cents; of medium, 42 8-10 cents; of coarse, 35½ cents. Fine wool averaged 15 per centum higher than medium, and medium 14 per centum higher than coarse.

The wools classed in the table as fine, included Saxon, grade Saxon, and choice lightish-fleece American Merino; the medium included American Merino and grade down, say to half blood; the coarse included wools one-fourth blood Merino and below. Each of these classes, of course, embraced wools of various qualities and prices.

The lessons to be derived from this table are most valuable to the wool grower. How very striking, for example, is the fact that during thirty-eight years — and with all the disturbing causes to the wool market which have been alluded to — there has not been a single year in which the average prices for the wools marked medium in the table would not *now* pay the actual cost of producing our heavy fleeced American Merino wools; and that there have not been more than half a dozen years, when those prices would not be decently remunerative! Of the production of how many other of our great staples of industry can as much be said?

STATEMENT

Exhibiting the value of Wool, and Manufactures of Wool, imported into and exported from the United States, from 1840 to 1861, both years inclusive.

YEARS ENDING—	WOOL UNMANUFACTURED. EXPORTS.			IMPORTS.	MANUFACTURES OF WOOL. EXPORTS.			IMPORTS.
	Foreign.	Domestic.	Total.		Foreign.	Domestic.	Total.	
September 30, 1840	$29,246	....	$29,246	$346,079	$418,309	....	$418,309	$9,071,164
do 30, 1841	44,226	....	44,226	1,001,063	171,814	....	171,814	11,001,939
do 30, 1842	90,866	....	90,866	797,382	145,123	....	145,123	8,375,725
June 30, 1843	34,661	....	34,661	248,679	61,907	....	61,907	2,472,154
do 30, 1844	22,153	....	22,153	861,461	67,483	....	67,483	9,475,782
do 30, 1845	41,571	$203,096	246,667	1,689,704	166,646	....	166,646	10,666,176
do 30, 1846	37,302	89,460	126,762	1,134,226	147,804	....	147,804	10,083,819
do 30, 1847	1,840	....	1,840	656,822	316,894	....	316,894	10,908,853
do 30, 1848	6,991	....	6,991	857,034	179,781	....	179,781	16,240,863
do 30, 1849	....	....	....	1,177,347	201,404	....	201,404	13,704,606
do 30, 1850	....	....	....	1,681,691	174,934	....	174,934	17,151,509
do 30, 1851	7,966	....	7,966	3,833,157	267,379	....	267,379	19,607,309
do 30, 1852	64,286	....	64,286	1,930,711	266,678	....	266,678	17,573,064
do 30, 1853	61,387	....	61,387	2,669,718	343,989	....	343,989	27,621,911
do 30, 1854	41,668	....	41,668	2,822,185	1,262,887	....	1,262,887	32,382,864
do 30, 1855	131,442	27,802	159,244	2,072,130	2,327,701	....	2,327,701	24,404,149
do 30, 1856	14,907	42,462	57,466	1,666,064	1,266,632	....	1,266,632	31,961,703
do 30, 1857	920	19,007	19,927	2,125,744	437,498	....	437,498	31,286,118
do 30, 1858	824,908	211,861	1,036,769	4,022,636	197,902	....	197,902	20,486,060
do 30, 1859	32,141	355,563	387,704	4,444,964	220,447	....	220,447	33,621,666
do 30, 1860	37,280	389,612	426,792	4,842,162	201,376	....	201,376	37,957,190
do 30, 1861	43,299	237,846	280,145	4,717,860	317,340	....	317,340	25,487,166
	$1,661,023	$1,562,602	$3,113,630	$46,077,273	$9,131,408	....	$9,131,408	$429,422,061

TREASURY DEPARTMENT, REGISTER'S OFFICE, Feb. 12, 1862.

J. A. GRAHAM, *Acting Register.*

Will this steady demand and these remunerating prices last? Here again the facts and figures of the past afford the most trustworthy answer. The table on preceding page was prepared for me in 1862, by the acting Register of the Treasury.

It is thus made to appear that during the twenty-two years which preceded the present war, our imports of unmanufactured wool exceeded our exports of the home-grown article in the value of $44,514,771, or upwards of two millions a year; and that during the same period, our imports of manufactured wool exceeded our exports of domestic manufactured wool in the value of $429,422,951, or upwards of nineteen millions a year!

There have been during the above period several "manias," as they have been termed, as strong as that of 1862–'63, to increase wool production in our country; yet, in spite of all contemporary predictions to the contrary, we see how utterly they failed in every instance to bring up, even temporarily, the supply to the demand. When every circumstance is taken into account, there cannot be a reasonable doubt entertained, that the United States can permanently furnish its own markets with a full supply of wool more cheaply than other countries can furnish it. I have not space here for the numerous facts and statistics which go to prove this assertion; nor is there need of it, they have been so fully set forth and discussed in a multitude of popular publications, particularly in those invaluable disseminators of information, our Agricultural Journals. Indeed, we might even compete with other countries in supplying wool to Europe. And yet, with such facts staring us in the face, there are so many other demands for capital, labor and enterprise in our country, that we continue and are likely to continue, no one can say how long, vast importers of one of the prime necessaries of life!

Sheep are not only the most profitable animals to depasture the cheap lands of our country—the mountain ranges of the South, and the vast plains of the West and South-west—but they are also justly beginning to be considered an absolute necessity of good farming on our choice grain-growing soils, where wheat, clover seed, etc., are staples.

I may be permitted to quote the two following paragraphs from my Report on Fine-Wool Husbandry, 1862:—"Sheep would be more profitable than cows on a multitude of the high, thin-soiled dairy farms of New York; and every person who has kept the two animals ought to know that sheep will enrich such lands far more rapidly than cows. On the imperfectly cleared and briery lands of our grazing regions, sheep will more than pay for their summer keep, for several years, merely in clearing and cleaning up the land. They effectually exterminate the blackberry (*Rubus villosus et trivialis*,) and raspberry (*Rubus strigosus et occidentalis*,) the common pests in such situations, and they banish or prevent the spread of many other troublesome shrubs and weeds. They also, unlike any other of our valuable domestic animals, exert a direct and observable influence in banishing coarse, wild, poor grasses from their pastures and bringing in the sweeter and more nutritious ones." It was a proverb of the Spaniards:—" Wherever the foot of the sheep touches, the land is turned into gold."

"And the growth of wool is peculiarly adapted to the pecuniary means and the circumstances of a portion of our rural population. Their capital is mostly in land. Hired labor is costly. Sheep husbandry will render all their cleared land profitably productive at a less annual expenditure for labor than any other branch of farming. By reason of the rapid increase of sheep, and the great facility of promptly improving inferior ones, they will stock a farm well, more expeditiously, and with far less outlay, than other animals. And, lastly, the ordinary processes and manipulations of sheep husbandry are simple and readily acquired. On no other domestic animal is the hazard of loss by death so small. It is as healthy and hardy as other animals, and unlike all the others, if decently managed, a good sheep can never die in the debt of man. If it dies at birth, it has consumed nothing. If it dies the first winter, its wool will pay for its consumption up to that period. If it lives to be sheared once, it brings its owner into debt to it, and if the ordinary and natural course of wool production and breeding goes on, that indebtedness will increase uniformly and with accelerating rapidity until the day of its death. If the horse or the steer die at three or four years old, or the cow before breeding, the loss is almost a total one."

The cost of producing wool depends upon that of keeping sheep, and this necessarily varies greatly in different situations. On the highest priced lands in New York and New England on which sheep are now usually kept for wool growing purposes, it, under judicious systems of winter

5

management, reaches about $2 a head per annum. In extensive regions of the South and South-west it is mainly comprised in the expense of herding, salting, and shearing, and where the number of sheep kept is large, does not exceed 25 cents a head. But it would be more profitable in those regions to provide some kind of shelter and give a little feed in the height of winter, and this would increase the cost of keeping to 50 cents a head. In some of our Western and North-western States, where sheep can have the run of lands belonging to the Government or to non-resident owners, in addition to those owned by the flock-master, the cost of keeping, including winter shelter, ranges from, say, 75 cents to $1 a head. In intermediate situations, between the densely populated and high-priced lands of the East and the broad, sparsely inhabited prairies of the West and Southwest, (open without price to the temporary occupant,) and between the warm South where vegetation flourishes almost throughout the year, and the cold North where winter feeding lasts from five to five and a half months, the cost of keeping will occupy every intermediate place between these extremes. Every experienced and sensible man acquainted with all the special circumstances, is the best judge of that cost in his own locality.

Improved Merino flocks of breeding ewes should average five pounds of washed wool per head in large flocks. Medium wool has sold on an average for 42 8-10 cents per pound for the thirty-five years preceding the high prices of the present war. This gives $2.14 to the fleece, which should pay for the cost of keeping, anywhere, and leave the owner the lambs and manure for his profit.* The increase of lambs will average about eighty per centum on the whole number of the breeding ewes.† The value of the manure would greatly vary in different situations. It may interest many to know how it is estimated in England. Mr. Spooner says:

"Four hundred South Down sheep are sufficient to fold twenty perches per day, or forty-five acres per year, the

---

* If he keeps wethers, he has for his profit their growth and about a dollar from each fleece. Wethers' fleeces should be worth about a dollar a piece more than ewes' fleeces.

† I gave this as the average fifteen years ago. With the improvement in sheep shelters, etc., it ought now to be higher. But a few usually fail to get with lamb, and occasionally there comes a "dying year" for lambs — when they are born feeble, goitred, rheumatic, or subject to some other maladies, so that they perish in extraordinary numbers. This was quite generally the case in New York in the spring of 1862. Taking a term of years together, I doubt whether, under average management, the increase by lambs yet exceeds 80 per cent.

value of which is therefore about £90 per year, or 4s. 6d. per sheep. * * Three hundred sheep have in this manner (with 'a standing fold on some dry and convenient spot, well littered with straw or stubble,') produced eighty large cart-loads of dung between October and March, and in this manner, after the expenses have been deducted, each sheep has earned 3d. per week."

A hundred Merino sheep, given abundance of bedding, will, between December 1st and May 1st, make at least forty two-horse loads of manure—and if fed roots, considerably more. I scarcely need to say that both the summer and winter manure of the sheep is far more valuable than that of the horse or cow.* Its manure on high-priced land which requires fertilizers, cannot be estimated at less than 50 cents per head per annum, and I should be inclined to put it still higher.

The value of the lambs and manure is the minimum of profit. That profit increases just as the market value of land and the cost of keeping decreases. On the rich plains of the West and South-west, manure is not yet reckoned among the appreciable profits, and the cost of transporting wool to market is from one to two cents per pound. The Western grower, then, gets the lamb and about half the fleece, as the profit on each sheep. The Texan grower gets the lamb and about three-quarters of the fleece, and so on. I do not deduct the extra prices paid from time to time for rams, because each good one vastly more than pays for himself in increasing the value of the flock.

The prices of lambs of different blood and in different places, vary too much to admit of even an approximately uniform rate of estimating them. But it does not anywhere cost more to raise a full-blood than a grade Merino lamb.

---

* Horses are not used as depasturing animals in any of the older States. The following remarks appeared in my Report on Fine-Wool Husbandry, 1862:—"If milch cows are not returned to their pastures at night in summer, or the manure made in the night is not returned to the pastures, the difference in the two animals in the particular named in the text, is still greater. Even grazing cattle kept constantly in the pastures, and whose manure is much better than that of dairy cows, are still greatly inferior to the sheep in enriching land. The manure of sheep is stronger, better distributed, and distributed in a way that admits of little loss. The small round pellets soon work down among the roots of the grass, and are in a great measure protected from sun and wind. Each pellet has a coat of mucus which still further protects it. On taking one of these out of the grass, it will be found the moisture is gradually dissolving it on the lower side, directly among the roots, while the upper coated surface remains entire. Finally, if there are hill tops, dry knolls, or elevations of any kind in the pasture, the sheep almost invariably lie on them nights, thus depositing an extra portion of manure on the least fertile part of the land, and where the wash of it will be less wasted. The manure of the milch cow, apart from its intrinsic inferiority, is deposited in masses which give up their best contents to the atmosphere before they are dry enough to be beaten to pieces and distributed over the soil."

Good grades have averaged about $2 per head in the fall for a number of years and the increasing demand for them by the butchers is steadily raising the price. Estimating 80 per cent. of lambs and 50 cents a head for manure, each sheep would thus average in both products $2.10—just about the equivalent of the fleece; so that it would be equally well, on high-priced lands requiring fertilizers, to say that the lambs and manure pay the cost of keeping, and the fleece is to be reckoned as the profit. According to the first computation, lands worth $50 per acre would give their owner a profit of seven per cent. if they would support a little over one and three-fifths sheep to the acre; and that would be indifferent grazing land, where the domesticated grasses are grown, and under proper systems of winter keeping, which would not support three sheep to the acre. It would be a very moderate estimate, taking a term of years together, to put full blood American Merino lambs — even from flocks of no especial reputation and not kept for what is technically designated "breeding purposes"— at double the price of grade lambs. They are now worth at least three times as much.

The prospect of the future demand for mutton has been sufficiently considered. I had hoped to be able to present an exhibit, in details, of the cost and profits of its production based on actual experiments. But I have been disappointed; and I will only reiterate the statement that the experience of England, and of portions of our own country, has clearly demonstrated that in regions appropriate for its production, it is a more profitable leading object of production than wool.

CHAPTER XI.

PRINCIPLES AND PRACTICE OF BREEDING.

BREEDING, in its technical sense, as applied to the reproduction of domesticated animals under the direction of man, is the art of selecting such males and females to procreate together as are best adapted, in conjunction, to produce an improved and uniform offspring. The first and most important fact to be kept in view, in pursuing the object of breeding, is that result of a fixed natural law which is expressed in the phrase, "like produces like." The painted oriole now flashing among the apple blossoms before my window wears the same bright dyes that were worn by the oriole ages ago. But the breeding maxim just quoted, is understood to assert more than that species and varieties continue to reproduce themselves: it implies that the special individual characteristics of parents are also transmitted to progeny. This is the prevailing rule, but it has a broad margin of exceptions and variations. Animals are oftentimes more or less unlike their parents, yet inherit a very distinct resemblance to remoter ancestors — sometimes to those several generations back. This is termed "breeding back." And, moreover, where the resemblance is to the immediate progenitors, the mode of its transmission is not uniform. Sometimes the progeny is strongly like one parent and sometimes like the other; sometimes, and perhaps oftenest, it bears a modified resemblance to both.

The physiological causes or laws which control the hereditary transmission of physical forms and properties — which determine the precise structure which the embryo shall assume in the womb, and give to each animal a distinct individuality which will accompany it through life and distinguish it from every other animal of the same breed and family — have not yet been, and probably never will be, fully understood. Nor can we, by the closest study of analogies or precedents, learn to anticipate their action with absolute certainty. Yet, by a proper course of breeding, we can control that action to a considerable degree; we can generally keep it in channels which are favorable to our wishes; we can avoid manifold evils which arise from promiscuous procreation; and a few, more gifted or more zealous in the attainment of their objects than the rest of us, can make permanent improvements in the forms and properties of our domestic animals, and thus confer important benefits on society.

If the male and female parent possess the same given peculiarity of structure, or in breeders' phrase, the same good or bad "point," the chances are very strong that the progeny will also possess it, because the progeny is most likely to inherit the structure of its immediate progenitors; and whether it receives that portion of the structure from one or the other of them, or partly from both, it still receives the same peculiar form. If all the remoter ancestors also possessed the same point, then the progeny must, in the ordinary course of nature, be sure to inherit it, for let it breed back to whatever ancestor it may, it must inherit the same conformation. This law applies to properties as well as forms. Hence it is that in breeding between pure blood animals of the same breed and family, we find like producing like, so far as the family likeness is concerned, in steady and endless order, and this necessarily includes a good deal of individual likeness. Indeed, it is this long continued preservation and transmission to descendants of the same properties by one family that constitutes "blood," in its technical sense — and its "purity" is its utter isolation from the blood of all other families. The full blood, or pure blood, or thorough-bred animal — for all these terms imply the same thing* — can inherit from its parents, or take from its remoter ancestors by breeding back, only the same family characteristics.

But in breeding between mongrels — animals produced by the crossing of different breeds — the closest resemblance of the parents in any point not common to both breeds, does not insure the transmission of their characteristics in that point to their offspring; for the offspring may obtain different ones by breeding back to either of the ancestors with which the cross commenced, or to some intermediate and partially

---

* At least, as they are used in this volume. An effort has been made in some quarters to introduce a distinction between these significations, but, in my judgment, without any authority.

assimilated ancestors. This occasional breeding back and consequent divergence from the existing type, is liable to continue for a great number of generations; and it can only be repressed by a long and uniform course of breeding, and by a rigorous "weeding out"—that is, exclusion from breeding—of every animal exhibiting a tendency toward such divergence.

We cannot always, among either pure bloods or mongrels, breed from perfect or approximately perfect individuals, or those which are alike in their structure and properties. Necessity sometimes, and economy frequently, requires us to make use of materials which we would not voluntarily select for the purpose. In such cases, it should always be the aim of the breeder to counteract the imperfection of one parent by the marked excellence of the other parent in the same point. If, for example, a portion of the ewes of a flock are too short-wooled, they should, other things being equal, be coupled with a particularly long-wooled ram.

The hereditary predispositions of breeding animals are also to be regarded, as well as their actual existing characteristics. In the case just given, if the long-wooled ram was descended from uniformly short-wooled ancestors, his length of wool would be what is termed an "accidental" trait or property; and there would be little probability of his transmitting it with uniformity and force to his offspring out of short-wooled ewes. There would be no certainty of his doing so, even among long-wooled ewes.

What are considered accidental characteristics are themselves generally the result of breeding back to a forgotten ancestor, but sometimes they are purely spontaneous. In such cases, they are exceptions, not to be accounted for by any of the known laws of reproduction. As a general thing they are not transmitted to posterity. In other cases they are feebly transmitted to the first generation and then disappear. But occasionally they are very vigorously reproduced, and if cultivated by inter-breeding, the related animals possessing them soon become fixed in their descendants apparently as firmly as the old and long-established peculiarities of breed.* The following is an instance of this,

---

* It is claimed that artificial peculiarities even—those produced by external causes after birth—are sometimes inherited, as for example, a limb distorted by accident. To this extent, I suspect the genuine cases of inheritance, are very rare. But *habitual* artificial properties, and to some extent, structures, marks etc., not unfrequently become hereditary. If, for example, men or brutes are kept healthy and vigorous for several generations, by proper food and exercise, they will have more vigorous offspring than the descendants of the same ancestors improperly fed and

which, so far as the facts occurred in the United States, fell under my own observation. A ram having ears of not more than a quarter the usual size appeared in a flock of Saxon sheep, in Germany. He was a superior animal, and got valuable stock. These were inter-bred and a "little-cared" sub-family created.* Some of these found their way into the United States, between 1824 and 1828. One of the rams came into Onondaga County, New York. He was a choice animal, and his owner, David Ely, valued his small ears as a distinctive mark of his blood. He bred a flock by him, and gradually almost bred off their ears entirely. His flock enjoyed great celebrity and popularity in its day, but has long been broken up, and many years have doubtless elapsed since any of the surrounding sheep owners have used a "little-eared" ram. Yet nearly every flock that retains a drop of that blood — even coarse mutton sheep bred away from it, probably for ten or fifteen generations, insomuch that all Saxon characteristics have totally disappeared — still continue to throw out an occasional lamb as distinctly marked with the precise peculiarity under consideration, as Mr. Ely's original stock.

Another much more important *alledged* case in point, is that of the Mauchamp family of Merinos in France. The published accounts of them declare that, in 1828, "a Merino ewe produced a peculiar ram lamb having a different shape from the usual Merino, and possessing a long, straight, silky character of wool," "similar to mohair," and "remarkable for its qualities as a combing wool." Mons. J. L. Graux, the owner of this lamb, bred from him others which resembled him. "In each subsequent year," the account continues, "the lambs were of two kinds, one possessing the curled, elastic wool of the old Merinos, only a little longer and finer; the other like the new breed. At last the skillful breeder obtained a flock combining the fine, silky fleece, with a smaller head, broader flanks and more capacious chest." This, excepting in the matter of being "finer" than the Merino, (and I am unable to say what Mons. Graux considers fine,) is a pretty good description of a mongrel between a Merino and some long-wooled variety,— and such I have no

---

enervated by idleness. And as vigor depends upon the volume of the muscle and upon the conformation of both the muscles and general frame, it follows that the shape is measurably controlled by the properties, and that artificial shapes become hereditary.

* This was the explanation given me of the origin of these sheep by my lamented friend, the late Henry D. Grove.

doubt it is. The "accidental" traits which are developed in breeding from pure animals of the same blood never, I suspect, at one bound, embrace quite such comprehensive particulars as a change, not only in the essential characteristics of the wool, but also in the general form of the carcass.*

But trustworthy cases of the vigorous transmission of accidental properties, involving visible changes, are sufficiently numerous. Involving slight changes or variations, not recognized as such by casual observers, they are more numerous. It is by noting these last, and cultivating the good ones, that the judicious breeder makes some of his best improvements. How otherwise can he possibly raise the progeny, in any given point, above the plane of its parents, and of *all* its ancestors? But while the breeder should avail himself of every opportunity of this kind to attempt to perpetuate accidental improvements on the pre-existing type, he must be prepared to meet with more disappointments than successes. My Merino ram "Premium"—mentioned particularly in "Sheep Husbandry in the South," and in some other publications, for his extraordinary individual qualities†—perhaps the finest wooled sheep then on record for one of equal weight of fleece, and ranking in the former particular with the choicest Saxons—did not get progeny peculiar for fineness. His own ancestors had been fine for the breed, but not remarkable in that particular. One of the showiest Merino rams now in New England does not inherit his showy traits, and he utterly fails to transmit them to his progeny. Exceptional good qualities are not, according to my observation, as likely to become hereditary, as indifferent or bad ones.

Accidental characteristics are less likely to be perpetuated where they are opposed to the special characteristics of the breed. For example, the Merino wool has had a peculiar curled or spiral form of the fiber, for ages—a fixed, marked trait, never wanting, and as much a characteristic of the wool as its fineness. Mons. Graux's first straight-wooled "Mauchamp Merino" ram, if an accidental instead of a mongrel animal, brought only his own individual power to transmit that peculiarity to his progeny (out of full blood Merino ewes)

---

* It will be seen that I have not introduced the case of these sheep with any view of illustrating the transmission of actual "accidental" qualities—but to caution my readers against what I have not a shadow of doubt is either an amusing case of credulity or a gross attempt at imposition.

† Sheep Husbandry in the South, p. 135. American Quarterly Journal of Agriculture, 1845; Ib, 1846, p. 290. Report on Fine-Wool Husbandry, 1862, pp. 65, 97.

against a hereditary power which had been acquiring force for ages.* His success therefore was the more marvelous. But in merely giving a smaller head, etc., to his progeny, he did not necessarily run counter to any special and fixed peculiarity of breed.† The heads of Merino sheep vary in size. Some of them are small. A malformation consisting of small ears, or of the want of any ears, or of one or more imperfect legs, or of having six legs, or any other *deformity*, does not impinge the special characteristics of a breed, or of one breed more than another. In all breeds alike, whether pure or impure, there is a tendency in nature to preserve and restore the normal form in the progeny; but occasionally, as in the case of Mr. Ely's sheep, that tendency is not strong enough to resist the tendency of like to produce like.

In all instances, pains should be taken to avoid breeding between males and females possessing the same defect, and particularly the same hereditary defect. In the first case, the individual force of hereditary transmission in both parents unites to reproduce the defect: in the second, both the individual and family hereditary force unite to reproduce it, and to escape from their combined effects would, of itself, be one of the strongest cases of "accidental" breeding.

When the same individual or family defects are thus transmitted by both parents to their offspring, the latter are apt to inherit them to a greater degree or extent than they are possessed by either parent. Such an increase or aggravation may be regarded as inevitable where the common defect is of the nature of an organic disease. If two human parents are affected by scrofula, and especially by hereditary scrofula, in a slight degree, their progeny may be expected to exhibit it in a much more malignant and destructive form. And the same law, in transmitting diseases, or morbific conditions, pertains equally to brutes. Relationship between parents also exerts a strong influence in such cases, but this will be more appropriately considered in the next Chapter.

The relative influence of the sire and dam in transmitting their own individual forms and other properties to the progeny, has been the theme of much observation and discussion. The prevalent opinion formerly was that each

---

* But if he was a mongrel, he brought the hereditary influence of straight-wooled and probably pure blood ancestors to bear against that of his Merino ancestors, and by breeding in-and-in, and by selection, he was made to give the preponderance to the former in the particular under consideration.

† I have no definite or reliable information in regard to the *form* of head in the Mauchamp Merino.

parent transmitted a portion of all the properties, or a trait here and a trait there, as chance or some special and independent power in each animal to "mark" its offspring, might dictate. An English gentleman by the name of Orton, broached the theory that the animal organization is transmitted by halves, the sire giving to the progeny the external organs and locomotive powers, and the dam the internal organs and vital functions. By this division, the general form, the bones, the external muscles, the legs, skin and wool would be like those of the male parent, while the heart, lungs and other viscera, and consequently those functions on which the integrity of the constitution mainly rests, would be like those of the female parent. But each parent was supposed by him to exert a degree of influence on the parts and functions chiefly inherited from the other parent; and this law " of limitations " he considered "scarcely less important to be understood than the fundamental law itself."

Mr. Walker, in his work on Intermarriage, presents the same theory, substantially, except that he denies that the series of organs inherited from one parent are modified or influenced by the other parent; and he assumes that between parents of the *same breed*, "either the male or the female parent may give either series of organs."*

Mr. Spooner, in an article on Cross-Breeding, which appeared in the Journal of the Royal Agricultural Society of England some years since the publication of his well known work on Sheep, adopts the Ortonian theory with some slight modifications. He says:—"The most probable supposition is that propagation is done by halves, each parent giving to the offspring the shape of one-half of the body. Thus the back, loins, hind quarters, general shape, skin and size follow one parent; and the fore quarters, head, vital and nervous system, the other; and we may go so far as to add, that the former, in the great majority of cases, go with the male parent and the latter with the female."†

The Ortonian theory, or either of the above modifications of it, if actually carried into practice, would lead to singular results. According to Mr. Orton, the effects of cross-breeding would, comparatively speaking, stop with the first cross, for each succeeding generation of cross-bred males and females would continue to transmit to their descendants substantially

---

* Vide pp. 142, 145.

† Journal of Royal Agricultural Society of England, 1859.

the same halves, in the same order, both with respect to form and general properties.*

According to Mr. Walker the effects of crossing, among animals of different breeds, would generally absolutely stop and become unchangeable with the first cross, for every generation of descendants would receive the same half of the organization without any modification! And on the other hand, between animals of the same breed, the descendants might either permanently exhibit the same relative paternal and maternal halves, or they might by in-and-in breeding, in the second generation, become exactly like their sire in both halves!†

The theory of propagation by halves appears to have considerable support from facts when it is applied to hybrids—animals derived from inter-breeding distinct species,—as for instance the male ass with the mare, the horse with the female ass, the goat with the sheep, etc. But as applied to sheep, every observing breeder ought to know that it is essentially unfounded and chimerical. The Merino ram crossed with a ewe of some thin and coarse-wooled family, does not, either fully or approximately, transmit the weight, fineness or other

---

* If this were so, half bloods, when bred together, would reproduce their own essential qualities about as uniformly as full bloods when bred together; and the attempt to form them into permanent families, occupying the same relative place they do between the original breeds of which they are composed, should result in as splendid success as it does, in point of fact, in complete and uniform failure. And by this theory, it would seem the half blood ram ought always to be used to perpetuate half bloods—yet experience shows that half blood rams are worthless for that object. I never have seen anything more than extracts from Mr. Orton's paper on this subject. I do not therefore know what exceptions he made for breeding back. He must of course have regarded it as only the exception, or else he could not have assumed any set of facts opposed to it to be the rule. Then, in his view, a majority at least of the descendants of half bloods, bred to half bloods, or to mongrels of their own degree, would continue uniformly to produce their own essential characteristics,—which every observing breeder knows they do *not* do.

† Mr. Walker says:—"Let the example be that in which, of the animals subjected to in-and-in breeding, the father breeds with the daughter, and again with the grand-daughter. Now, it is certain the father gives half his organization to the daughter, (suppose the anterior series of organs,) and so far they are identical; but, in breeding with the daughter, he may give the other half of his organization to the grand-daughter, (namely, the posterior series of organs,) and as the grand-daughter will then have both his series of organs—the former from the mother and the latter from himself—it is evident that there exists between the male and his grand-daughter a quasi identity. [p. 210.]

Mr. Spooner does not develop his views very fully, but so far as he states them, he would appear to adopt Mr. Walker's theory of a strict propagation by halves, and at the same time to assume, by implication, that either parent may give either series of organs, in all cases, as Mr. Walker only assumes they may among animals of the *same breed*. If these are Mr. Spooner's real opinions, he must be prepared to believe that results like the following may ensue:—If a Merino ram was put to a Leicester ewe he would transmit half of his organization to their common progeny. If the same ram was put to his own half-blood daughter of that cross, he might give the other half of his organization to the progeny, so that it would be, *de facto*, a pure Merino. This would be a very summary process of creating pure Merinos out of Leicesters! If the same rule held good in regard to horses, an Arabian stallion might in two generations produce pure Arabian stock from cart mares! Is Mr. Spooner prepared to adopt such a *sequitur* to his theory?

qualities of his fleece to his progeny. He, it is true, transmits a fleece which is much heavier and finer than that of the ewe; and if again crossed with the half-blood, he transmits additional weight and fineness. Each ascending grade toward the Merino will continue more and more to resemble the Merino in these particulars. But the process is gradual, not immediate; the properties are transmitted *by degrees*, not *by halves*.

The Ortonian theory, as applied to the transmission of form, in sheep, has a little more apparent foundation. The ram does, much oftener than the ewe, transmit his general external structure to the progeny. But the hypothesis that he does so as invariably as Mr. Orton contends, or as Mr. Walker contends in the case of crosses between different breeds, or even as generally as Mr. Spooner supposes,* will fall to the ground at once when examined in the light of actual facts. In any and every flock of lambs, whether pure blood or crossed, there will be found entirely too many to be classed as mere exceptions, which, without breeding back of their immediate parents, do take the general form of the dam, and not that of the sire. And it will also be found that the instances which, even by the most liberal resort to imagination, can be adduced as proofs of the theory of a strict transmission by halves, and of such a division of those halves as the advocates of the theory have agreed on, do not comprise a majority of cases. In my judgment, they do not include a fourth of them; and could scarcely be shown conclusively to include any. As a general thing we see distinct resemblances to each parent, or modified resemblances to both parents, existing in different proportions in the form, the fleece and the skin. One lamb has a carcass mostly like that of its sire and a fleece mostly like that of its dam.† Another takes a middle place between its parents in one or both particulars. Another actually, to some degree, divides the form, taking, for example, the shoulders of the dam with the hind quarters of the sire, or *vice versa*. I have a specific case in view of a ram ("21 per cent.,") which has a shoulder obviously defective in being too thin. He transmits most of his form, his fleece, etc., to his progeny, with marked force. But not one in thirty of them exhibits a thin shoulder. By

---

* I mean making all due allowance for breeding back, or for an exceptional want of relative vigor in the male, &c., &c.

† I think it is *not* common to see *these* two characteristics quite so broadly divided; and probably never, when the pure blood ram is coupled with the cross-bred ewe. But with both those pure and cross-breeds which most resemble their sires in form, it is common to see the fleece at least *equally* partaking of the characteristics of the dam.

the half-and-half theory, all this would be impossible. According to that theory, all these characteristics belong to the *same* half of the organization, which is always transmitted as an entirety by one parent or the other.

But it is easier to defend the half-and-half theory, so far as it pertains to the viscera and internal organization, because it is very difficult to follow it there! I do not see how a really reliable decision can be arrived at except by a practical ocular examination of the parts, and it is not easy to understand how even the dissecting knife would let in much light on the subject. In healthy animals, it is not probable that any particular and persistent differences could be discovered in the viscera, except in the mere particular of size, and in this, the theory would not be likely to derive any support from a comparison of facts.* If it be contended that internal structure is to be judged or inferred by certain effects — such as constitution, strength, appetite, etc., I undertake to say, from abundant experience, that the progeny as often and as fully inherit these qualities from the sire as from the dam, even when they most distinctly inherit the general form of the sire.

I have pursued this subject at greater length, because I have observed that too many men who have the word "practical" ever on their lips (who seem to consider themselves *practical* on all agricultural subjects, because they *work* practically with their own hands on a farm!) are always ready to adopt the most baseless theories: and I consider the Ortonian theory as mischievous as it is baseless.

I have said that the ram much the oftenest gives the leading characteristics of the form; and I will now add, that he much the oftenest gives the size, and several of the leading properties of the fleece, particularly its length, density, and yolkiness. Its fineness and general style are probably usually, other things being equal, as much controlled by the dam as by the sire. But I do not believe the superior power of the ram to transmit his own qualities is purely an incident of sex. I believe co-operating causes are equally potential, and that the chief of these are superiority of blood, and superiority of individual vigor.

---

* I suppose that if a large ram were put to a small ewe, and as usual gave his size (comparatively) to the progeny, the *size* of the viscera would necessarily follow the size of the sires', because the viscera always correspond with the size of the external structures and of the cavity to be filled. If, on the other hand, the ewe gave the size of carcass, she would also give the size of the viscera. This is exactly at variance with the Ortonian theory, if the size of the intestines is one of those properties said to be given by *that* parent which does *not* give the size and form.

The ram is generally "higher bred" than the ewes, even in full blood flocks. As pure blood is only separate family blood which has been kept distinct until it transmits but one set of family characteristics, so higher blood is produced by the selection of pure blood animals of choicer qualities and breeding them together separate and distinct from all others, until they form a smaller improved sub-family, alike possessing a permanent hereditary character. The thin-chined, low fore-ended, roach-backed, black-faced sheep which formerly depastured the downs of Sussex, were of as pure blood as the superb South Downs which Mr. Ellman created out of them — but they were not so highly or well bred. The improved South Down ram of to-day does not transmit the same properties to his progeny which the unimproved animal of eighty years ago did. He not only transmits better ones, but he transmits them with more force and uniformity. This last is occasioned by two circumstances. The restriction of the sub-family for a number of generations to one fixed standard, gives greater force of hereditary transmission to the fewer properties — that is, fewer in kind — which that standard admits of, because by that law on which "blood" or "species" rests, the oftener *the same* quality is reproduced, the stronger becomes its tendency to continued reproduction. The improved South Down breeds, so to speak, to one uniform pattern. The unimproved one breeds to a dozen different varieties of a family pattern. The second circumstance which gives a stronger power of strict hereditary transmission to the high-bred animal, consists (after the improved family becomes thoroughly established) in the restriction placed on the limits of breeding back. The unimproved South Down could breed back to fifty different ancestors, all differing quite widely; the improved one, unless he casually goes far back of the ordinary limits of breeding back, can only breed back to ancestors of very close resemblance.

If the pure blood ram is put to grade ewes of different and no determinate blood, his strong power of hereditary transmission is encountered by no corresponding power on the other side, and the resemblance of the progeny to himself is unexpectedly striking, considering that they are but half of the same breed. If put to full blood ewes of his own breed, but lower bred than himself, the resemblance to himself is much less marked, though it is still very perceptible. If put to ewes of the same breed and as high bred as himself, the resemblance to himself is still fainter and considerably less uniform. In these last, he has encountered a force of hereditary transmission equal to his own, except in so far as he is aided by superior power of sex.

Persons who buy rams, generally buy from flocks better bred than their own, and hence is witnessed that assimilation of the progeny to the sire, and consequently that improvement, which is by some referred exclusively to sex, and by others to some inherent property to "mark" his offspring supposed to be peculiar to the sire. This hypothesis is not overthrown by the notorious fact that rams from the same flock exhibit the power of hereditary transmission in essentially different degrees, any more than is the hypothesis of the superior influence of the male sex overthrown by the same fact. Every flock has separate and better strains of blood within itself — even where all are descended from the same stock. Not only better males occasionally present themselves, but also better females. If the latter are found to transmit their own properties in a special degree to their offspring, they are highly prized and carefully reserved from all sales. Each female descendant is prized and reserved in the same way, and a sub-family is thus created. A touch of in- and-in breeding (by using a ram from the same sub-family on his relatives, as well as on the rest of the flock,) frequently aids to confer an identity on this little group of sheep which preserves itself for generations — as long as the flock is kept together. I am not acquainted with a celebrated breeding flock which has not within it several such recognized groups or sub-families of different value, but all better than the body of the flock. This explains how rams of the same blood and flock, and perhaps general appearance, may differ materially in their qualities as sires, without imagining the existence of an independent faculty based on no physical properties.

There is still another circumstance which affects the power of hereditary transmission, viz., vigor, — general physical vigor, and also special sexual vigor. A very strong, powerfully developed ram, full of power and vital energy — and full of untiring sexual ardor — will get stronger and better lambs and impress his own qualities on them more strongly than an ill, or feeble, or flaccid ram, with naturally weak or exhausted sexual powers. The ram should be essentially masculine in every organ and function.* He

---

* Large testicles, and large, firm spermatic cords connecting these with the body, are regarded as indications of sexual vigor in the ram. The capacity to "bear heavy feed" has also much to do with a ram's endurance in this particular.

should not even have what is termed a "ewe's fleece," but a longer, thicker and coarser one.*

The Merino ram produces strong, healthy lambs from the age of seven or eight months to that of eight or ten years, and sometimes later, if he has never been over-worked. He does not attain his full maturity of vigor until he is three, and he usually begins to decline at seven or eight. A ram lamb ought not, for his own good, to be used to over ten or fifteen ewes — merely enough to test his qualities as a sire; and to fit him properly for even this amount of work, he should be large, strong, and fleshy. A yearling can, without injury, do one-third and a two-year-old two-thirds the work of a mature ram. Strong, mature rams will, on the average, properly serve about two hundred ewes a year. I speak in all the above cases of but a single service to each ewe, and of a coupling season extending from forty to forty-five days. Rams have often exceeded these numbers. An Infantado ram lamb owned by Loyal C. Wright, of Cornwall, Vermont, got one hundred and three lambs in the fall of 1862. The "Wooster Ram," so celebrated throughout Vermont, served three hundred ewes when a year old.† Some strong rams, in their prime, have served four hundred. The "Old Robinson Ram" is believed to have got nearly three thousand lambs during his life of thirteen or fourteen years. The Merino ewe breeds from her second to her tenth or twelfth year, and sometimes considerably longer, if carefully nursed after she begins to decline.‡ It is better for her, however, not to breed until her third year. Some, however, who have valuable ewes,

---

* A ram of the same blood and breeding does not require to be as fine as a ewe, to get female progeny equal to her in fineness; and an over-fine ram generally gets too light-fleeced progeny. His own fineness, unless an exceptional quality, shows that he has been bred too far in the direction of fineness, and, consequently, away from the proper standard of weight, for the maximum of these two qualities in the same fleece is not even approximately attainable. If the over-fine ram has himself a fleece of good weight, it is to be apprehended—in the absence of a full knowledge of antecedents—that the latter quality is exceptional, and that he may breed too much in the opposite direction.

† So I am informed by Mr. Abel J. Wooster, of West Cornwall, Vermont. He purchased the ram of Mr. Hammond when a lamb—and hence the name of "Wooster Ram," or rather, according to a prevailing Americanism, "Wooster *Buck*." Some Merino breeders who find this name in the pedigrees of their sheep may be interested to learn the following particulars communicated to me by Mr. Wooster. The ram never exceeded about 100 lbs. weight with his fleece off. His first fleece weighed 12½ lbs., his second 19½ lbs., and "after that he began to run down," and died before the completion of his fourth year. "He would bear heavy feed, and that and hard service shortened his life."

‡ I stated in my Report on Fine-Wool Husbanry, 1862, that I had been informed that the dam of the "Old Robinson Ram" produced a lamb in her twenty-second year. I have since ascertained that I was misinformed on the subject.

put them to breeding at two, but take off their lambs and give them to foster-mothers. If the young ewe is carefully dried off her milk, she will experience no injury and no loss of growth. The increase of growth during pregnancy will make up for the slight falling off after yeaning. The English breeds both mature and decline considerably earlier in life.

A theory of considerable importance to the breeder, if true, has recently been started, viz., that the male which first impregnates a female, continues to exert an influence on some of the qualities of her subsequent offspring, or at least is liable to do so. I have not, in my own experience, observed any proofs of this.*

It has been a prevailing opinion among American breeders that it is much better to breed between a small male and large female, than in the contrary direction. The reason assigned by Mr. Cline, of England, who first, I think, publicly advanced this view, was that the fœtus begotten by the larger male has not room to expand and develop itself properly in the womb of the small female; that it does not obtain sufficient nutrition from stores intended for a smaller fœtus; and that, in consequence of these things, it can not obtain its normal size and proportions anterior to birth: secondly, that it is liable on account of its extra size to cause difficulty, if not danger to its dam in yeaning; and finally, that the opposite course, by giving the fœtus unusual room and extra nutriment, tends to its most perfect development. This is probably true as between different breeds, where the disparity in size is extreme, as, for instance, between the Saxon Merino ewe and the Cotswold ram. I would not expect a greatly overgrown ram to get as good stock as a more moderate sized one, even on ewes of the same breed, but it would be quite as much for another reason as for any of the preceding ones, viz., that these overgrown animals never possess the highest attainable amount of vigor and general excellence themselves, and are not therefore fitted for sires, irrespective of relative size. But the rule should not be extended to the exclusion of large rams of the breed, if good in other particulars. Nature adapts herself unexpectedly to circumstances, in the face of all theories. Constant and recent experiments, in England,

---

* Those who wish to see the facts and arguments which are set forth to support this theory will find them in Mr. S. L. Goodale's interesting work on the Principles of Breeding, published in 1861.

in crossing ewes with the rams of much larger breeds (to obtain large lambs for the butcher) demonstrate, as has been already seen, that the prevailing fears on this subject have been somewhat exaggerated.*

---

* The Down or New Leicester ram is coupled with almost any of the smaller sized local varieties for the purpose of getting larger and earlier maturing lambs for the market. The very small and hornless heads of the Down and New Leicester lambs, it is true, peculiarly fit them for easy and safe parturition; but in other respects, they are exposed to all the disadvantages of disproportioned size before and after birth, and these are not found sufficient, in practice, to prevent the crosses from proving highly profitable for the objects in view.

CHAPTER XII.

BREEDING IN-AND-IN.

Breeding in-and-in is ordinarily understood, in our country, to mean breeding between relatives, without reference to the degree of consanguinity; and I shall therefore use it in that sense in this work, specifying, when there is occasion, whether the degree of consanguinity is *close* or *remote*. But this is not the sense in which it has been used by those eminent European writers who have done so much to plant an inveterate prejudice against its very name in the public mind. Sir John Sebright ranks among the highest of these, and he did not consider procreation between father and daughter, and mother and son, to be breeding in-and-in! Breeding between brother and sister he thought might "be called a little close," but "should they both be very good, and particularly should the same defects not predominate in both, but the perfections of the one promise to correct in the produce the imperfections of the other, he did not think it objectionable!" And again, he says breeding in-and in "may be beneficial, if not carried too far, particularly in fixing any variety which may be thought valuable." It is to be regretted that Sir John does not define what he considers to be in-and-in breeding. I apprehend that he means by it breeding the father with the daughter and again with the grand-daughter, or the mother with the son and again with the grand-son. In all the distinguished British works I have ever perused on the subject, I have found the same lack of definitions. The authors evidently vary in the meaning they attach to the term, but I think I can confidently say that none of them make it include breeding between all relatives, or object to breeding, when there is occasion for it, between relatives not of near consanguinity.

It is a very prevalent impression in the United States, particularly among those who have no personal experience on the subject, that the inter-breeding of the most remote relatives is fatal — fatal not only to the physical organization, but to the mind among human beings, and even to the instinct among brutes.

It was stated in the preceding Chapter that when hereditary disease or a predisposition toward *it*, exists in either parent, there is always danger that it will be transmitted to offspring, and that if the disease or predisposition exists in both parents, that danger is greatly increased. If the parents be nearly related to each other, the danger of transmission is virtually converted into certainty, with an aggravation of the conditions and increased incurableness in the malady. Consequently when mankind degenerated from their original physical perfection — when disease entered the world and predispositions to it became engrafted in the human system — the Divine Lawgiver made cohabitation within certain degrees of affinity a crime by prohibition. But if it was evil in itself (*malum in se*) why was it not prohibited to the immediate descendants of our first parents, and why were not unrelated human beings created to avoid its necessity? The peopling of the world in the second generation at least, was *necessarily* carried on between brothers and sisters, the closest possible relations. Can it be supposed that, under the direct ordination of Omnipotence, the human race originated in a crime against nature — in an extreme violation of the fundamental laws which regulate physical and mental well being?

The brute, it is fair to assume, was started in its course of procreation equally unrestricted, for it would understand no prohibition; and it was created with habits which must constantly and necessarily lead to cohabitation and breeding between the nearest relatives. Some varieties of birds, like the dove, are hatched in pairs, one of each sex, and with habits which would render the separation of those pairs, for procreation, the exception instead of the rule. Some varieties of quadrupeds, like the lion, are born and brought up in isolated families; and having no aversion to breeding between relatives, it would be most natural that those who thus live together should at maturity pair together. In herds of elephants, wild horses, buffaloes, etc., particular males dominate over the same herd for years, and make it their harem until they become enfeebled and are conquered by some more youthful and more vigorous rival — probably a son — who in turn dominates, decays and gives place to a successor. In this course of things, the father must be constantly breeding with his own daughters, and, if he lives long enough, with his grand-daughters; and his male successors must commence breeding with sisters and continue it with their descendants. All these animals are, *de facto*, paired together by that Being who created their instincts and gave them their habits. Is there any visible proof that their races have become physically degenerate on this account? Are not the lion and the elephant as large, healthy and powerful as they were ages ago?

No one pretends to the contrary. But we are told — and this was Sebright's argument — that a natural provision was also made to prevent animals from degenerating from the effects of in-and-in breeding. "A severe winter, or a scarcity of food, by destroying the weak and the unhealthy, has all the good effects of the most skillful selection." And he might have added, that the strong male kills the weak male, the herd trample down the sick and the feeble, and gore to death the wounded. Such causes, undoubtedly, combine to extirpate what may be termed accidental degeneracy. But these facts do not go far enough to sustain the position of those who believe that in-and-in breeding necessarily results in degeneracy. If it did, instead of a few, the whole or nearly the whole flock or herd or family, in such cases as I have mentioned, would perish; and whole races would long since have become extinct.

The moment we step from the domain of nature to the domain of man, the scene changes. We have treated our domesticated animals as we have treated ourselves. By artificial surroundings — by changing the natural habits in regard to nutrition, exercise, etc. — by cruelty or kindness — by breeding the diseased with the healthy — we have brought malformation, infirmity, disease and premature death among all of them; and we have continued the causes until we have made the effects a part of the physical systems, and thoroughly hereditary among them. Therefore no longer, like the free normal denizens of the forest and the air, can they follow their natural instincts with impunity; and the inter-breeding of the infirm and diseased, and especially of infirm and diseased relatives, must, as in the case of man, be prevented.

But all the facts I have ever seen or ascertained from entirely reliable sources, go to show that the inter-breeding of relatives, and even near ones, is innocuous when both parents are free from all defects and infirmities which tend to impair the normal physical organization. It is difficult to improve animals, give them a marked family uniformity, and give their peculiar excellencies a permanent hereditary character, without in-and-in breeding. Consequently a great majority of the ablest breeders of domestic animals of every description in England — such as Bakewell among long-wooled sheep; Ellman among short-wooled sheep; the Collings, Mason, Maynard, Wetherell, Knightly, Bates and the Booths among Short-Horn cattle;* Price among the Herefords,† and a multitude of others of nearly equal celebrity — have been *close* in-and-in breeders. The Stud Book abounds in examples of celebrated horses produced by this course of breeding. The same is true of nearly all the improved English varieties of smaller animals, such as pigs, rabbits, fowls, pigeons, etc.

But we need not go abroad for examples. The Paular sheep of the Rich family were first crossed in 1842. They were then pre-eminently hardy. No one claims that they have gained either in hardiness or size by the cross. Yet for thirty years preceding that period, they had been bred strictly in-and-in, to say nothing of their previous in-and-in breeding in Spain. Whether and how far the Spaniards aimed to avoid breeding from *very close* individual relationships I am not informed. I have never learned that they paid any attention to them one way or the other; and their general course of breeding was certainly in-and-in. Each Cabana, or permanent flock, was kept entirely free from admixture with

---

* I quote the following from a note in my Report on Fine-Wool Husbandry, 1862: "In the first volume of American Short-Horn Herd Book (edited by Lewis F. Allen, Esq.,) are diagrams showing the continuous and *close* in-and-in breeding which produced the bull Comet, by far the most superb and celebrated animal of his day, and which sold, at Charles Colling's sale for the then unprecedented price of $5,000. His pedigree cannot be *stated* so as to make the extent of the in-and-in breeding, of which he was the result, fully apparent, except to persons familiar with such things, and such persons probably need no information on the subject. But this much all will see the force of: the bull Bolingbroke and the cow Phenix, which were more closely related to each other than half-brother and sister, were coupled and produced the bull Favorite. Favorite was then coupled with *his own dam* and produced the cow Young Phenix. He was then coupled with *his own daughter* (Young Phenix) and their produce was the world-famed Comet. One of the best breeding cows in Sir C. Knightly's herd (Restless) was the result of still more continuous in-and-in breeding. I will state a part of the pedigree. The bull Favorite was put to his own daughter, and then to his own grand-daughter, and *so on to the produce of his produce in regular succession for six generations*. The cow which was the result of the sixth inter-breeding, was then put to the bull Wellington, "deeply inter-bred on the side *of both sire and dam in the blood of Favorite*, and the produce was the cow Clarissa, an admirable animal and the mother of Restless. Mr. Bates, whose Short-Horns were never excelled (if equaled) in England, put sire to daughter and grand-daughter, son to dam and grand-dam, and brother to sister, indifferently, his rule being 'always to put the best animals together, regardless of any affinity of blood,' as A. B. Allen informs me he distinctly declared to him, and indeed as his recorded practice in the Herd Book fully proves."

† Mr. Price, whose Herefords were the best in England in his day, declared, in an article published in the British Farmer's Magazine, that he had not gone beyond his own herd for a bull or a cow for forty years.

others, and its stock rams were selected from its own number. Consequently fathers and daughters, and brothers and sisters must have constantly bred with each other. Mr. Chamberlain's Silesians have not received any cross, or any fresh blood from either of the original families, within half a century; yet they are 50 per cent. larger than the sheep they originated from and are entirely healthy. Mr. Hammond's Infantados present a still stronger case. They were bred in-and-in by Col. Humphreys up to the period of Mr. Atwood's purchase; Mr. Atwood bred his entire flock from *one ewe*, and never used any but pure Humphreys rams; Mr. Hammond has preserved the same blood entirely intact — and thus, after being drawn beyond all doubt from an unmixed Spanish Cabana, they have been bred in-and-in, in the United States, for upwards of sixty years. Fortunately Mr. Hammond has preserved some of his leading individual pedigrees, and I will give one of these as a most forcible illustration of the subject under examination. For that purpose I will select the pedigree of Gold-Drop, one of his present stock rams. It includes that of Sweepstakes — the ram figured in the frontispiece — and has the advantage of exhibiting the course of breeding for two generations later. The pedigree is given on next page.

PEDIGREE OF GOLD-DROP AND SWEEPSTAKES.

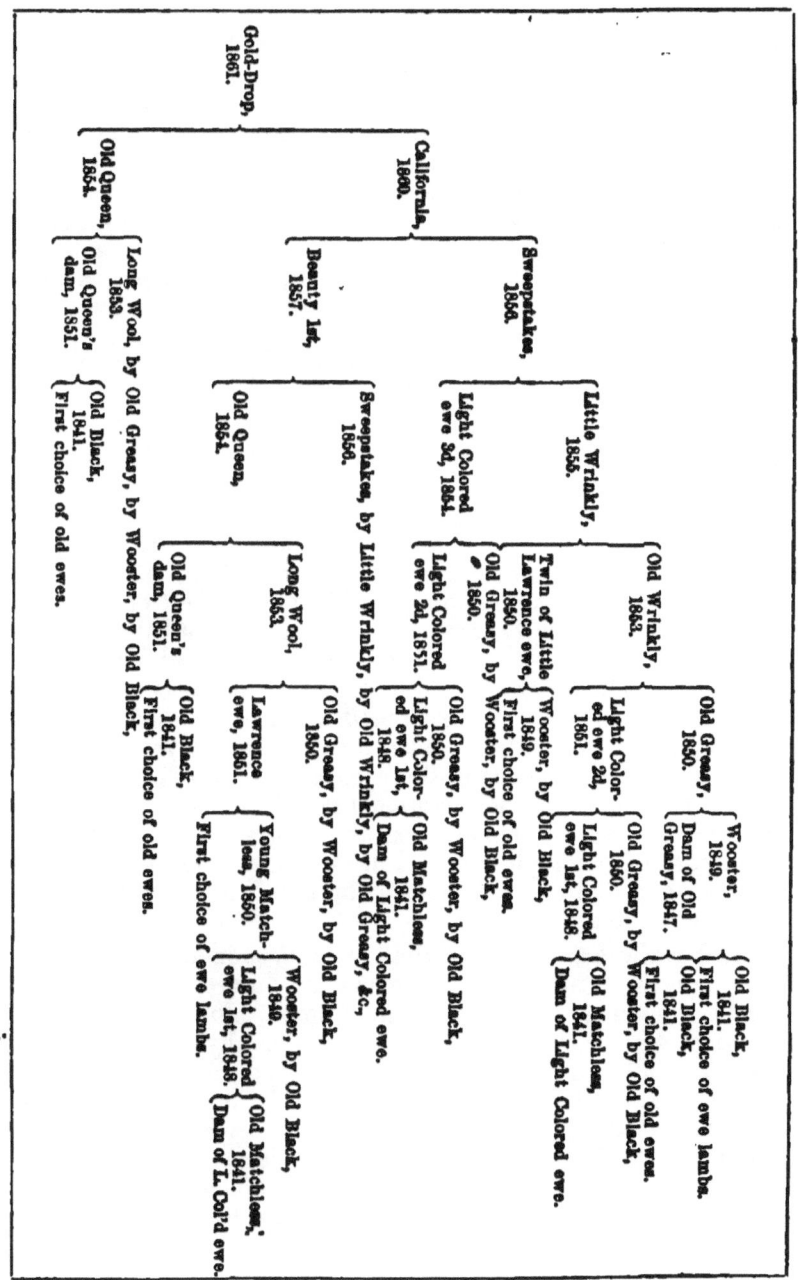

It will be seen that Gold-Drop, after the recurrence of seven generations, traces every drop of his blood to two rams and three ewes, purchased of Mr. Atwood! A careful study of this pedigree will disclose a closeness of in-and-in breeding which will surprise most persons, and will surprise a portion of them the more in view of the fact that Mr. Hammond's whole flock has been bred with the same disregard of consanguinity, and yet all the time since his purchase of its foundation, has been increasing, not only in amount of wool, but in size, bone, spread of rib, compactness, easiness of keep; in short, in all those things which indicate improved constitution. Nor has there been the least tendency toward that barrenness which has been thought by some to be one of the results of in-and-in breeding.*

Every one who draws rams from his own flock and breeds from the best, will inevitably find himself a *close* in-and-in breeder. The best beget the best. If a ram of surpassing excellence as a sire arises and makes a decided improvement in the flock, he is of course coupled with the best ewes, and all the choicest young animals in the flock are soon of his get — and consequently, leaving out of view all previous consanguinity, are as nearly related as half brothers and sisters. These must be bred with each other, or the best of one sex sold, or the highest grade of perfection, on one side, prevented from being joined with the highest grade of perfection on the other. The latter alternatives are most discouraging hindrances in the progress of breeding improvement; and how can we assume that they are necessary, in the face of such facts as those above given? I could add hundreds of examples, both in Europe and the United States, to prove that in-and-in breeding does not, *per se*, produce degeneracy.

But while I am satisfied that even *close* in-and-in breeding is one of the most powerful levers of improvement in the hands of such men as Bakewell, Ellman, and Hammond — breeders who thoroughly understand the physiology of their art — I shall not claim that it is so, or even that it is *safe*, in the hands of those who do not fully and clearly know what is perfect and imperfect in structure; who cannot detect every visible indication of hereditary disease; and who are not familiar by long experience with the effects of combining different forms, qualities and conditions by inter-breeding.

---

* See APPENDIX A.

With such notable instances of successful in-and-in breeders as I have given, and with the hundreds that might be added to the list, it is equally true that the instances of those who have failed have been vastly more numerous. When the masterly hand of Bakewell no longer guided his improved Leicesters, but a very small number among all the prominent breeders of them were found able to preserve them without some admixture of fresh blood. When not ruined entirely, they became delicate and inclined to sterility. And so the pinnacle of success is often but one step from the final overthrow. In view of all the facts, therefore, the great majority of sheep farmers, who do not make breeding a study and an art, had better continue to avoid anything like close in-and-in breeding — though there is no occasion for those exaggerated fears which many entertain on the subject, in respect to remote relatives, where the animals to be coupled are obviously robust and well formed.

Some persons believe that the dangers of in-and-in breeding are less between animals of pure blood than between mongrels or grade animals.* I can see no reason for this, if the latter are equally perfect in that structural organization on which health depends.

---

* See Goodale on the Principles of Breeding.

CHAPTER XIII.

CROSS-BREEDING.

CROSS-BREEDING THE MERINO AND COARSE BREEDS — CROSSING DIFFERENT FAMILIES OF MERINOS — CROSSING BETWEEN ENGLISH BREEDS AND FAMILIES — RECAPITULATION.

CROSS-BREEDING, as I shall use the term, signifies breeding between animals of different breeds, varieties, or families; but it is not applicable to breeding between animals of the same family, though they belong to different and unrelated flocks.

CROSS-BREEDING BETWEEN THE MERINO AND COARSE BREEDS.—The range of cross-breeding between fine and coarse-wooled sheep is comparatively limited, because there is but one breed of the former of any recognized importance, viz., the Merino. And no intelligent man, at the present day, would any more think of crossing the Merino with another breed to improve the characteristics sought in the Merino, than he would of alloying gold with copper to improve the qualities of the gold.

When the object of such crossing has been to improve coarse inferior races, it has succeeded for certain purposes. The coarse common sheep of our country, for example, are always rendered more valuable by an infusion of Merino blood. They gain materially in fleece, and lose in no other particular. But all crosses between the Merino and the large, early-maturing improved English breeds and families, such as the Leicesters, Cotswolds, and the different families of Downs, have uniformly resulted in failure, and must always do so, as long as the characteristics of the respective breeds remain the same. The largest and heaviest fleeced Merinos would probably increase the weight of fleece of even the heaviest fleeced English long-wools, but the wool loses by the cross its present specific adaptation to a demand always great in England and now rapidly increasing in the United States.* The mutton is not injured, nay, for American tastes, it is decidedly improved by the cross; but the long-wool sheep loses its size, its early maturity, its propensity to fatten, and its great prolificacy in breeding. It loses the faultless form of the English sheep, without even acquiring the knotty compactness of the Merino. In short, in the expressive common phrase, it becomes "neither one thing nor the other," but only a comparatively valueless mongrel between two — for their own separate objects — unimprovable breeds!†

The cross between the Merino and the Down materially increases and improves the fleece of the latter. But it is held to detract from the value of the mutton, and it seriously impairs the value of the Down in all the same particulars in which it impairs the value of long-wools.

All attempts to establish *permanent intermediate varieties* of value by crosses between the Merino and any family of mutton sheep, with a view of combining the especial excellencies of each, have ended in utter failure. Those with the Down and the Ryeland seemed to promise best,‡ yet they not only resulted in disappointment, but produced mongrels incapable of being bred back to either of the English types.

The Merino, owing doubtless to its greater purity of blood compared with most other breeds, and to its vastly greater antiquity of blood compared with any of them,§ possesses a force and tenacity of hereditary transmission which renders it a most unmanageable material in any cross aiming at middle results. Its distinctive peculiarities are

---

* The combination of a wool so pre-eminent for certain necessary objects with such valuable mutton properties, render these sheep one of those great gifts to mankind which it would seem almost *wicked* to tamper with!

† I made some experiments in this cross — quite enough to satisfy me — in the earlier part of my life.

‡ I bred a few hundred South Down and Merino cross-breeds, many years ago, and they made a very pretty sheep. They were not much larger than the largest sized Infantados of the present day — because, filled with Mr. Cline's ideas, I selected a very small and excessively high-bred ram for the cross. He was bred by Francis Rotch, Esq., and got by a prize ram of Mr. Ellman's out of an Ellman ewe.

§ The fine-wooled sheep of Spain are clearly traceable to a period anterior to the Christian Era, on the authority of Strabo, Pliny and other Roman writers of conceded veracity. Pliny was himself the Roman Procurator in Spain in the opening part of the first century, and could speak from the result of his own observations. The often re-published statement — that the breed was formed and subsequently perfected by crossing these fine-wooled sheep with coarse, hairy, long-wooled Barbary rams, introduced for that purpose by Columella, Pedro IV, of Castile, and Cardinal Ximenes — rests on no sound historical proof, and is not credited by any recent intelligent writer on sheep. It never was credited by men who were practically acquainted with the breeding of Merino sheep. If these Barbary crosses are not altogether mythical, they undoubtedly were made with, or first formed, the Chunahs, a long, coarse-wooled breed of sheep which have existed for ages in Spain

made to give way with difficulty, and its tendency to breed back is almost unconquerable. But if the Merino fuses with reluctance, it absorbs other breeds with rapidity. A cross between it and a coarse breed is always legitimate and successful, where the object is to merge that coarse breed entirely in the Merino. This is accomplished by putting the ewes of such breed, and every new generation of their cross-bred descendants, steadily to pure blood Merino rams.

Many grade flocks were commenced in this way, a few years since, in the Southern States, and particularly in Texas,—not a few of them under my advice, and to some extent under my direction. The pasture lands in those regions were limitless and their market value only nominal. They were generally yielding no returns to their owners. If they could be stocked speedily with any kind of sheep, the gain would be immense. But wool would be the main object, as there was little or no market for mutton. To stock such large tracts with pure blood Merinos was out of the question, both on the score of expense, and because they could not be obtained rapidly enough at any cost. I therefore counseled the purchase of the common ewes of the country where there were any, and where there were none, those most readily to be obtained,— even though, as it often happened in Western Texas, none could be obtained better than the small, coarse, thin-wooled, miserable Mexican ewes. These and their progeny being bred steadily to Merino rams, the result was in every instance a decided success. The first generation of cross-breeds, even from Mexican sheep, were signally improved in weight and quality of wool, and when from a mediocre Merino ram, would sell for more than twice the price of their dams; and each ascending grade toward the Merino continued to increase steadily in value.*

---

* George W. Kendall, Esq., by far the largest and most experienced wool grower in Texas, who started a portion of his flock with Mexican ewes, in a letter published in the Texas Almanac, 1858, says:

"The produce of the old Mexican ewes gave evident signs of great improvement, not only in form and apparent vigor of constitution, but particularly in the quantity and quality of the wool. Here I might state that a Mexican ewe, shearing one pound of coarse wool, if bred to a Merino buck of pure and approved good blood, will produce a lamb, which, when one year old, will shear at least three pounds of much finer wool; and the produce of this lamb, again, if a ewe, will go up to four and a half or five pounds of still finer wool. I can now show wethers in my flock of the third remove from the original coarse Mexican stock which last May sheared seven pounds of wool— unwashed, it is true, but of exceeding fine quality, and worth 30 cents per pound at this time in New York, or $2.10 for the fleece. This is a rapid improvement. Had the old ewe and her produce been bred constantly to Mexican bucks, the wether would have sheared about 35 cents worth of coarse wool—not more than 40 cents worth at the outside." (These facts further show the *nonsense* of the half-and-half theory of propagation!)

In such crosses the high qualities of choice rams render themselves eminently conspicuous — even more so, relatively, than in breeding among full-bloods. The descendants of such rams in the second cross (¾ blood) are frequently more valuable than those of mediocre rams in the fourth or fifth cross (15/16 or 31/32 blood.)

In the matter of profit — for the mere purposes of wool growing for our American market — these grades approach the full-blood rapidly. But there never was a more preposterous delusion than that entertained by the early French breeders, that "a Merino in the fourth generation [15/16 blood] from even the worst wooled ewes, was *in every respect* equal to the stock of the sire." Chancellor Livingston, who asserts this to have been the opinion of the French breeders, further says:—"No difference is now [1809] made in Europe in the choice of a ram, whether he is a full-blood or fifteen-sixteenths."* This undoubtedly solves problems in relation to a portion of the French Merinos, which otherwise would be quite inexplicable. They are, undoubtedly, *grade* sheep. The Germans, on the other hand, refuse to the highest bred grade sheep any other designation than "improved half-bloods." They found, says Mr. Fleichmann, that their original coarse sheep had 5,500 fibers of wool on a square inch of skin; that grades of the third or fourth Merino cross have about 8,000; the twentieth cross 27,000; the perfect pure blood from 40,000 to 48,000.† I do not apprehend that there is any thing like an equal difference between the number of fibers on a given surface of the American Merino and its grades; but in thirty years observation of such grades of every rank — some of them higher than the tenth cross, where there is but one part of the blood of the coarse sheep to 1,023 parts of Merino blood‡ — I never have yet seen one which, in every particular, equaled a full blood of the highest class.

CROSSING DIFFERENT FAMILIES OF MERINOS. — This has resulted more or less favorably under different circumstances. The Spaniards did not practice it. The French were the first who undertook it on a comprehensive scale. They selected, as we have seen, from all the Spanish families indiscriminately

---

* Livingston's Essay on Sheep, p. 131.

† See Mr. Fleichmann's article on German sheep in the Patent Office Report, 1847.

‡ Probably most persons are familiar with reckoning the degrees of blood in ascending crosses — but for those who are not, I will say that the first cross has 1-2 improved blood; 2d, 3-4; 3d, 7-8; 4th, 15-16; 5th, 31-32; 6th, 63-64; 7th, 127-128; 8th, 255-256; 9th, 511-512; 10th, 1023-1024, and so on.

where they could find animals which presented desirable qualities, and mixed these families indiscriminately together. To this cause, in a very considerable measure, is to be attributed the remarkably unhomogeneous character of the French flocks. Breeding back, in the hands of persons entertaining different views, has separated them into almost as many families as they started from; and the new families lack within themselves the uniformity and permanent hereditary character of the original ones. Mr. Jarvis, in the United States, crossed several families — all prime Leonese, and not widely variant in character. The cross was guided by a single intelligent will, and always toward a definite and consistent end. Therefore a much greater degree of uniformity was obtained.

The present highly popular Paular family in Vermont is, as has been already seen, dashed with Infantado and mixed Leonese (Jarvis) strains of blood.* Crosses between the present Paulars and Infantados are now common throughout Vermont, and the produce is held in high estimation. The Paular ewe in such cases is usually bred to the Infantado ram. It should be borne in mind that the widest of these crosses do not go beyond six original cabanas of prime Leonese sheep,—among the best and most uniform of Spain.

The cross began in Germany by Ferdinand Fischer,

---

* I gave an account of the origin of this cross in my Report on Fine-Wool Husbandry, 1862, from the information of those who ought to have known the facts; but on fuller investigation it proves to have been erroneous in some particulars. The Rich (Paular) and Jarvis (mixed Leonese) sheep had been crossed somewhat anterior to 1844. Judge M. W. C. Wright, of Shoreham, Vermont, having conceived the idea of crossing the produce with the Infantado or Atwood family, purchased a ram for that purpose of Mr. Atwood at the New York State Fair in the fall of the last named year. Judge Wright sold the ram, immediately after his return to Vermont, to Prosper Elithorp, of Bridport, and Loyal C. Remelee, of Shoreham, but used him himself more or less for three years. This, the "Atwood ram," got the "Elithorp ram" out of a ewe bred by Mr. Remelee, and sold by him to Mr. Elithorp. The dam of the Elithorp ram was got by Judge Wright's "Black Hawk" out of a pure Jarvis ewe, purchased by Mr. Remelee of Mr. Jarvis. Black Hawk was got by "Fortune," out of a pure Jarvis ewe purchased by Judge Wright of Mr. Jarvis. Fortune was bred by Tyler Stickney, and got by "Consul" out of a pure Paular (Rich) ewe. Consul was a pure Jarvis ram, purchased by Mr. Stickney of Mr. Jarvis. Mr. Elithorp sold the Elithorp ram, then a lamb, in the fall of 1845, to Erastus Robinson, of Shoreham. The Elithorp ram got the "Old Robinson ram" out of a ewe bred by Mr. Elithorp, and sold by him, with twenty-nine others, to Mr. Robinson in 1848. The dam of the Old Robinson ram was got by the Atwood ram, above mentioned, out of a pure Paular (Rich) ewe bred by Mr. Robinson, and sold by him to Mr. Elithorp in 1843. The Atwood, Elithorp and Old Robinson rams, and particularly the last named, were the founders of the crossed family. The Old Robinson ram in the hands of Mr. Robinson and his brother-in-law, Mr. Stickney, (who subsequently purchased him of the former,) begot an immense number of lambs, which were very strongly marked with his own characteristics, and which, in turn, generally transmitted them with great force to their posterity. They were generally smallish, short, exceedingly round and compact, with fine, yolky, and for those times and for the size of the sheep, heavy fleeces. Messrs. Robinson and Stickney spread rams of this family far and wide. See APPENDIX B.

between the Negretti and Infantado families, and continued in the United States by Mr. Chamberlain, and its results have already been described.

The cross between the French and American Merino has been well spoken of in some quarters, but it has not yet, so far as my individual observation has extended, justified those expectations which, it would seem, might reasonably be based on the character of the materials. The best French ewe, or the French and American Merino ewe (with a sufficient infusion of French blood to have large size,) has few superiors as a pure wool-producing animal. But the wool lacks yolk to give it weight. The full-blood French sheep also lacks in hardiness*. Both it and its cross-breeds are excellent nurses. The American Merino ram has a super-abundance of the desired yolkiness of fleece and of hardiness. As the smaller animal, his progeny have especial advantages for an excellent development before parturition, and they receive abundant nutrition afterwards. Here then, seemingly, are all the requisite conditions for an excellent cross; and I cannot but believe that such a cross will be made with decided success, as soon as precisely the fitting individual materials are brought together and managed with the requisite skill.†

The cross between the American and Saxon Merino results proverbially well — better in almost every instance than it would be considered reasonable to anticipate. I gave a

---

* It lacks very materially in hardiness if from a pampered flock, or immediately descended from pampered ancestors. The early crosses between French and American Merino sheep require extra attention when young, but when fully grown are, on fair keep, a healthy and hardy animal.

† I tried this cross a few years since, and the following statement of the results appeared in my Report on Fine Wool Husbandry, 1862:—"My own experiments in this cross, candor requires me to say, have been less successful. Some of them were made with a ram bred by Col. F. M. Rotch and pure-blood American Merino ewes; some were purchased of gentlemen who started with such ewes and bred them to first-rate French rams obtained of Messrs. Taintor and Patterson; and some were got by pure American rams on high grade French and American ewes (averaging say fifteen-sixteenths or more French, and the remainder American Merino blood.) From this last cross I expected much. The ewes were compact and noble looking animals. The produce was obviously better than the get of French rams on the same ewes, but after watching it for two years, I have recently come rather reluctantly to the conclusion that, in this climate, even these grades are not intrinsically as valuable as pure American Merinos. But the Merino ram which got them, though apparently presenting the most admirable combination of points for such a cross, has not proved himself a superior sire with other ewes; and I do not therefore regard this experiment as conclusive. (This ram weighed about 140 lbs., was compact and symmetrical, and his fleece weighed 14 lbs. washed. He was a very dark, yolky sheep. He was bred in Vermont; and though undoubtedly full blood, probably did not spring from ancestors as good as himself, or in other words, he was an "accidental" animal.) Some well-managed experiments of both these kinds have been tried by the Messrs. Baker, of Lafayette, and the Messrs. Clapp, of Pompey, N. Y. They bred toward the French until they obtained about fifteen-sixteenths of that blood, and now find the cross best the other way. One of the last of these crosses now appears to promise extremely well."

6*

striking instance, in my Report on Fine-Wool Husbandry, 1862, of the good results of a Paular and Saxon cross. I will now give one of an Infantado and Saxon cross. Capt. Davis Cossit (U. S. V.) of Onondaga, New York, had in 1859 a flock of Saxon ewes with sufficient American Merino blood to yield, on ordinary keep, about four pounds of washed wool per head. In that and the two succeeding years he put his ewes to the Infantado ram "21 per cent.," (named in connection with Petri's table of the dimensions, etc., of Spanish sheep in Chapter 1st of this volume.) In 1862 the fleeces of the young sheep produced by this cross were first weighed separately. Eighty-three two-year old ewes yielded 552 lbs., and eighty yearling ewes 504 lbs. of washed wool — within a fraction of 6½ lbs. per head, and an advance of about 2½ lbs. per head over the fleeces of their dams. Each lot was the entire one (of ewes) of its year: not one having been excluded on account of inferiority. I saw them several times before shearing, and them and their wool immediately after shearing. The wool was in good condition; and the sheep obviously had not been pampered. They were very uniform in size and shape, and bore a strong resemblance to their sire. Not one of the whole number had short or thin wool.

In 1863, sixty-five two-year olds (the portion remaining on hand of the eighty yearlings of the preceding year) and ninety-two yearlings (the third crop of lambs got by "21 per cent.") yielded 1,119½ lbs. of washed wool, or an average of 7 lbs. 2 oz. per head. All these sheep had been heavily tagged and the tags, which would not have averaged less than 2 oz. of washed wool per head, were not weighed with the fleeces.*

Notwithstanding these brilliant and rather frequent successes in crossing different Merino families, (especially where the object is to merge an inferior in a superior family,) the failures, or comparative failures, have been far more numerous. To cross different families of any breed merely for the sake of crossing, under the impression that it is *in itself* beneficial to health, or in any other particular — or with

---

* I do not give the weight of the three-year olds' fleeces in 1863, because they were put in with the fleeces of other breeding ewes, and not weighed separately. About fifteen of the yearling ewes were out of some young ewes of a previous cross, then just come into breeding, which yielded about 5 lbs. of wool per head. The two-year olds were sheared on the 24th of May in 1862, and on the 8th and 9th of June in 1863, so that their fleeces were of 12½ months' growth. The yearlings were dropped between the 6th of April and 1st of June, 1862, and sheared at the same time with the preceding in 1863, so that their fleeces did not average over fourteen months' growth—the usual one at the first shearing. Neither lot was pampered.

a vague hope that some improvement of a character which cannot be anticipated may result from it, is the height of folly and weakness. Even uniform mediocrity is far preferable to mediocrity without uniformity; and he who has the former should not break it up by crossing, without having a definite purpose, a definite plan for attaining that purpose, and enough knowledge and experience on the subject to afford a decent prospect of success. It is always safer and better in seeking any improvement, to adhere strictly to the same breed and family, if that family contains within itself all the requisite elements of the desired improvement, or as good ones as can be found elsewhere. The most splendid successes, among all classes of domestic animals have been won in this way.* Successful crossing generally requires as much skill as successful in-and-in breeding. And as it is vastly more common, so vastly more flocks in this country have been impaired in value by it, or at least hindered from making any important and permanent improvement. They are not permitted to become established in any improvement, before it is upset by a new cross; and these rapid crosses finally so destroy the family character of the flock — infuse into it so many family and individual strains of blood to be bred back to — that it sometimes becomes a mere medley *which has lost the benefit that blood confers* — viz., family likeness and the power to transmit family likeness to posterity.

Every breeder or flockmaster should, after due observation and reflection, fix upon a standard for his flock — a standard

---

* The English race-horse and the Short-Horned family of cattle are both frequently cited as instances of choice breeds originating from a mixed origin. In regard to the origin of the race-horse, the weight of proof and intelligent opinion is the other way. In regard to that of the Short-Horn, the matter is involved in much doubt. (Those who wish to see the facts on both sides of the question stated, will find them in Stevens' edition of Youatt and Martin on Cattle 1851.) But conceding, for the sake of the argument, that both breeds were originally the result of crosses, can any one show that they owed such merit as they first possessed to the cross? And have either of them been *improved* up to their present matchless character, *by the aid of any new crosses?* Mr. Youatt says:—"In the descent of almost every modern racer, not the slightest flaw can be discovered; or when, with the splendid exception of Sampson and Bay Malton, one drop of common blood has mingled with the pure stream, it has been immediately detected in the inferiority of form, and deficiency of bottom, and it has required two or three generations to wipe away the stain and get rid of its consequences." The Short-Horns have been bred pure, with an equally jealous exclusiveness; and no breeder of them would admit a cross in his pedigrees sooner than he would a bar-sinister on his family escutcheon, except in the single case of the descendents of a polled Galloway cow, to which Charles Colling resorted for a cross with some of his Short-Horns. He took but a single cross and bred back ever after to the Short-Horns, so that there is not probably a thousandth, or perhaps five thousandth part of the blood of that Galloway cow in any of the *Alloy* (as the descendants of the cross are called,) now living. Yet the English breeders think one of the Alloy can now be distinguished from a pure Short-Horn, by its appearance! This cross once enjoyed — perhaps was *written* into — great popularity; but its reputation has waned; and there are many leading breeders in England who would not on any consideration have a valuable cow bulled by the best sire of the family.

of form, of size, of length of wool, of quality of wool, etc., etc.; and on this he should keep his eyes as steadily as the mariner keeps his eyes on the light house, in the darkness, when on a dangerous coast. Even in using a fresh ram from an unrelated flock of the same family, (which is not crossing,) he should use one which conforms as nearly as possible to his standard. If he disregards this; if he uses rams now tall and long bodied, and now low and short; now short and yolky wooled, and now long and dry wooled; now fine, and now coarse — in a word, each varying from its predecessor in some essential quality — he will not, perhaps, break up his flock quite as much as he would by *crossing* equally at random, but he will do the next thing to it; he will give it an unsettled and unhomogenous character and materially retard, if not altogether prevent essential improvement.

CROSSING BETWEEN ENGLISH BREEDS AND FAMILIES. — If we assume, with Mr. Youatt, that the long and short-wooled sheep of England are each respectively descended from common ancestors, they form but two breeds of sheep, according to the mode of classification adopted in this volume. There have been but a very few successful crosses between these two breeds. The Hampshire and Shropshire Downs, however, both ranked as first class sheep, and both officially classed as short-wools, have usually a dip of long-wool blood. The Oxfordshire Downs are the result of a direct cross between the Down and the Cotswold, and they are already claimed to be an "established variety."* But the instances of failure in blending the breeds have been so much more numerous than the successes, that the balance of intelligent opinion seems to be decidedly against such attempts. With them, as with the Merino, the successes in crossing between the different families of the same breed, have been numerous and signal. Mr. Bakewell, there is little doubt, was the first great improver in this direction, though we are scarcely authorized to cite his example, because, with a spirit much better befitting

---

* In this and all similar instances, we should not forget that a breed regarded as "established" in England, might not prove so, literally, elsewhere. The English breeders, as a class, are men of education, and of ample wealth and leisure to choose materials for their experiments, devote time to those experiments, and sacrifice by weeding out, without regard to time or money. And by devoting themselves to the pursuit, and constantly comparing their opinions with other opinions, and their stock with other stock, among a whole nation of breeders striving to excel each other, they acquire a degree of knowledge, taste and skill on the subject which is professional, and which far exceeds that (within their own particular circle of breeding,) of any other people. And in no place has English breeding skill manifested itself more than in creating, moulding and "establishing" mutton breeds of sheep.

a nostrum vender than a reputable breeder, he veiled all his proceedings in the closest mystery, and even permitted the knowledge of them to die with him. Some therefore have affected to believe that he resorted to different breeds, as he is known to have done to different families, in selecting his materials. But there are no proofs of the fact, and all the probabilities favor the conclusion that he adhered strictly to the long-wooled families.* Among the facts which would seem, by analogy, to favor the latter conclusion, was his own rigid in-and-in line of breeding, after his materials were selected. If he deemed such quasi-identity both in blood and structure necessary or favorable to the completion of his object, it can scarcely be supposed that he would have voluntarily, and wholly *unnecessarily*, disregarded so great a discrepancy as that of a total difference in breed, in its outset; or, even that he would have spread his selection over any unnecessary number of families within the same breed.

Mr. Bakewell's improved Leicesters have, since his death, again been improved by a dip of Cotswold blood. It is found to invigorate their constitutions, and to render them better in the hind quarters. The Cotswolds of the present day have generally been rendered a little more disposed to take on fat rapidly, and to mature earlier, by a Leicester cross. The New Oxfordshire sheep, as has been seen, is but a Cotswold improved by Leicester blood.

The Hampshire and Shropshire Downs may be cited as conspicuous examples of successful crossing between the short-wooled families — for it is, in my opinion, mainly to these families they owe their peculiar excellence, and not to any strain of long-wool blood, where it exists in them. Various of the minor British short-wooled families have also been improved by crosses with the Down, and with each other.

For another and merely temporary purpose, viz., to obtain larger and earlier lambs or sheep for the butcher, it is legitimate to cross between different breeds or families indiscriminately, where the object in view can be effected in the first cross. The nature of the soil, food or climate may be unfavorable to the large, early-maturing mutton families, but sufficiently favorable to some smaller and hardier sheep; indeed, many such localities in all old countries have families, grown on them for many generations, which have gradually

---

* This is decidedly Mr. Youatt's opinion, though, like other British writers, he uses the word *breed* to classify the different *families* (as they are termed in this volume) of the long-wooled breed.

become so adapted to their surroundings, that conditions highly unfavorable to other sheep have become innocuous, if not actually favorable to them. Yet these local families may be ill adapted to meet the requisitions of the most accessible mutton markets, or, indeed, of any mutton market. They may be too small, too late in maturing, too indisposed to take on flesh, fat, etc. In such cases, rams of an improved mutton family — the family being selected with especial reference to the demands of the particular market and the defects to be counteracted in the local family — are put to the ewes of the local family, and the produce, as is usual with half-bloods, partakes strongly of the physical properties of the sire and yet retains enough of the hardiness and local adaptation of the dam to thrive and mature where the full-blood or high bred grade of the superior family could not do so. But in all such instances, the grower should stop with the first cross. If, seduced by the beauty of that cross, he makes a second one between the full-blood ram and the half-blood females, he obtains animals very little better than their dams for the purposes of mutton sheep, and decidedly less adapted to the local circumstances. Accordingly, some portions of the local family should always also be bred pure by themselves, to furnish females for the cross. This last course is generally pursued among the breeders of England who make such crosses.

It is wonderful that, with the highly successful example of the English constantly before us, in the mode of cross-breeding last described, it has not been more extensively resorted to in the United States. In the heart of the mutton-growing region on our Atlantic sea-board, there are very many localities which, by the poverty of the soil, by the severity of the climate and the want of proper winter conveniencies, or by these causes combined, are rendered unfit to sustain the large English mutton breeds. But they sustain local varieties, or in default of these, would sustain the coarse, hardy "common sheep" of the country; and these bred to Down or Leicester rams would produce lambs which, with a little better keep, would sell, at four or five months old, for as much as the cost of their dams, so that, if the fleece and manure would pay for keeping, and if the number of lambs equaled that of the ewes (always practicable with such sheep when not kept in large numbers,) the net profit of 100 per centum would be annually made on the flock.*

---

* Mr. Thorne, whose superb South Downs have been described, finds his lands well adapted to the pure South Down, but his sheep of that family are too valuable

An analagous course of crossing might be resorted to with great profit by those farmers in our Western States, who *prefer* to make mutton production the leading object of their sheep husbandry, and who now grow those immense flocks of "common sheep," which are annually driven eastward to find a market. A single proper cross of English blood on these sheep would produce a stock which it would cost little more to raise than it now costs to raise common sheep *in the most profitable way*, and which would habitually command 50 per cent. more in market and be ready for market a year earlier than the common sheep. They would require good feed and consequently not overstocked ranges in summer, and comfortable sheds and an abundance of corn in winter. In regions where the latter can be grown more cheaply than its equivalent in meadow hay in the Atlantic States, nay, more cheaply than an equivalent of prairie hay can be cut and stored on the same farm, it is a sufficiently cheap feed; and no one will fatten sheep more rapidly or produce more wool.* The value of the wool would not be lessened by any of the proper English crosses, and would be considerably increased by some of them.

The selection of the English family for the purposes of the above cross should be made with strict reference to local circumstances. On rich, sufficiently moist lands, unsubject to summer drouth, bearing an abundance of the domesticated grasses, and near good local mutton markets, the unrivalled earliness of maturity in the Leicester would give it great advantages; but it would bear no even partial deprivation of feed, no hardships of any kind, and no long drives to distant markets. The Cotswold is a hardier, better working and

---

for breeding purposes, to be sold as mutton; and, living in the mutton-growing region and having more land than is necessary for his breeding flock, he pursues the following course. He purchases the common sheep of the Western States—say, one part Merino to three parts of coarse-wooled varieties—as soon as they begin to be driven eastward, about mid-summer or a little later. He has generally, in past years, bought good ones from $2.50 to $3.00 a head. It is necessary that they have some Merino blood or they will not take the ram early enough. He puts them to a South Down ram as near as practicable to the first of September. The ewes are kept on hay in winter until just before lambing, when they get turnips, and after lambing, meal or bran slop in addition. The lambs are also fed separately. They are sold when they reach 40 lbs. weight, and all are generally disposed of by first of June. They have always brought $5 a head on the average. The ewes having only to provide for themselves during summer get into good condition, and a little grain fed to them after frost has touched the grass ripens them for the butcher. They, too, have sold for $5 a head, on the average. If the fleece, manure, and one dollar a head in addition, will pay for the keeping, this leaves 200 per cent. net profit. One hundred and fifty per cent. ought to leave a margin wide enough for all casualties. See Mr. Thorne's letter to me in my Report on Fine-Wool Husbandry, 1862, p. 104.

* I mean corn cut up and cured with all the ears on, and fed out in that state. The system of Western keeping and corn feeding will be fully examined in Chapter XXI of this volume.

driving sheep, inferior to the Leicester in no particular, which would be very essential in such situations; and I cannot but think that, for the object under consideration, those sub-families of it which have not been too deeply infused with Leicester blood, offer excellent materials for a cross. The different Down families will bear shorter keep than the preceding, and will range over larger surfaces to obtain it. They are considerably hardier than the Leicesters, or those families of the improved Cotswolds which have much Leicester blood. They can endure slight and temporary deprivation of food better than the long-wools; but it is a mistake to suppose that any mutton breed or family will fully, or profitably, attain the objects of its production, without abundance of suitable food being the rule, and deprivations of it any more than the occasional exception.* The Downs also produce better mutton; and the dark legs and faces of the half-bloods always gives them a readier and better market. But the half-blood Downs would generally carry less wool than the half-blood long-wools.

In hardiness, patience of short keep, and adaptability to driving long distances, any of the half-bloods would surpass their English ancestors, and would, under the conditions already stated, generally flourish vigorously in our Western States. If the views here expressed of the value of such a cross are even approximately correct, the utility of embarking in it at once, and the immense advantages which would thereby accrue to individuals and to our whole country, must be apparent to all eyes.

Though the crossing of mutton breeds has, in many instances, entirely different objects from those sought in crossing sheep kept specially for the production of wool, and though, consequently, the proper modes of crossing in the two cases often vary essentially, still the general views expressed at page 130 in regard to unmeaning, aimless and unnecessary crossing, are as applicable to the English mutton sheep as to the Merino.

RECAPITULATION.—I will now, for greater convenience of reference, recapitulate the principal positions taken in this chapter.

I. That it is wholly inexpedient to cross Merino sheep with

---

* I speak of course of sheep which are grown only for the butcher, the leading objects of whose production is high condition and early maturity.

any other breed to improve the Merino in any of the characteristics now sought in that breed.

II. That while an infusion of Merino blood is highly beneficial to unimproved coarse families, to increase the fineness and quality of their wool, it injures the improved mutton races more in size, early maturity, propensity to fatten and prolificacy in breeding than it benefits them in respect to the fleece, or otherwise.

III. That no valuable intermediate family of permanent hereditary character has yet been formed, or is likely to be formed, by crossing between Merinos and coarse sheep; and that the only successful *continuous* cross between them is when the object is to merge a coarse-wooled family wholly in the Merino, and when the breeding is steadily continued toward the Merino (i. e., when no ram is ever used but the full-blood Merino.)

IV. That an infusion of the blood of one coarse-wooled breed has been supposed, in a very few instances, to benefit another coarse-wooled breed, but that as a general thing it is much safer to avoid all crossing between *distinct breeds*.

V. That crossing between different *families of the same breed*, for the purpose of obtaining permanent sub-families, has, both among the Merinos and English sheep, resulted highly favorably in many instances; but that, nevertheless, the instances of failure have been much more numerous; that it is not expedient to cross even different families of the same breed for this object, except in pursuance of a well-digested and definite plan, founded on some experimental knowledge of the subject; and finally, that such crosses (like all others) should only be made when the necessary materials for the desired improvement cannot be found within one of the families (in other cases breeds) which it is proposed to cross together.

VI. That crossing between different families of the same breed for the purpose of *merging one family in another* is still more likely to prove successful: but that, in attaining either this or the preceding object, it is desirable to unite families presenting the fewest differences, and to limit the cross to as few families as the circumstances admit of.

VII. That for the *purposes of mutton production* it is highly expedient to breed rams of the best mutton families with ewes of hardier and more easily kept local families — but that, in such cases, it is almost uniformly advisable to stop with the first cross. That such a system to produce early lambs for the butcher on sterile and exposed situations of the mutton region proper, or to produce earlier and better mutton on the natural pastures and corn-producing soils of the West, where its production as a leading object is *preferred* to the production of wool, would redound enormously to individual profit and to public utility.

VIII. That with all breeds and families, crossing for the sake of crossing, without a definite and well understood object — under the vague impression that it is in itself beneficial to health or thrift, or that some benefit, the character of which cannot be anticipated, is likely to spring from it—is in the highest degree improper and absurd. That in using rams of *the same breed and family* taken from different and not directly related flocks, the utmost care should be used to select such only as conform as nearly as practicable to a uniform standard of qualities, which the owner should have previously adopted as the settled one of his flock.

CHAPTER XIV.

SPRING MANAGEMENT.

CATCHING AND HANDLING — TURNING OUT TO GRASS — TAGGING — BURS — LAMBING — PROPER PLACE FOR LAMBING — MECHANICAL ASSISTANCE IN LAMBING — INVERTED WOMB — MANAGEMENT OF NEW-BORN LAMBS — ARTIFICIAL BREEDING — CHILLED LAMBS — CONSTIPATION — CUTTING TEETH — PINNING — DIARRHEA OR PURGING.

CATCHING AND HANDLING SHEEP.—As nearly every operation of practical sheep husbandry is necessarily attended with the catching and handling of sheep, I will make these the first of those practical manipulations which I am now to describe. A sheep should always be caught by throwing the hands *about* the neck; or by seizing one hind leg immediately above the hock with the hand; or by hooking the crook round it at the same place. When thus caught by the hand, the sheep should be drawn gently back until the disengaged hand can be placed in front of its neck. The crook is very convenient to reach out and draw a sheep from a number huddled by a dog or in a corner, without the shepherd's making a spring for it and thus putting the rest to flight; and a person accustomed to its use will catch moderately tame sheep almost anywhere with this implement. But it must be handled with care. It should be used with a quick but gentle motion — and the caught sheep immediately drawn back rapidly enough to prevent it from springing to one side or the other, and thus wrenching the leg, or throwing itself down, by exerting its force at an angle with the line of draft in the

SHEPHERD'S CROOK*.

---

* The cut represents the crook with but a small portion of the handle. This is made seven or eight feet long, of light, strong wood.

crook. Care must be taken not to hook the crook to a sheep when it is so deep in a huddle with others that they are liable to spring against the caught one, or against the handle of the crook, either of which may occasion a severe lateral strain on the leg. When the sheep is drawn within reach, the leg held by the crook should at once be seized by the hand, and the crook removed.

A sheep should be lifted either by placing both arms around its body, immediately back of the fore-legs; or by standing sideways to it and placing one arm before the fore-legs and the other behind the hind-legs; or by throwing one arm round the fore parts and taking up the sheep between the arm and the hip; or by lifting it with the left arm under the brisket, the right hand grasping the thigh on the other side, so that the sheep lays on the left arm with its back against the catcher's body.. The two first modes are handiest and safest with large sheep; the third mode is very convenient with small sheep or lambs; and a change between them all operates as a relief to the catcher who has a large number to handle.

Under no circumstances whatever should a sheep be seized, and much less lifted, by the wool. The skin is thus sometimes literally torn from the flesh, and even where this extent of injury is not inflicted, killing and skinning would invariably disclose more or less congestion occasioned by lacerating the cellular tissue between the skin and flesh, and thus prove how much purely unnecessary pain and injury has been inflicted on an unoffending and valuable animal, by the ignorance or brutality of its attendant. *

It cannot be too strongly enforced that gentleness in every manipulation and movement connected with sheep is the first and one of the main conditions of success in managing them. They should be taught to fear no injury from man. They should be made tame and even affectionate — so that they will follow their keeper about the field — and so that, in the stable, they will scarcely rise to get out of his way. Wild sheep are constantly suffering some loss or deprivation themselves, and constantly occasioning some annoyance or damage to their owner; and under the modern system of winter stable-management, it is difficult to get them through the yeaning season with safety to their lambs.

---

* Let him who doubts the impropriety of lifting a sheep by the wool, have himself lifted a few times by his hair! And let him who falls into a passion and kicks and thumps sheep because they crowd about him and impede his movements when feeding, or because they attempt to get away when he has occasion to hold them, &c., &c., test the comfort and utility of these processes in the same way — by having them tried on himself. Such a person *ought* not to lack this convincing kind of experience.

Turning out to Grass.—In northern regions, where sheep are yarded and fed only on dry feed in winter, they should be put upon their grass feed, in the spring, gradually. It is better to turn them out before the new grass has started much, and only during a portion of each day for the first few days, returning them to their yards at night and feeding them with dry hay. If this course is pursued, they make the change without that purging and sudden debility which ensues when they are kept up later, and abruptly changed from entire dry to entire green feed. This last is always a very perilous procedure in the case of poor or weak sheep, particularly if they are yearlings or pregnant ewes.

Tagging.—After the fresh grass starts vigorously in the spring, sheep are apt to purge or scour, notwithstanding the preceding precautions. The wool about and below the vent becomes covered with dung, which dries into hard knobs if the scouring ceases; otherwise, it accumulates in a filthy mass which is unsightly, unhealthy, and to a certain degree dangerous — for maggots are not unfrequently generated under it. In the case of a ewe, it is a great annoyance, and sometimes damage to her lamb, for the filth trickles down the udder and teats so that it mingles with the milk drawn by the lamb, and often miserably besmears its face. I have seen the lamb thus prevented from attempting to suck at all. Whether the dung is wet or dry it cannot be washed out by brook washing: it must sooner or later be cut from the fleece and at the waste of considerable wool.

Tagging sheep before they are let out to grass, prevents this. This is cutting away the wool around the vent and from the roots of the tail down the inside of the thigh, (as shown in cut,) in a strip wide enough so that the dung will fall to the ground without touching any wool. Wool on or about the udder which is liable to impede the lamb in sucking, should also be cut away — but not to an unnecessay degree during cold weather, so as to denude this delicate part of adequate protection. Tagging is sometimes performed by an attendant holding the sheep on its rump with its legs drawn apart for the convenience of the shearer. But it is best done by the attendant holding the sheep on its side on a table, or on a large box, covered, except at one end, and the breech of the sheep is placed at the opening, so that the tags will drop into it as they are cut away. This is the only safe position in which to place a breeding ewe for the operation, when near to lambing, unless it be on her feet — and tagging on the feet is excessively inconvenient. If a ewe is handled with violence, there is danger of so changing the position of the fœtus in the womb as to render its presentation at birth more or less irregular and dangerous. But if the operation is performed as last described, and the catching and handling are done with proper care, there is no danger whatever.

BURS.— Pastures containing dry weeds of the previous year, which bear burs or prickles liable to get into the fleece, should be carefully looked over before sheep are turned on them in the spring, and all such weeds brought together and burned. The common Burdock (*Arctium lappa*,) the large and small Hounds-tongue, or Tory-weed (*Cynoglossum officinale et Virginicum*;*) and the wild Bur-marigold, Beggar-ticks, or Cuckold, (*Bidens frondosa*,) are peculiarly injurious to wool. The damage that a large quantity of them would do to half a dozen fleeces, would exceed the cost of exterminating them from a large field. The dry prickles of thistles are also hurtful to wool, and they render it excessively disagreeable to wash and shear the sheep. They readily snap off in the fleece, when sheep are grazing about and among them in early spring.

LAMBING.— It used to be the aim of flock-masters in the Northern States, to have their lambs yeaned from about the 1st to the 15th of May — particularly when Saxon and grade Saxon sheep were in vogue. Small flocks with abundant range would grow up their lambs, born even at this season, large and strong enough to winter well; but in the case of large flocks they were not sure, or very likely to do so, except under highly favorable circumstances. The least scarcity of good fall feed told very destructively on them — and if there were those which were dropped as late as June, they generally perished before the close of winter.

From the 15th of April to the 15th of May is now the preferred yeaning season among a majority of Northern flock-masters. Some, however, have it commence as early

---

* The first named variety grows at the roots of stumps and by the sides of decaying logs, etc., along road-sides, and in new cleared and other fields—the other grows more particularly in woods and thickets. The last variety has finer stems, and its burs are considerably smaller, but I think more difficult to remove from wool.

as the 1st of April, and those who breed rams for sale, as early as the 10th or 15th of March. These very early lambs, if properly fed and kept growing, are about as much matured at their first, as late dropped ones are at their second shearing.*

It is understood, of course, that lambs yeaned earlier than May, in the Northern States, must, as a general thing, be yeaned in stables. But this in reality diminishes instead of increasing the labors of the shepherd. The yeaning flock is thus kept together, and no time is spent traversing pastures to see if any ewe or lamb requires assistance, or in getting a weak lamb and its dam to shelter, or in driving in the flock at night and before storms. And the yeaning season may thus be got through with before it is time for the farmer to commence his summer work in the fields.

PROPER PLACE FOR LAMBING.—Stable yeaning, too, is safest, (though I once thought otherwise,) even in quite pleasant weather, provided the stables are roomy, properly littered down and ventilated, and provided the sheep are sufficiently docile to allow themselves to be handled and their keeper to pass round among them, without crowding from side to side and running over their lambs. While the stables should not be kept hot and tight, they should be capable of being closed all round; and they should be so close that in a cold night the heat of the sheep will preserve a moderate temperature. On the other hand, they should be provided with movable windows, or ventilators, so that excess of heat, or impure air, can always be avoided.

Excessive care is not requisite with hardy sheep in lambing, and too much interference is not beneficial. It is well to look into the sheep-house at night, the last thing before going to bed, to see that all is well; but then if all is well, many even of the best Merino shepherds leave their flocks undisturbed until morning, holding that the lamb which cannot get up, suck, and take care of itself until morning in a clean, well-strawed, comfortable stable, is not worth raising. Our English shepherds, who have charge of choice breeding flocks, usually go round once in two hours through the night

---

* We have seen that Mr. Chamberlain, the importer and leading breeder of the Silesian Merinos in this country, has his lambs dropped from November to February. Under the admirable arrangements of Mr. C.; and under the admirable handling of his German shepherd, this works well, and a lamb is rarely lost: and being early taught to eat roots, &c., separate from their dams, they attain a remarkable earliness of maturity. Such a system would not, of course, succeed with ordinary arrangements and handling, nor would it be profitable for ordinary purposes.

during the height of the lambing season. This may be rather more necessary among breeds which are accustomed to bring forth twins — for one of a pair is less likely to be missed and cared for by the mother, if it accidentally gets separated from her. But unless the sheep are extremely tame, more harm than good, even in this particular, would result from disturbing them in the night.

MECHANICAL ASSISTANCE IN LAMBING.—The Merino ewe rarely requires mechanical assistance in lambing. The high-kept English ewe requires it oftener. But in neither case should it be rendered, if the presentation of the lamb is proper, until nature has exhausted her own energies in the effort, and prostration begins to supervene. The labors are often protracted, or renewed at intervals, through many hours, and finally terminate successfully without the slightest interference. But if the ewe ceases to rise, if her efforts to expel the fœtus are less vigorous, and her strength is obviously beginning to fail, the shepherd should approach her, without alarming or disturbing her, if possible, and at once render his aid. The natural presentation of the lamb is with the nose first and the fore-feet on each side of it. The shepherd with every throe of the sheep should draw very gently on each fore-leg, alternately. If this does not suffice, he should attempt to assist the passage of the head with his finger, proceding slowly and with extreme caution. If the head is too large to be drawn out thus gently, both the fore-legs must be grasped, the fingers (after being greased or oiled) introduced into the vagina, and the head and legs drawn forward together with as much force as is safe. But haste or violence will destroy the lamb, if not the dam also. If the former cannot be drawn forth by the application of considerable force, it is better to dissect it away. In these operations the ewe must be held by an assistant.

If the fore-legs do not protrude far enough to be grasped, the head of the lamb is to be pushed back and down, which will generally bring them into place — or they may be felt for by the hand and brought into place. If the fore-legs protrude and the head is turned back, then the fœtus must be pushed back into the womb, and the head brought along with the legs into natural position. There are several other false presentations, such as having the crown of the head, the side, back or rump come first to the mouth of the womb. The only directions which I can render intelligible in all such cases is to say that the lamb should be *pushed back into the womb*, and either placed in natural position or its hinder legs allowed to come first into the vagina. A lamb is born perfectly safely with its hind feet first. In applying force to pull away the lamb, it should always be exerted if practicable simultaneously with the efforts of nature toward the same end, provided the throes are continued and are of reasonably frequent occurrence. But on the other hand, if a throe occurs while the hand of the operator is *in the womb*, he should at once suspend every movement until the throe is over, or else there will be great danger of his rupturing the womb — a calamity always fatal. But if the throes are suspended, or only recur faintly and at long intervals, and the strength is failing, the operator should, as a dernier resort, attempt to get away the lamb independently of them; and he may even, where death is certain without it, use a degree of force that would be justifiable under no other circumstances.

The English shepherds administer cordials to their ewes during protracted labors to increase their efforts or to keep up their strength. In some cases, they give ginger and the ergot of rye * — in others oatmeal gruel and linseed.† They also sometimes administer restoratives after long and exhausting parturition. One of these is thus compounded: — To half a pint of oatmeal gruel is added a gill of sound beer warmed, and from two to four drachms of laudanum. This is given and repeated at intervals of three or four hours, as the case may require; the same quantities of nitric ether being substituted for the laudanum if the pain is less violent and the animal seems to rally a little. ‡ The diseases occurring after parturition, will be mentioned among the general diseases of sheep.

INVERTED WOMB.— The womb is sometimes inverted and appears externally — especially when parturition has been severe, and force applied for the extraction of the fœtus. It should be very carefully cleansed of any dirt with tepid water — washed with strong alum-water — or a decoction of oak bark — and then returned. If again protruded, its return should be followed by taking a stitch (rather deep, to prevent tearing out,) with small twine, through the lips of the vagina,

---

* Youatt on Sheep, 502. Amounts not stated.
† Spooner on Sheep, 360. Amounts not stated.
‡ See W. C. Sibbald's prize report "On the Diseases occurring after Parturition in Cows and Sheep, and their Remedies," Jour. of Royal Ag'l Soc. of England, Vol. 12, p. 554.

by means of a curved needle, and tying those lips loosely enough together to permit the passage of the urine. The parts should be washed frequently with alum-water or decoction of oak bark, and some of the fluid be often injected with moderate force into the vagina. If this fails to effect a cure and the protrusion of the womb becomes habitual, it should be strongly corded close to the vagina (or the back of the sheep) and allowed to slough off. The ewe will not, of course, breed after this operation, but she will fatten for the butcher.

MANAGEMENT OF NEW-BORN LAMBS.—If a lamb can help itself from the outset, it is better not to interfere in any way to assist it. If the weather is mild, if the ewe apparently has abundance of milk, and stands kindly for her lamb, and if the latter is strong and disposed to help itself, there is usually little danger. But if the lamb is weak and makes no successful efforts to suck, and particularly when this occurs in cold or raw weather, the attendant—the "lamber," as he is called in England—should at once render his aid. The ewe should not be thrown down, if it can be avoided, but the lamb assisted, if necessary, to stand in the natural posture of sucking, a teat placed in its mouth, and its back and particularly the rump about the roots of its tail lightly and rapidly rubbed with a finger, which it mistakes for the licking of its dam. This last generally produces an immediate effort to suck. If it does not, a little milk should be milked from the teat into its mouth, and the licking motion of the finger continued. These efforts will generally succeed speedily—but occasionally a lamb is very stupid or very obstinate. In that case, gentleness and perseverance are the only remedies, and they will always in the end triumph. Too speedy resort to the spoon or sucking-bottle frequently causes a lamb to rely on this kind of aid, and a number of days may pass by before it can be taught to help itself properly, even from a full udder of milk.

ARTIFICIAL FEEDING.—If the dam of a new-born lamb has not good milk ready for it, it is better to allow it to fill itself *the first time* from another ewe, or from a couple of ewes, which can spare the milk from their own lambs. And it is well to continue the same supply two or three days, if there is a prospect that the dam will in that time have milk—for ewes' milk is better for young lambs than cows' milk. If cows' milk must be resorted to, it should by all means be that of a new-milch cow. This is generally fed from a bottle having on its nose an artificial India-rubber lambs' nipple — now manufactured and sold for the express purpose. But milk flows less freely from a bottle than from a vessel having two vents, and accordingly tea-pots; or other vessels manufactured for the purpose, with spouts so constructed as to hold the artificial nipple, are now more used.* Milk should be fed at about its natural temperature — but when cold, never be heated rapidly enough to scald it, which renders it costive in its effects. A new-born lamb fed on other ewes', or on cows' milk, should be fed about six times, at equal intervals between sun-rise and ten o'clock at night, and allowed each time to take all it wants.† After two or three days it need not be fed so often.

Some farmers feed from a spoon instead of a nipple — others milk directly from a cow's teat into the mouth of the lamb. By neither mode is the habit and disposition to *suck* as well preserved — and by both modes, and especially by the last, there is great danger of the milk entering the throat so rapidly that a portion of it will be forced into the lungs. If the strangulation of the weak little animal at the time passes unnoticed by the careless "lamber," a rattling sound will soon be heard in the lungs, accompanying each respiration; and it is a death-rattle. I never knew one to recover.

A farrow cow's milk is unsuited to young lambs, and it is very difficult to raise them on it. When it must be used, it is generally mixed with a little "sale" molasses, as that made from the cane is familiarly termed, to distinguish it from domestic or maple molasses, which is not supposed to be equally purgative in its effects. Others do not mix molasses with the milk, but in lieu of it, administer a teaspoonful of lard to the lamb every other day.‡ A farmer of my acquaintance who is very successful in raising lambs, feeds in such cases beaten eggs with, or in the place of, milk. This is a highly nutritious food, and he informs me that it is quite as

---

* My friend, Mr. Rich, has devised a good substitute by winding cloth around the spout of a lamp-filler, so that it will hold the artificial nipple.

† Some persons do not allow lambs thus to fill themselves at first. If the lamb is fed soon after birth, and then as often as above recommended, it is decidedly best. But if a lamb has been for some hours deprived of food at birth — or is subsequently kept on very scanty feed — a sudden admission to an unbounded supply is undoubtedly hurtful and dangerous.

‡ Some persons mix molasses, and others molasses and water, with new milch cows milk. I used to do this, but have come to the conclusion that it is inexpedient.

good for the lamb as new milk, and that it passes the bowels freely, without being too laxative.

CHILLED LAMBS.— When a lamb is found "chilled" in cold weather, i. e., unable to move, or swallow, and perhaps with its jaws "set," no time is to be lost. It can not be restored by mere friction; and if only wrapped in a blanket and put in a warm room, it will inevitably die. It should at once be placed in a *heated oven,* or in a bath of water about *as hot as can be comfortably borne by the hand.* The restoration must be immediate, and to effect this the degree of warmth applied greater than an inexperienced person would suppose a lamb capable of enduring. Where neither oven nor water are ready, (one of these always ought to be ready at such times in the farm house,) the lamb should be held over a fire or over coals, constantly turning it, rubbing it with the hands, bending its joints, &c. On taking it from the water it should be rubbed thoroughly dry. If sufficient animation is restored for it to suck, and it at once fills itself, the danger is over. But if it revives slowly, or remains too weak or languid to suck, it should, as soon as it can swallow,* receive from half to a full teaspoonful of gin, whiskey or other spirits, mixed with enough milk for a feed — the amount of the spirits being proportioned to the size and apparent necessities of the lamb.

If taken to the stable to suck it should be wrapped in a woolen blanket while on the way, if the cold is severe; and the temperature of the stable will decide whether it is safe to leave it there, or whether it should be returned to the house for a few hours longer. If returned, it should not be placed in a room heated above the common temperature of those occupied by a family. It is astonishing from how near a point to death lambs can be restored by the above means. It often appears literally like a re-animation of the dead.

If a lamb is found beginning to be chilled — inactive, stupid, but still able to swallow — the dose of spirits above recommended acts on it like a charm. If it will not drink the mixture from the sucking bottle — which is scarcely to be expected — it must be poured down it carefully with a spoon, giving ample time to swallow. Some administer ground black pepper in the place of spirits. It is not so prompt or so decided in its effects, and its effects do not so rapidly pass away, leaving the restored functions to their natural action.

---

* Under no possible circumstances should fluid be poured down the throat before the lamb can swallow.

But, in emergency, *any* stimulus should be resorted to which is not likely to be followed with directly injurious results. One of the most skillful shepherds in the United States administers strong tea in such cases—in extreme ones, tea laced with gin.

All lambs which get an insufficient supply of milk from their dams, or from other ewes, should regularly be fed cows' milk from the sucking bottle two or three times a day, until the amount given by the dam can be increased by better keeping. They will learn to come for it as regularly as lambs brought up entirely by hand. If the sheep are not yet let out to grass, those deficient in milk should, with their lambs, be separated from the flock and fed the choicest of hay and roots, oatmeal, bran-slop or the like. Some persons partition off a little place with slats which stop the sheep, but which allow the free ingress and egress of the lambs; and in this they put a rack of hay for the lambs, and a trough into which is daily sprinkled a little meal. The lambs soon learn to eat hay and meal, and it benefits them as much in proportion as grown sheep.*

CONSTIPATION OR COSTIVENESS.—Lambs fed on cows' milk, or fed on any milk artificially, are quite subject to constipation. The first milk of the mother, too, sometimes produces this effect.† A lamb that gets strayed from its dam for several hours and then surfeits itself on a full udder of milk—or one that is changed, after it is several days old, from one ewe to another—is subject to constipation. In all these cases the evacuations cease, or they are hard and are expelled with great difficulty. The lamb becomes dull, drooping, disinclined to move about, and lies down most of the time. Its belly or sides usually appear a little more distended than usual. It becomes torpid—sleeps most of the

---

* Mr. Chamberlain's Silesian lambs, yeaned in early winter, are thus fed separately all winter—but they, according to the German custom, are caught out of the flock, and *confined* in a separate place during most of each day. They eat at their racks and troughs as regularly as the old sheep. This undoubtedly materially contributes to the extraordinary size they obtain the first year. The poet Burns had a good idea of a shepherd's duties! Among the "Dying words of Poor Mailie," to be borne to her "Master dear," are the following, in respect to her "helpless lambs" left to his care:

"O bid him save their harmless lives
Frae dogs, 'an tods, an' butchers' knives!
But gie them guid cow milk their fill,
Till they be fit to fend themsel';
An' tent them duly, e'en an' morn,
Wi' teats o' hay an' rips o' corn."

† While the ewes are in the yards and before they are let out to grass. After being let out to grass, I think the milk of the mother very rarely produces this effect.

time — and if not relieved speedily dies. This not unfrequently happens when the lamb is a number of days old and had previously appeared healthy. Constipation is liable to attack the same lamb several times if the exciting causes are continued. Cathartics are not rapid enough in their action to meet the case at the stage when it is generally first observed. An injection of milk warmed to blood heat, with a sufficient infusion of molasses to give it a chocolate color, should at once be administered with a small syringe — say two ounces at a time for a small lamb, and three for a larger one.* The lamb is held up perpendicularly by the hind-legs, so that the fore-feet but just touch the floor, during and for a moment after the injection. If hardened dung is not discharged with the fluid, or soon afterwards, the injection is to be repeated. This process generally gives prompt and entire relief, but if the lamb continues inactive and dull, the tonic contained in half a dozen teaspoonfulls of strong boneset or thoroughwort (*Eupatorium perfoliatum*) tea, has an excellent effect. And where, as it often happens, the urinary action is also insufficient, pumpkin seed tea is the readiest and safest remedy in the hands of most farmers. The syringe and the injection constitute the very sheet-anchor of artificial lamb raising. The flock-master had better be without all other remedies than these.

There is another form of constipation occurring to very young lambs, with their first evacuations. The dung (yet of a bright yellow color) is so pasty and sticky that it is voided with great effort, and the lamb sometimes utters short bleats, expressive of considerable pain, in the process. The injection is here also the most rapid remedy; but two or three spoonfuls of hogs' lard administered as a purgative, will usually answer the same purpose.

CUTTING TEETH. — Sometimes a healthy looking lamb seems strangely disinclined to suck. It seizes the teat as if very hungry, but soon relinquishes it. It repeats this perhaps once or twice, and then gives up the attempt. On examining its mouth it will be found that the front teeth are not through the gums, and that the latter, over the edges of the teeth, are sufficiently inflamed to be very tender. Drawing the back of the thumb nail across the teeth with sufficient force to press

---

* It is not necessary to be exact. There are about eight ounces in half a pint of fluid; and the ordinary teacup or water-tumbler hold half a pint.

them up through the gums, is the usual resort; but a keen-edged knife or lancet inflicts less pain and leaves the inflammation to subside more rapidly. It generally, however, subsides in either case in a few hours; but it is well enough to watch both the lamb and the ewe to see that the former does not suffer for food, and that the udder of the latter is properly drawn.

PINNING. — The first yellow, gummy excrements of the lamb often adhere to the tail and about the vent, and if suffered to harden there, *pin down* the tail to the breech and hinder or entirely prevent later evacuations. The dung should be carefully removed and the parts rubbed with pulverized dry clay, chalk, or, in the absence of anything better, dirt. If there is a tendency to a recurrence of the pinning, docking the tail lessens the danger.

DIARRHEA OR PURGING.— Lambs which suck their dams, very rarely purge, and if they do, they usually scarcely require attention. If a fed lamb purges, the cause should be ascertained and discontinued— and a spoonful of prepared chalk given in milk, and the dose repeated after a few hours, if necessary.

CHAPTER XV.

SPRING MANAGEMENT — CONTINUED.

CONGENITAL GOITRE — IMPERFECTLY DEVELOPED LAMBS — RHEUMATISM — TREATMENT OF THE EWE AFTER LAMBING — CLOSED TEATS — UNEASINESS — INFLAMED UDDER — DRYING OFF — DISOWNING LAMBS — FOSTER LAMBS — DOCKING LAMBS — CASTRATION.

CONGENITAL GOITRE, OR SWELLED NECK.—The thyroid glands are small, soft, spongy bodies on each side of the upper portion of the trachea, (wind-pipe.) Lambs are sometimes born with them enlarged to once or twice the size of an almond, and they then have the feeling of a firm, separate body, lying between the cellular tissue and the muscles of the neck. The lamb thus affected is generally small and lean, or if it is large and plump it has a soft, jelly-like feeling, as if its muscular tissues were imperfectly developed. In either case, the bones are unnaturally small. It is excessively weak — the plump, soft ones being often unable to stand, and usually dying soon after birth. The others perhaps linger a little longer — sometimes several days — but they perish on the least exposure. So far as my observations have extended this condition always, to a greater or lesser extent, accompanies the glandular enlargement under consideration; but it also appears without it, and, as I shall presently show, sometimes to a highly destructive extent.

Having early adopted the view that the preservation of the life of a lamb, which is incapable of attaining that full structural development on which the vigor of the constitution depends, is a loss instead of a gain — and being specially averse to tolerating in a breeding flock any animal even *suspected* of being capable of carrying along and transmitting a hereditary disease — I never have applied any remedy whatever for "swelled neck." I have seen very little of it for the last few years; but events in 1862, presently to be mentioned, have surrounded the subject with new interest, and I now regret that I have not experimented more fully in order to ascertain the precise nature of the malady.

I have learned some new facts in relation to it. Two or three lambs which I saw, in 1862, decidedly affected by it, but not as weak or as attenuated in the bony structures as usual, very rapidly threw off all appearance of the goitrous enlargement of the glands; and they thenceforth grew about as rapidly and appeared about as strong as ordinary lambs. I saw another such case in 1863. I made no memorandum of the facts at the time, but my impression is that in all these instances the enlargement of the thyroid glands disappeared within the space of as short a period as a fortnight. An intelligent friend informed me that having some goitrous lambs in his flock, last spring, he placed a bandage round the neck of each over the thyroid glands, and wet it a few times a day with camphor (dissolved in alcohol.) The swelling, he thinks, disappeared in less time than a fortnight. Mr. Daniel Kelly, Jr., of Wheaton, Illinois, who is represented to be a highly successful flock-master, states in an article in the Rural New-Yorker, that the disease is frequent among his lambs; that he binds a woolen cloth about their necks and keeps it wet "with spirits of camphor or the tincture of iodine"— that "there is little, if any, difference in the effectiveness of these tinctures"—that either "is sure to cure them."*

These facts would seem to add to the number of anomalous features of the malady, when they are compared with those which appear in the human subject of goitre, if indeed it is the same malady;† and they suggest some doubts of the latter fact. But fortunately no question affecting the practical treatment of the disease is to be settled by the determination of that identity. It would now seem that mere evaporants and external stimulants rapidly control it. Should the fact be found otherwise, in the case of a lamb worth saving, the application of iodine would undoubtedly remove the glandular

---

* I should rather say the article is published under the head of Western Editorial Notes, Mr. C. D. Bragdon giving the statements as he received them from Mr. Kelly.

† I was the first public writer, so far as I know, who classified the "swelled neck" of lambs as goitre or bronchocele, (in Sheep Husbandry in the South,)— though conscious then that some of its conditions were very different from those generally exhibited in the human subject of that disease. These exceptional conditions were:—1. That it was so often congenital; 2. That it so frequently affected the progeny of parents that were not themselves subjects of the disease or known ever to have been subjects of it; and 3. That it should so often affect young animals, and so comparatively rarely affect grown ones. The additional anomalies disclosed by the facts stated in the text (if they are facts,) are the following:—4. The very sudden and spontaneous disappearance of the supposed goitrous enlargement. 5. Its sudden disappearance on the application of camphor, and the apparent equal power possessed by camphor and iodine to cause its absorption.

enlargement. It might be applied to the parts with a little less trouble in the form of an ointment, composed of one part by weight of hydriodate of potash to seven parts of lard.

IMPERFECTLY DEVELOPED LAMBS.—Aside from abortions and premature births, lambs are sometimes yeaned of the feeble and imperfect class described under the preceding head, but apparently exhibiting no specific form of disease. The plump, soft ones, and perhaps some of the others, are frequently so colorless about the nose, eyes and the skin generally, that they have the appearance of being nearly destitute of blood. The small ones are often almost destitute of the ordinary wooly coating. This, with their diminutive size, the smallness of their bones, the remarkable delicacy of their tissues, their general appearance of fragility, and their feeble, languid movements, gives them so much resemblance to prematurely born lambs, that the observer finds it difficult to believe they are not so, until dates and other circumstances are investigated.

Far more of these imperfect lambs were produced in 1862 than in any other year within my recollection. Some counties in New York lost twenty-five and others probably thirty-three per cent. of their entire number, and the mortality is said to have extended to a greater or lesser degree further west. I saw large numbers of these imperfect and perishing lambs. A few, in some of the flocks, were affected by goitre, but in others there was not an instance of it; and taking all I saw together, not five per cent. of them were affected by that, or, so far as I could discover, any other specific disease.

Any mode of treating lambs which are in the condition I have described, so that they will, in more than an occasional instance, ultimately attain the average size and the average integrity of structures and functions possessed by good sheep, is, according to my experience, wholly out of the question; and the bestowal of excessive care merely to preserve the life of an animal essentially lacking in the above particulars, is, as remarked under the preceding head, labor thrown away: indeed, it is much worse than thrown away if the animal is suffered to remain in a breeding flock. No good sheep breeder would permit this. And even if the subsequent structural development appeared to become about as complete as usual, I confess I should still feel decidedly averse to breeding from such an animal. In the case of a ram, I should regard it as inexcusable. We cannot too jealously guard our flocks from the remotest predispositions to hereditary defect, especially in the cardinal point of constitution. I fully concur in this particular with Mr. George W. Kendall, of Texas, who, on ordering some rams of me for the use of his flock, sent the following "particular description" of the points which he wished to have regarded in their selection: he said they must have, "1st, constitution; 2d, constitution; 3d, constitution." And a *congenital* defect of any kind, whether ostensibly removed or unremoved, should be a subject of peculiar apprehension, from the stronger probability which exists of its being hereditary. Acting under these views, my directions in regard to my own flocks have always been to give all lambs of the class under consideration merely good care, and if that prove insufficient, to let them die. If they live until fall, they are sold for any trifle they will fetch as avowedly imperfect lambs, or are given away.

The causes which lead to the production of these imperfectly developed lambs will receive some attention when I treat of the winter management of breeding ewes.

RHEUMATISM.— Lambs on being first turned out of warm, dry, and well-littered yards and stables into the pastures where they lie on the damp ground, and where they are for the first time exposed to cold rains and chilly winds, sometimes exhibit symptoms which, with the present limited information which I possess on the subject, I can only classify as rheumatism. The lamb suddenly becomes unable to walk except with difficulty. It is lame in the loins, and the hind quarters are nearly powerless; or it partly loses the use of all the legs, without the back appearing to be particularly affected; the legs, either from pain or weakness, are unable to support the weight of the body; the lamb hobbles about, and occasionally becomes wholly unable to walk. The neck sometimes becomes stiff, is firmly drawn down, and is perhaps drawn to one side.* Usually there is not much appearance of constitutional disease. The lamb seems to be bright and feeds well. But in some cases, a hollowness and heaving at the flank indicate a degree of fever. Those unable to rise, and those whose necks are so drawn down that they cannot reach the teat, would soon perish without assistance; but in no other way do any of the forms of the disease, as a general thing, very strongly tend to fatal results.

---

* I was not at first disposed to consider this the result of the same disease—but I now have very little doubt of this fact.

So far as my information extends, this malady is new, infrequent, and in any other form than "stiff neck" is yet limited to comparatively few localities in our country. Warmth, dryness, non-exposure to the damp ground, etc., and the careful feeding (from the teats of their dams) of those unable to suck, are conditions necessary to recovery; and as the weather becomes warm and settled it generally disappears without other remedies. In a few cases, however, it has proved quite destructive. Mr. Luther Baker, of Lafayette, New York, had a very valuable flock of Merino ewes, about 20 per cent. of the lambs of which died one year, and 50 per cent. another, of this malady — though his sheep were very carefully and judiciously managed. This is by far the severest mortality which has come to my knowledge. Mr. Baker then put his ewes to ram so the lambs would not come until the flock began to be turned to grass, and the malady almost entirely disappeared. The present year (1863) he had but two or three cases, and these were promptly cured by administering three spoonfuls of lard and one spoonful of turpentine, once or twice, as required, to each lamb. Some of Mr. Baker's neighbors who had one or two diseased lambs apiece, made use of the same remedy with equal success. The dose above mentioned may prove rather large for a very young lamb. Its constituents render it an appropriate internal remedy for rheumatism. The cathartic, and the stimulating and diuretic properties of the turpentine, are called for. Mr. Spooner recommends (for a grown sheep) two ounces epsom salts, one drachm of ginger and half an ounce of spirit of nitrous ether — rubbing the affected parts with stimulants, like hartshorn or opodeldoc; and he says if the disease assumes a chronic form, a seaton should be inserted near the part. Rheumatism in grown sheep, or chronic rheumatism in lambs, appears to be yet unknown in the United States.

TREATMENT OF THE EWE AFTER LAMBING.— Every sound principle of physiology goes to show that the ewe, like every other domestic animal, and like the female human being, should be suffered to remain as quiet as possible for some time after parturition. To drive her for any considerable distance immediately after her lamb drops, when exhausted with her labors, and when her womb remains fully distended, is cruel and injurious; "hounding" her with a shepherd's dog, in that situation, as is sometimes done in driving, because she lingers behind the flock, is to the last degree brutal.

As already said, there should be no hasty interference with a new-born lamb, if it appears to be doing well. But if, on making the usual effort, it fails to obtain a supply of milk, the ewe should at once be examined. The natural flow of milk does not always, particularly in young ewes, commence immediately after lambing, though in a few hours it may be abundant. In this case the lamb should be fed, in the meantime, artificially. If from the smallness of the udder or other indications, there is a prospect that the supply of milk will be permanently small, the ewe should be separated from the flock and nursed with better feed, as mentioned in preceding Chapter. Some careful flock-masters separate from the flock all the two-year-old breeding ewes, and all the old and weak ones, either a few days before, or immediately after lambing, and give them feed especially intended to promote the secretion of milk.

CLOSED TEATS.—Sometimes when a ewe has a full udder of milk the opening of the teats are so firmly closed that the lamb can not force them open. The pressure of the human fingers, lubricated with spittle to prevent chafing or straining the skin, will readily remove the difficulty. If the teat has been cut off by the shearer and has healed up so as to leave no opening, it should be re-opened with a needle, and this followed by inserting a small, smooth, round-ended wire, heated sufficiently to cauterize the parts very moderately. Neither of these should enter the teat but a little way— barely sufficient to permit the milk to flow out. The sucking of the lamb will generally keep the orifice open—but it may require a little looking to and the application of something calculated to allay inflammation.

UNEASINESS.—A young ewe, owing partly, perhaps, to the novelty of her situation, and partly sometimes either to her excessive fondness for, or indifference toward her lamb, will not stand for it to suck. As soon as it makes the attempt, she will turn about to caress it, or will step a little away. In cold weather, she may thus interpose a dangerous delay to its feeding. If she is caught and held by the neck until the udder is once well drawn out, she will generally require no further attention.

INFLAMED UDDER.—But a ewe that refuses thus to stand will sometimes be found to have a hot, hard, inflamed or "caked" udder — particularly if she is in high condition, and lambs late in the season. In this case, the udder should be fomented frequently for some time with *hot* water containing a slight infusion of opium, obtained from the crude article, from laudanum or from steeped poppy leaves. The oftener the fomentation is repeated the sooner the inflammation will subside and the proper flow of milk ensue. Repeated washings with cold water will produce the same effect, but less rapidly, and I think with a less favorable influence on the subsequent secretions of milk. If a ewe has lost her lamb, and from neglect the udder has become swollen and indurations have formed in it, the iodine ointment is one of the best applications. (For further particulars, see Garget, among Diseases of Sheep.)

DRYING OFF.—If a grown ewe having a full udder of milk loses her lamb, she should receive a foster lamb, or be reserved to give temporary supplies of milk to the new-born lambs requiring it. But if it becomes necessary to dry off a ewe, even a young one not having much milk, she should, if convenient, be fed on dry feed, and care taken to milk out the udder as often as once a day for several days, and a few times afterwards, as may appear necessary, at intervals of increasing length. The daily application of an evaporant — say water with 15 grains of sugar of lead dissolved in a pint — would facilitate the process. I am satisfied that many of the troubles shepherds experience in raising lambs are produced or greatly increased by the very careless manner in which ewes are habitually dried off.

DISOWNING LAMBS.—Ewes, and especially young or very poor ones, or those which have been prostrated by difficult parturition, occasionally refuse to own their lambs or are exceedingly neglectful of them. When, notwithstanding, it is advisable to compel the ewe to raise her lamb, both should immediately be separated from the flock and placed in a small, *dark* inclosure together, and if convenient out of hearing of other sheep—care being taken to hold the ewe, at first, as often as five or six times a day for the lamb to suck. As soon as she takes to it, she may be let out; but for a few days she should be let out only with her lamb, and be closely watched, for when she mixes with other sheep as soon as she regains her liberty, her indifference sometimes returns. It is very convenient to attach some peculiar paint mark both to the ewe and lamb, so that they can be readily recognized. If a ewe is obstinate about accepting her lamb, frightening her sometimes aids to arouse her maternal instincts. Some shepherds show her a strange dog, a child wearing a bright colored mantle, or the like. I never chanced to suffer inconvenience by it, but I am informed by good shepherds that on driving flocks of ewes with new-born lambs, *when they are wet*, into a crowded barn, and keeping them there for some time, it produces great confusion in the recognition of lambs, particularly by the young ewes: and my informants attributed this to the lambs rubbing together, and thus blending or disguising those odors by which each ewe is supposed alone to distinguish her own lamb, until she becomes accustomed to recognize it by sight and by its voice. If a ewe exhibits the least indifference to her lamb when it is first born — or if it is quite weak, or in a crowded stable, or requires help of any kind, a pen should be immediately brought and placed around them.

PENS.— Every breeding barn should be provided with a dozen or two of pens, ready made, and hung up on pegs overhead. They should be about three by three and a half, or three and a half by four feet in dimensions, very light but strong; and in field lambing, canvas covers on top and one canvas side cover to a few of them would be highly convenient to keep off rain and cold winds.

FOSTER LAMBS.— If a ewe having a good udder of milk loses her lamb, and a young or feeble ewe disowns hers, or is unable to raise it properly, the lamb of the latter should be transferred to the former. This can usually be readily effected. If the skin of the foster dam's lamb can be taken off soon after death, and fastened on the lamb she is required to adopt, she will generally take to it at once or after only a moment's hesitation. Neither the head, legs nor tail of the skin need to be retained. It should be fastened by strings (sewed through the edges of it,) tied under the neck and body — the labor of a moment — and that is all that is required. Those persons, already mentioned, who transfer all the lambs of their two-year old ewes to foster dams, in some instances put good-milking coarse ewes to ram at the same time with their young ewes, or a trifle later. These are watched and when one yeans, her lamb is immediately taken away, if practicable, before she sees it. The foster lamb is rubbed about in "the waters," (amniotic fluid,) blood, etc., which accompanies the "cleanings," (placenta,) and then is left with her in a pen. She generally does not suspect the substitution, or if she does, after a short delay the adoption on both sides becomes complete. When neither of the above modes is available, the ewe required to adopt a lamb is treated like one which disowns her own. Some take to them pretty readily; others exhibit great obstinacy. If the ewe is confined long in a pen, she should be given feed calculated to produce milk, or should, after a little, be let out daily in a small, green paddock alone with the lamb.

DOCKING LAMBS.—This is most safely performed when the lamb is not over two or three weeks old. Some experienced shepherds do it well, on simply having the lamb lifted by an attendant and its breech held toward them — the lamb being held with its back uppermost and in about the same position as if it was standing on the ground. The shepherd seizes the tail with one hand, places the knife *under* and cuts *up* and toward himself, with a swift, firm motion. But an inexperienced person attempting this, will cut the tails of different lengths, cut off some of them obliquely, and will occasionally leave the bone projecting half an inch outside of the skin, to heal over slowly and cause a vast deal of unnecessary pain. This last is sure to occur in a good share of cases if an unfeeling booby performs the operation, without an attendant, holding the lamb by the tail as it stands on the ground pulling with all its might to escape.* A flock of choice sheep owe too much to the neat and uniform appearance of their tails — especially among the Merinos, where it has become a "fancy point"— not to have the process well performed. The safest mode is to have an attendant hold the lamb, upright but leaning back, with its rump resting on a block, and the hind-legs drawn up out of the way. The shepherd with his right hand fore-finger and thumb slides the skin of the tail toward the body, places a two or three inch chisel across the tail, with his left hand — pressing it down enough to keep the skin slidden toward the body; and taking a mallet in the right hand he severs the tail at a blow. The tail of the Merino should be left barely long enough to cover the anus and

---

* I knew a brutal fellow who, cutting thus, with all his strength, severed not only the tail but one of the hind-legs of a lamb.

vagina. The breeders of English sheep usually leave it three or four inches long.

Docking is best performed in cool, dry weather, and the lambs should not be previously heated by chasing or even driving them fast. The flock should be driven into a stable, the lambs caught out, one by one, and as they are docked placed in another apartment. The tails of the rams should be thrown into one pile and those of the ewes into another, so that when the docking is done, a count of each pile will give the number of each sex; and this should then and there be recorded in the "Sheep Book" of the farm. It is well, also, to mark those of one sex with a brand, or a dot made by the end of a cob dipped in paint, to facilitate later separations. Sometimes, though very rarely, a lamb bleeds to death from docking. This generally can be stopped by a tightly drawn ligature. If this fails, resort should at once be had to actual cautery—the red-hot iron. If lambs are docked after the weather becomes quite hot, it is advisable to apply a mixture of tar, butter and turpentine to the parts. I this year saw eighty lambs, docked on the 7th of July, with their tails swollen and covered with small maggots, for the want of some such application to keep away the fly. The scrotums of the castrated ones were also filled with maggots. Docking is necessary to guard against filthiness. Maggots, too, are liable to be produced under that filth, and to cause the death of the animal. And, finally, habit has rendered a long tail an unsightly appendage to the sheep.

Castration—Is usually performed at the same time with docking—but it is rather severe on the young lamb to do both at the same time. Some, therefore, put off castration a few days later. It should be performed with still more care in regard to the weather, heating the lamb in advance, etc. An attendant holds the lamb (with a fore and hind-leg grasped in each hand,) in an upright position, with its back placed against his own body. He draws the hind-legs up and apart, and presses against the lamb's body with sufficient force to cause the lower part of the belly to protrude between the thighs and the scrotum to be well exposed. The operator then cuts off about one-third of the scrotum; takes each testicle in turn between the thumb and fore-finger, and after sliding down the loose enveloping membrane to the spermatic chord, pulls out the testicle with a moderately quick but not violently jerking motion. The connecting tissues (of the spermatic cord) snap with very little bleeding.* If they snap so that a portion of the nerve adhering to the body remains exposed, it should be cut off. Tar, butter and turpentine should be applied to the parts.

---

* Some foreign shepherds have various absurd processes of severing the last attachments, before the entire spermatic cord snaps asunder. Some chew them off—others cut them off by rubbing the thumb nail across them. Mr. Spooner recommends, even in the case of a *young lamb*, to put iron clams on the spermatic cords and to divide them with a hot iron.

I have given the process, in the text, as it is generally performed, and as it is always performed among my own sheep. But there is no denying that pulling out the testicle in this way often draws out the spermatic nerves (*plexus testiculares*) so that they do not snap within *three or even four inches* of the testicles. The remaining part, of course, retracts within the abdominal ring, which must certainly be injurious, and might, with an animal less capable of enduring all sorts of mistreatment, have serious consequences. I have tolerated the practice because thus tearing the spermatic cord asunder, prevents bleeding; and the hot iron, etc., are inconvenient. Pulling out the testicle far enough and severing it with a hot iron (without using the clams) might also sufficiently prevent bleeding.

CHAPTER XVI.

SUMMER MANAGEMENT.

MODE OF WASHING SHEEP — UTILITY OF WASHING CONSIDERED — CUTTING THE HOOFS — TIME BETWEEN WASHING AND SHEARING — SHEARING — STUBBLE SHEARING AND TRIMMING — SHEARING LAMBS AND SHEARING SHEEP SEMI-ANNUALLY — DOING UP WOOL — FRAUDS IN DOING UP WOOL — STORING WOOL — PLACE FOR SELLING WOOL — WOOL DEPOTS AND COMMISSION STORES — SACKING WOOL.

MODES OF WASHING SHEEP.—Sheep are now washed, in the Northern States, somewhat earlier than formerly — usually between the first and fifteenth of June — as early as the warmth of the streams will admit. When it used to be considered an object to sell clean wool, it was the common practice to wash fine-wooled sheep under the fall of a mill-dam; or to make an artificial fall by damming up a small stream, conducting its water a few feet in a race, and having it fall thence a couple of feet into a tub or washing vat. The vat was a strong box, large enough to hold four sheep at a time. It was from three and a half to four feet deep, about two and a half feet of it rising above the surrounding platform for the washers, and the remaining portion being sunk in the ground. The sheep were penned close at hand, and the lambs immediately taken out to prevent their being trampled under foot. Two washers generally worked together, and a catcher brought the sheep to them. If the sheep were dry, four were usually placed in the vat together, so that two were soaking while two were being washed. Every part of each fleece was exposed for a short time to the full force of the descending current. The dirtier parts, the breech, belly and neck, were thoroughly squeezed, (by pressing the wool together in masses between the palms of the hands,) and these operations continued until the water ran entirely clear from the fleece. The animal was then grasped by the fore parts, plunged down deep into the water and the re-bound taken advantage of to lift it over the edges of the vat without touching them. It was set carefully on its feet, and, if old or weak, a portion of the water was pressed from the fleece. Washing under a mill-dam was performed in substantially the same manner, except that the washers were compelled to stand in the water.

These modes rendered wool quite too clean for the fashion of the present day. The reasons for the change have been elsewhere adverted to. The object now is, with a large proportion of the growers, to see how little they can wash their wool and yet have it sell as "washed wool." It would be difficult, if indeed desirable, to give any instructions on this head! English sheep require very little washing compared with the Merino, and it can be done with sufficient expedition and thoroughness in any clear, running water of proper depth.

UTILITY OF WASHING CONSIDERED. — The utility of washing sheep before shearing is now the subject of a good deal of discussion. One class of producers advocate it on the ground that it prevents a useless transportation of dirt to market, that it improves the saleableness of wool, and that it avoids the operation of an unequal rule of shrinkage applied by buyers indiscriminately to all unwashed wools. Another class of producers contend that it is injurious to the health of sheep; that it renders shearing impracticable at that period which best tends both to the comfort and productiveness of the animal, and which enables the producer to avail himself of the early wool markets; that it subjects sheep to the danger of contracting contagious diseases; and, finally, that any custom of buying, or conventional rule of shrinkage, which is found unfair in itself or opposed to public utility, should be promptly abandoned.

The objection to transporting dirt is a good one, unless it secures some advantage which counterbalances its cost. I am satisfied that washing, properly conducted, in water of suitable temperature, is not in the least injurious to decently hardy sheep — not any more so than an hour's rain any time within a month after shearing — the rain being of the same temperature with brook water when fit for washing. But if it can be shown that shearing before about the 25th of June is better for the sheep, or gives the grower a better chance to sell, there is a weighty and perfectly legitimate reason against washing in many portions of the Northern States — for the streams are not warm enough usually for washing sheep without injury until about the second week of June. This is true among the high lands of New York[*] and Northern Pennsylvania, and certainly ought to be still more so in Vermont, New Hampshire, etc., where the snows which feed the streams lie later on the mountains.

Highly intelligent and candid flock-masters who have tried the experiment, (I have never myself done so,) assure me that Merino sheep sheared a month before the usual period — say from 20th of May to 1st of June — get sooner into condition if they are lacking in that particular; that the wool obtains a better start before the opening of hot weather, and retains it through the year; and that the sheep have better protection from inclemencies of weather during those periods when they most require it — that is, in the winter — and still more particularly during the cold storms of autumn. Whatever may be thought of the two first of these propositions — and they certainly are not unreasonable ones — the last is undeniably true; and the additional autumn protection alone would be a sufficient reason for earlier shearing, in the absence of any special reason to the contrary. The apprehension of contagious diseases, too, from using the same washing yards, from temporarily occupying the same fields during the process, and even from driving sheep over the same roads, is, as I know from bitter experience,[†] perfectly well founded; and it is often highly inconvenient, if not altogether impracticable, for the farmer to wash his sheep without using the same washing pens, or at least the same roads, with the public.

And what sound objection can the *buyer* have to the

---

[*] My residence is less than 1,200 feet above tide-water, surrounded by no lofty hills, and I know that *here* it is generally difficult to find the water as warm as it *ought to be* to wash sheep, before about the time specified in the text.

[†] I have had four different visitations of hoof-rot in my flocks — all clearly and distinctly traceable to contagion. The third case occurred from some wethers affected by that disease, getting *once* among a flock of my breeding ewes. The wethers were found with the ewes at 9 o'clock, A. M., and were not with them at night-fall the preceding day. They might therefore have been with them a few hours, or only a few moments. In the fourth case, half a dozen of my lambs and sheep jumped into the road when a lame flock was passing, and remained with them half an hour. Both lots of animals were thus exposed when I was not aware there was a sheep having hoof-rot in the town! The diseased sheep had just been brought in by drovers, and the farmer who took them to pasture, in the lot adjoining mine, in the third case, did not dream of their being thus affected; and they had mixed with mine before I knew there was a new flock in the neighborhood. I mention these facts to show how readily sheep contract the disease, and how idle it would be for any man to lay aside all fears of contagion in going to and occupying a public washing pen — because he supposed he knew there were no diseased flocks in his neighborhood. There could be no better place for contracting hoof-rot or scab, than a washing-pen.

farmer's shearing his sheep a month earlier and unwashed, if he chooses to do so, even if we should admit, for the sake of argument, that all the reasons assigned for it have no real weight? If the farmer sends *dirt* to market, he, not the buyer, pays for the transportation. Washed or unwashed, the wool must go through the same cleansing process. Am I asked if the buyer has not the right to judge of the conditions in which *he* shall voluntarily purchase a commodity with his own money? By no sound principle, either of morals or commerce, have any class of buyers a right to establish rules of purchasing, not necessary to protect their own legitimate interests, which are calculated to injure the legitimate interests of producers.

The rule that all wools shall be washed or subjected to a deduction of one-third to put them on a par with brook-washed wools, operates very unequally. A large, highly yolky ram, housed in the summer, will have at least two pounds, and a ewe one pound, more yolk in its fleece than would the same animal if unhoused in the summer. Should the unwashed wool then sell at the same rate of shrinkage in both cases? If we were to admit that one-third is a fair *average* rate of shrinkage on all unwashed wools, is there any justice in making the producer of the cleaner ones suffer for the benefit of the person who chooses to grow yolkier wools, or who houses his sheep in summer to preserve all their yolk? Does the manufacturer wish to pay a premium on the production and preservation of yolk in the wool?

No manufacturer claims that the present rule of shrinkage operates strictly equitably in all cases; but some manufacturers contend that a discrimination in unwashed wools would be impracticable, or at least inconvenient, and that if the present rule injures the interest of the producer, all he has to do is to wash his wool. It would be difficult for any one to show that there is any greater practical inconvenience in deciding between the different amounts of yolk in unwashed wool than there is in deciding between the different amounts of foul seed in wheat and other varieties of grain, of useless weeds in hay, or even of *yolk in washed wool;* yet who thinks of buying these impure commodities at a fixed rate of shrinkage? Still less excuse is there for preserving an arbitrary and unequal rule, as a quasi *punishment* on growers who only believe themselves consulting their own legitimate interests, and who certainly are not invading those of others.

The ground directly or impliedly assumed by some

growers, that a reduction of the present rate of shrinkage is all that is now called for — leaving it as fixed in its rate as at present — must be a pleasing one to those who grow and preserve the largest amount of yolk, for this would increase the present premium on yolk precisely in proportion to that reduction. But it would do it at the expense either of the producer of cleaner wools, or of the manufacturer. Equally fallacious and interested is the pretence that unwashed wools come nearer to a uniform standard in respect to cleanliness than washed ones, and therefore that, as a matter of right or mutual protection, all wool growers ought to combine to omit washing for the purpose of forcing all wools on the market in that situation.

The only sound and equitable course is to abolish any fixed rule in the premises — to buy unwashed wool as wheat, other grain, hay, and washed wool containing impurities are now bought, viz., subject to a deduction proportioned to the amount of impurity in each particular case — *clean wool* being made the standard. It is as easy for the buyer and seller to agree on the amount of deduction as to agree on the quality. Indeed, they have no especial occasion to agree in terms on either; nor do they now, in the case of washed wools of different qualities and degrees of cleanliness. They simply agree or disagree on *price*, each basing his estimates on such data as he pleases. The moment this mode of purchasing is adopted and put fairly into operation, its propriety will commend it to all. It will equally promote the legitimate interests of both buyer and seller. But one leading purchaser has to adopt it rapidly to procure its general adoption — because those who bought thus would secure the decided advantage of acting without competition in the rapidly increasing market of unwashed wools, while they still could compete on equal terms in the market of washed wools.

Two sets of persons have taken what I esteem to be very uncalled for positions on this subject. Those who assume that manufacturers should, at the first intimation and without understanding the reasons, abandon any established custom of their calling, or submit to the imputation of laboring to take advantage of the wool producer, and of "combining" to secure that advantage, assume positions which are equally unsupported by proof and at war with good sense. The manufacturers have been at least as much sinned against as sinning. There is no more intelligent, honorable, public-spirited and liberal class of business men in our country.

The one-third rule of shrinkage was adopted by them at an early day, when but very little domestic wool came unwashed into the market. It was brought in usually by owners of small lots, who took no care of their sheep. The wool was not only frequently filled with wood-dirt, sand and dung, but it was also frequently out of condition — here a fleece cotted, there one jointed, and anon one filled with burs. It was not convenient to classify these with good washed wools, nor was it obligatory on anybody to encourage their continued production. Under such circumstances, the one-third rule of shrinkage met the case fairly enough.

Very few persons are the first to discover that *their* customs have survived their original causes. Even sensible men surrender old ones with reluctance, and are quite apt to suspect the motives of proposed innovators. Weak and prejudiced men mistake them for principles and support them with bigotry and fury. As soon as the manufacturers become convinced that the present feeling among flock-masters against the washing of wool springs from legitimate motives, and indicates a settled purpose instead of a mere freak, they will meet it, not by a suspension of purchases or by holding on to any unequal and unjust rules, but in a fair and business-like way. But if the grower errs in denouncing and "passing resolutions" against the manufacturer who does not at once accede to his precise terms, not less does the manufacturer err in assuming, in a matter where his own real interests are not at stake, to dictate modes and times of preparing a commodity for market to the producer of it; and especially in assuming that the reasons offered by the latter for the change under consideration are either false or frivolous.

I have in this connection spoken only of the manufacturers as buyers, though, directly, other classes of buyers are equally concerned in the question. But I have done this on the supposition that as all wools go ultimately into the hands of the former to be prepared for consumption, their action in the premises would be the controlling one among all classes of purchasers.

CUTTING THE HOOFS.— The hoofs of the improved English mutton breeds usually retain nearly their natural size and form. The hoofs of the Merino often continue growing to twice their natural length, and their horny crusts turn up in front and curl under at the sides. There is some difference between individuals in this particular, and considerable is made between flocks, by the nature of their summer pastures.

Moist, low grounds encourage the growth of the horn; and it is also highly increased by the presence of hoof-rot. But all Merino flocks require examination, at least once a year, in this particular, or else a considerable portion of the sheep will have their hoofs grown out to an extent which is highly unsightly, which gives them a hobbling, "groggy" gait, and which, when the hoof turns under at the sides, confines between it and the sole a mass of mud or filth which remains there constantly. Occasionally, the hoof turns under so far that these impurities are also kept confined between the toes. This situation of things greatly increases the tendency to fouls, and aggravates hoof-rot where it exists. In England it would probably be thought to originate both.

Where no disease is present, and the hoofs only require their usual annual shortening, the time of washing is often a very convenient one to attend to it. The hoofs are then freed from dirt and softened by soaking. When the sheep is removed from the washing-vat, the washer, or an attendant, holds it sitting on its rump with its back resting against his legs. He then, with a thin-bladed, strong, *sharp* knife, cuts away the horn underneath the foot so as to restore it to a level with the sole; and some of the sole should be pared off too, if it has become unnaturally thick. Care should be taken to preserve the natural bearing of the foot—not lowering the heel so much as to throw the weight on the toes, and not lowering the latter so much as to throw it on the heel. An experienced, firm, swift hand will perform this operation on each foot by one or two rapid strokes with the knife. The long toes are then to be cut off with a pair of nippers made for the purpose. As these are sometimes necessarily used when the hoofs are dry and tough, they should be made very strong, with handles eighteen or twenty inches long, the rivet being half an inch in diameter and confined with a nut, so they can be taken apart for sharpening. The cutting edge should descend upon a strip of copper inserted in the iron to prevent dulling. With this instrument, the largest hoofs are readily severed. All these operations should be performed in a little more time than it takes to read this description of them— or else deferred until some other occasion, because, both on account of the washers and the sheep, the washing process is one which ought not to brook much delay.

TOE-NIPPERS.

TIME BETWEEN WASHING AND SHEARING.—This should be determined by the weather. The fleece should become thoroughly dry, and be so far again lubricated with yolk as to have its natural silky feel and glossy appearance. The secretion of yolk depends much on temperature. More of it is secreted in one really hot day than in half a dozen dry, cool ones. Consequently the time of shearing should be controlled by the condition of the wool, and not by the lapse of any established period of time. The old-fashioned wool growers usually sheared within ten days of washing, if the weather was dry, without much respect to temperature. Their successors, for reasons which have been repeatedly alluded to, generally aim to let enough time elapse for the fleece to become well nigh as yolky as it was before washing.

SHEARING.—This should always be performed on smooth, clean floors or platforms, with the sheep penned close at hand. If the weather is fair, it is best to drive only enough sheep into the pen at once to employ the shearers three hours — the rest remaining in the pasture to keep themselves filled with feed. A hungry, empty sheep is more impatient, and the shears run round its collapsed belly and sides with more difficulty. The bottom of the pen should be kept clean with straw, saw-dust, or corn-cobs.* If there are any sheep in the pen dirty from purging, they should be the first taken out. They should be carried a little aside from the shearing floor and the dungy locks cut away. When the catcher catches a sheep in the pen he should lift it in his arms clear of the floor, instead of dragging it to the door and thus filling its feet with straw, manure, &c. At the door of the pen, he should hold it up with its back resting against his own body and its feet projecting toward the shearer, who should be there with a proper shaped stick to clear its feet of loose filth, and with a short broom to free its belly from any adhering straws, chaff or saw-dust — before the sheep is carried to the place of shearing.

It is difficult to give any practical directions for shearing which are of any use to the novice; and experienced shearers do not need them. The art can only be properly acquired by experience and observation. A few suggestions, however, may not be entirely thrown away. The first care of the

---

* These last, if spread on the bottom of the pen a few inches deep, answer the purpose admirably. They keep the feet clean and do not adhere to the wool if the sheep lie down.

shearer should be to clip off the wool evenly and smoothly, without breaking the fleece and without cutting the wool twice in two, or cutting the skin. It is difficult to avoid the last, occasionally, on the corrugated surfaces of the Merino: but repeated and severe cuts should always procure the shearer's dismissal. Especial pains are to be taken in this particular about the udders of ewes. There is perhaps less danger if these are large and in sight. In the case of a young Merino ewe having no lamb, and whose udder is small and mostly covered with wool, I have repeatedly seen a teat clipped off—thus rendering it forever after incapable of allowing the passage of milk, unless re-opened by the artificial process already described at page 157. The shearer who holds his sheep in the easiest manner for itself, who keeps it confined for the least period in one and especially an uncomfortable position, and who makes use of the least violence in case it attempts to escape, accomplishes more work, performs it better, and incurs far less labor and fatigue.

Wool should be cut off reasonably close, but not close enough to have the skin show naked and red—so as to expose it to sun-burn, or to have the sheep suffer severely from a moderate degree of cold. The English shepherds have a system of shearing their large sheep in uniform ridges or flutings, running in a particular way, which has a very pleasing appearance. I see no objections to it; and everything which tends to raise any process toward the dignity of art, and increase the *esprit du corps* of any class of laborers, is beneficial both to themselves and their employers.

Fair ordinary shearers will shear about twenty-five common Merinos in a day, and active ones from five to ten more. The highly corrugated sheep which are now becoming fashionable among a class, demand far more time. The comparatively open-fleeces, and smooth, round carcasses of the English breeds, admit of considerably more rapid shearing.

While sheep are being sheared, the catcher should always be at hand with shovel and broom to remove dung, pick up scattered locks, and keep the floor perfectly clean. When a sheep is sheared, he should catch another for the shearer and set it on a new place on the floor, before taking up the fleece of its predecessor. This done, he should bring the preceding fleece together as it lies with its inner side up, and then, pressing it between his hands and arms, lift it up, carry it to the folding table and *turn it over* as he lays it down. He next should go back, pick up every "frib," and sweep the place so that it will be ready for another sheep.*

STUBBLE SHEARING AND TRIMMING.—If wool is left half an inch long or more at shearing, it will, of course, (in the case of all varieties which do not annually shed their wool,) retain that extra length through the ensuing year. This is called "stubble shearing;" and is sometimes resorted to by the sellers of Merino sheep to deceive purchasers in relation to the actual length of the staple. The sellers are always ready to make or produce *affidavits*, if need be, of the *time* of shearing — but the *mode* of shearing is not stated in these interesting documents! Indeed, thousands of unsuspecting buyers never think to ask that question. "Stubbling" is particularly convenient to convert an unimproved Merino into an improved one in appearance, by doubling the length of wool about the head, legs, belly, etc., where the former is most deficient.

"Trimming" is a little higher branch of the same art. It is "cutting a sheep into form," by shortening the wool where there is over-fullness, and leaving it longer where there is a lack of fullness, so that the sheep takes many of its leading points — such as fullness in the crops, straightness of back, etc. — quite as much from the shears as from nature. This is practiced by exhibitors for prizes in the show yards of the Royal Agricultural Society of England!†

"Trimming" has entirely the advantage on the score of respectability of association, for "stubbling" in this country is not practiced by any but the acknowledged Bedouins "of the profession!" Both are disreputable frauds.

SHEARING LAMBS AND SHEARING SHEEP SEMI-ANNUALLY. — When lambs are yeaned, as Mr. Chamberlain's Silesians are, in the early part of winter, and fed up to a large size before shearing, there is no impropriety in shearing them in the spring with their dams; but there can be no good reason for shearing spring lambs when two or three months old.

---

* I once knew a powerful Englishman who would thus tend twelve good shearers, do up the wool beautifully, (this was when the fleeces were done up entirely by hand,) and bring out the sheep so fast that the shearers were constantly hurried by him! Most who both catch and do up the wool do not tend more than half a dozen shearers, and want a boy to pick up the fribs.

† So says the Editor of the Mark-Lane Express (by implication,) in his paper of January 19th, 1863, and he there entirely dissents from the opinion of a correspondent who asserts that the animals which take the prizes are those which are "*least* cut into form."

Sheep are sheared twice a year in portions of the Southern States. This may be a sort of necessity to save the wool, where they are suffered to run at large in forests or on lands infested by brambles. But where sheep are treated like domesticated animals, and kept on cleared and inclosed pastures, neither necessity nor utility can be pleaded for the practice.

DOING UP WOOL.—The fleece having been desposited on the folding table, with its inside ends downward, the wool-tyer

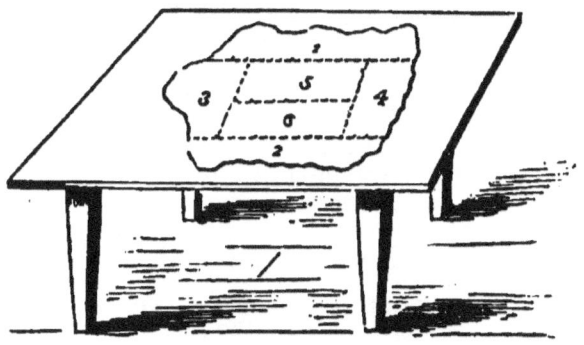

FOLDING TABLE.

first spreads it out to its full extent, restoring every part to its natural relative position. Dung and other impurities being removed, the fleece is pressed together in the same position as closely as practicable. One of the sides (1 in above cut,) is then folded directly over or inverted toward the middle of the fleece so that it covers 5. The opposite side (2) is then folded over and inward in the same way, covering 6, and leaving the fleece in a long strip, some twenty inches wide. The neck (3) is next folded toward the breech; and the breech (4) toward the neck. The fleece is now brought into the oblong square represented by 5 and 6. Having placed the clean

FLEECE READY FOR PRESS.

fribs belonging to the fleece in a bunch on top, and having folded 5 over on 6, so that it will take the form presented in the preceding cut, it is ready for the wool press. The wool-tyer then takes it carefully between his hands and arms, so as not to disturb its arrangement, and places it unbroken in the wool press, either on one side, as in the left hand cut annexed, or on what may be termed its edge, as in right hand cut annexed.

FLEECE IN PRESS.

The wool press I consider one of the most convenient minor agricultural inventions of the day. Combining some previous plans with my own, I furnished a plan of it substantially as it now is, except that it was worked by a lever instead of the crank arrangement described below, to Mr. James Geddes, of Fairmount, New York. Mr. Geddes perfected it by adding that arrangement. I am indebted to him for the following cut and description:

"The Press consists of a substantial and firmly made box, supported on legs of convenient height; the length of the box, four feet, and its width eleven inches, and its depth ten and one-half inches, both measured inside of the box.* One end or head of this box (a) is fixed, and strongly braced by a sort of iron bracket made for the purpose; the other or movable head (b,) has a horizontal support to which it is also firmly braced, and slides under the cleet nailed at f up to within any requisite distance of the other head, a. Through both the heads there are three perpendicular slits which render so many braces essential to their strength, and through which the strings are extended for the tying of the fleece. In oper-

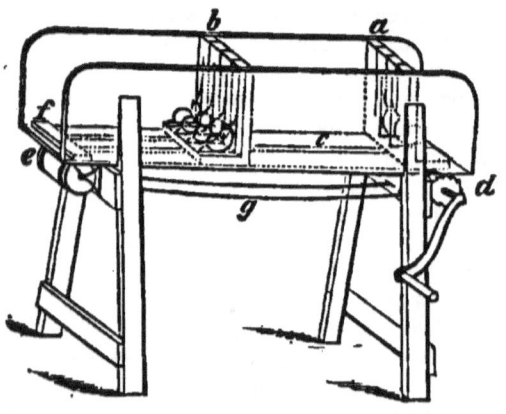

WOOL PRESS.

ation, these strings having been put in place, the fleece is folded to go into the box, but not rolled; the crank, turned by hand and prevented by a ratchet from springing back, moves the roller at d, which, by means of the strap, two inches wide, shown at c, pulls up the follower

---

* Large fleeces require a rather larger box.

b—the strings are tied; the catch lifted and crank reversed, when the straps, one inch wide each, at *g*, draw back the follower, and the fleece is released in perfect shape."

There are several other forms of wool presses, but they possess so little proportionable value that I do not regard them as worth describing.*

The fleece comes from the press in a nearly square mass, and if it is properly folded, and placed in the machine with respectable skill, not a black or outside end of a single lock is visible; and none but the best parts of the fleece are visible. This is expected by the buyer, and therefore has no odor of deception about it.

The twine used in tying should be of flax or hemp. If of cotton, particles of it are liable to be mixed with the wool and to become incorporated with the cloth. They receive different colors from wool in the process of dyeing, and might thus spot the surfaces of dark, fine cloths. Wool twine should be large enough not to render the continuous tying of it too painful to the fingers, but if over large, it looks unworkmanlike and also as if the seller was anxious to sell twine for wool. The three bands of twine placed on each fleece in the press is sufficient, unless it comes loose at the edges and requires an extra band placed round it, the other way, after being taken from the press.

FRAUDS IN DOING UP WOOL.—Some farmers have the habit, if they have a few sheep die in winter, of putting the wool pulled from them into the sheared wool, distributing a a handful or two into each fleece. If the pulled wool is unwashed and the fleeces are sold as washed, the practice is a serious fraud. If the pulled wool is washed, or is in the same condition in this respect with the fleece wool, then it is a petty fraud—for pulled wool is not as well adapted to some

---

* The only possible exception, I think, is the original of this press, worked by a lever. It is not so good an implement as the above, but is much more conveniently made with the rough tools usually found on a farm. One end of the lever passes through a hole in the middle of the cross-piece or brace, which is nailed on the left hand legs of the machine, near their bottom, as seen in the cut. The strap (*c*,) which is attached in above cut to the movable head (*b*,) is fastened to the lever under the front end of box (*d*.) The lever is a couple of feet longer than the box, so that a man can, if necessary, stand on the elevated end to press it down. That end is raised about half-way from the floor to the box, when the movable head (*b*) is slid back to *f*. Consequently when forced down by the foot, it draws forward the sliding head toward the stationary one, in the same manner as the crank does above. A strip of notched iron attached perpendicularly to the inside of one of the fore-legs with a piece of iron on the lever to catch into the notches, holds down the lever to any point to which it is pressed. The lever-press requires to be fastened to the floor by a hook and staple at the rear end, to prevent it tipping up when the weight of a man is put on the lever at the other end.

purposes as sheared wool, and "dead wool" is apt to be inferior in various particulars.* Putting unwashed tags into washed fleeces is also fraudulent. If as well washed as the wool, it is not fraudulent, for they are parts of the same fleeces.† Breech wool simply discolored by dung may enter the fleece, but all respectable flock-masters should take good care that no lumps or masses of dung are accidentally rolled up in it. Locks wet with urine should be dried in the sun before being done up in the fleece. It is not a fraud to put the hairy shank wool in the fleece, but it is unworkmanlike. It is fraudulent to sell fleeces burred to any extent, unless the buyer is distinctly put on his guard. All such fleeces, however much or little burred, should be put by themselves, and the buyer invited to open them.‡

STORING WOOL.—Wool should be stored in a clean, dry room, into which neither dust, vermin nor insects can obtain entrance. Both of the latter are very fond of building nests in it.§ A north light is the best one to show wool in. If there is room for it, the fleeces should be piled up neatly and regularly in walls, with alleys between, so that a large proportion of them can be seen by the purchaser without disturbing their arrangement. Fleeces of the same lot or flock should be piled promiscuously, or divided into lots according to quality. If the want of room or other circumstances require the wool to be piled in a large, compact mass, it is not only for the character but even often for the immediate interests of the seller to place a full proportion of the inferior fleeces in sight. Few persons buy without opening the pile somewhat, and he who opens it and finds that it has been "faced" with the best fleeces, is apt to overestimate the inferiority of that which remains unseen.

It is a common but erroneous idea, that wool continues to gain in weight for long periods after being stored. It does so for a short time: at any rate it has where I have seen the fact tested; but every wool merchant knows that in the course of a year it loses several per cent. by the evaporation of yolk and moisture.

---

* When the sheep die of diseases it is apt to be uneven, jointed, weak, harsh and unelastic.

† And the buyer is a gainer by their being washed separately, because, being severed from the sheep, they receive no yolk after washing.

‡ However badly wool is burred, not one is usually visible on the outside of fleece when it is well done up in a press.

§ Especially rats, mice and bumble-bees.

Place for Selling Wool.—My own experience and observation for more than thirty years, in regard to selling wool, has satisfied me that, on the whole, the best, and, to the farmer, by far the most satisfactory place for disposing of his clip, is at home in his own wool room. It shows better there than in the sack; and the bargain a man makes for himself, he is bound to rest contented with. The local competition, too, in places frequented by buyers, I think usually runs up prices to quite as high a point as the general market authorizes at the time of sale — not unfrequently quite as high as would be received directly from the manufacturer, after deducting freight and the other incidental charges which cluster round such transactions.

Wool Depots and Commission Stores.—The wool depot system, as it was called, was introduced by H. Blanchard, at Kinderhook, New York, in 1844. It was conducted on the same general principles with the ordinary commission establishments, but varied in its method of transacting business. Each lot of wool was graded and stapled and the owner credited with the amount; but his wool was no longer kept separate. The charges were for receiving, sorting and selling, one cent a pound; cartage, three cents a bale; and insurance, usually thirty cents on $100 for three months. The anticipated advantages of the system were that each owner would get the highest market value for his wool, and that the manufacturer could afford to pay a better price when he could buy the kind he wanted unmixed with others. T. C. Peters opened such an establishment at Buffalo, New York, in 1847,. Perkins & Brown one at Springfield, Massachusetts; and I think others were commenced. It was anticipated for a time that they would receive and sell most of the wool of the country, but, though conducted with acknowledged skill and probity, the system failed utterly. Americans generally prefer to do their own bargaining. Wool commission stores, however, still flourish in the important centers of commerce. For a class of sellers — those like the prairie wool growers, for example, who have large lots and no suitable place of storage, or those who are remote from regular markets and wish to realize at stated periods — they are indispensable.

Sacking Wool.—When wool is sold at the barn, the place of delivery is the subject of stipulation. The sacking, unless otherwise agreed, must be done by the purchaser. It is sacked in bales nine feet long, formed of two breadths of "burlaps" from 35 to 40 inches wide. The mouth of the sack is sowed with twine round a strongly iron-riveted hoop, and the body of it is let down through a circular aperture usually in the floor of the loft or room where the wool is stored, if it is in an upper story. If sacked on the farm, and the wool room is not in an upper story, a temporary platform is sometimes erected for that purpose, and the wool tossed up to a catcher. The hoop rests on the edges of the floor around the hole, and the suspended sack should swing clear of everything beneath. A man enters it, and another standing at the mouth passes down the fleeces to him. He arranges them as closely as possible in successive layers and tramples them down with his feet until they are as compact as they can be made. When the bale is filled, the top of it is sowed up with twine, and it is marked as the buyer wishes. It renders the bales more convenient for lifting, if handles are formed by tying up a little wool in their lower corners.

CHAPTER XVII.

SUMMER MANAGEMENT — CONTINUED.

DRAFTING AND SELECTION — REGISTRATION — MARKING AND NUMBERING — STORMS AFTER SHEARING — SUN-SCALD — TICKS — SHORTENING HORNS — MAGGOTS — CONFINING RAMS — TRAINING RAMS — FENCES — SALT — TAR, SULPHUR, ALUM, &C. — WATER IN PASTURES — SHADE IN PASTURES — HOUSING SHEEP IN SUMMER — PAMPERING.

DRAFTING AND SELECTION.—To secure constant improvement in a stock of sheep, as well as to remove all animals from it which have individual peculiarities which render them comparatively unprofitable, or troublesome, it is necessary annually to "draft" the flock, as it is termed, that is, exclude from it all animals which fall below a certain standard of excellence. The leading defects to be had in view in drafting are, first, the general ones of a want of the requisite degree of perfection in the form and fleece, judged by the existing standard of the flock. What satisfies the owner, in these respects, in one generation of sheep, ought not to in the next. However perfect the flock, there ought to be some degree of improvement visible in the get of every new stock ram, or that ram ought at once to give place to another. And as each year brings more perfect younger animals into breeding, the most defective old ones should be excluded, or drafted, to make place for them. If, however, the get of a new stock ram do not meet expectation — or if it is found that they bring some new prominent fault into the flock, or, what is still worse, restore an old one partly bred out and toward which a predisposition yet lingers in the flock — or if they present a type not uniform with the established type of the flock, even though, in itself, it may be an equally good one — it would be better to draft this entire get of lambs, and allow the year of their birth to be a stationary one in the progress of the flock.

The principal special and, in prime flocks, exceptional defects which call for drafting, are weakness of constitution, predispositions to particular diseases, poor qualities either as breeders or mothers, difficulties of any kind connected with lambing, tendencies to barrenness, or any important vices, such as wool-biting, jumping, untamable wildness, &c. Ewes which have attained an advanced age are usually excluded unless they are peculiar favorites. If crones are retained on account of their marked value as breeders, they ought, both on the score of utility and appearance, to be separated from the rest of the flock and fed and nursed by themselves.

The selection of the young stock to take the place of the drafted sheep, should not depend on one examination, however deliberate and careful. It is one of the most important operations of the sheep farm, and can only be properly performed by noting the characteristics of every animal in the young flock, from the time it is yeaned until that for selection arrives.

The best time for drafting is at shearing. There is no other one period during the entire year when all the characteristics of each individual are either so apparent to the eye, or so fresh in the recollection, as then. No person ever attains so perfect a knowledge of the fleece in any other way as by seeing it roll from the carcass under the shears, spread out on the folding table, handled into and out of the wool-press, and put to the last and crowning test of being separtely weighed. The least defect of form, too, is then laid most naked. And, finally, in the case of sheep not permanently numbered, if the drafting and selection are not then made, the removal of the fleece usually destroys all means of distinctly identifying the animal, and consequently of recalling its *past* history, unless in the case of a few very superior or otherwise peculiar animals.

REGISTRATION.— Some owners of small and very carefully managed flocks remember, or imagine they remember, the history of every sheep in them; but this is obviously impracticable in regard to flocks of any considerable size. A history of each individual sheep is by no means necessary in a flock kept mainly for wool-growing or mutton purposes, or in order to effect a good and even a rapid degree of general improvement in any flock; but it is indispensable to the *breeder* to enable him to make the greatest individual as well as general improvement — to preserve his pedigrees correctly — and to sell sheep with a full understanding of their particular qualities at periods of the year when those qualities cannot be determined solely by the eye. The careful breeder should invariably be on the shearing floor with his Register in his hand, minutely scrutinizing each sheep as its fleece is taken off, and noting down his observations on the spot. It is most convenient to have a prescribed form of record in which each particular can be stated by a figure; and it will, of course, include those particulars which each person is most desirous of preserving. I have always had my own include such facts as would give me a full general idea of the sheep without going beyond the record. I have changed the form several times, but that used for the last three or four years has been a blank book with each page ruled into columns, and headed as follows:

Number.	Age.	Size.	Form.	Lamb at heel.	Breeding qualities.	Weight of fleece.	Quality of wool.	Length of staple.	Thickness of fleece.	Yolkiness.	Covering of belly.	Covering of head.	Wrinkliness.	Constitution.	Remarks.
1	4	1	3	0	1	8¾	3	1	2	3	1	4	1	1	
*2	5	5	1	4	3	5	5	3	3	2	4	1	5	4	

Except in the columns for number, age, and weight of fleece, the figures imply relative degree or quality: and 1 is assumed as the maximum and 5 as the minimum of that degree or quality. Thus the first of the above records being translated reads thus: No. 1 is four years old, very large, of middling form, has no lamb, has hitherto exhibited first rate breeding qualities, yields 8½ lbs. of wool, the wool is of middling quality, and of the longest staple, its thickness is better than middling but not first rate, yolkiness medium, covering on belly excellent, the head badly covered, wrinkled in the highest degree, constitution excellent. The second would read thus: No. 2 is 5 years old, is of the smallest size, of the best form, has an inferior lamb, her breeding qualities are only middling, weight of fleece 5 lbs., quality of wool prime, length of staple middling, thickness of fleece middling, fleece of more than medium yolkiness, covering of belly below middling, covering of head first rate, no wrinkles, constitution quite defective. The star at the left of No. 2, signifies that she is to be drafted from the flock. If I had a ram exceedingly strong in the points where No. 1 was most defective viz., in form, quality of wool and covering of head, I should be likely to write his name opposite in the column of "Remarks," to signify the propriety of coupling them the ensuing fall. If any sheep had any special defect not included in the record, I would place that fact in the same column.*

The above system of registration may appear to many persons to be attended with a good deal of labor and trouble. I know by abundant experience that there is not the slightest difficulty in recording these memoranda with the utmost care and accuracy, and at the same time keeping up with five or six shearers. To prevent any confusion, where there is alone a chance for it, namely, in crediting fleeces to the wrong sheep, I throw down a card by each sheep which is being sheared, marked with its number as entered in the Register, in connection with its other qualities. The card is taken up with the fleece, and kept with it until the latter is done up and weighed. Habit soon renders the eye prompt to decide, and at least *as accurate here as under any other circumstances.* I had as lief sell sheep, or select them for coupling, by my Register, as to give them a new examination at the time; and I certainly could do so far more understandingly than by examination without the Register at any period within five or six months after shearing.

MARKING AND NUMBERING.—Sheep should be marked immediately after shearing with the mark of ownership—usually two of the owner's initials stamped on the side by an iron brand dipped in paint. Whether they need additional marks, so that each can at any time be distinguished from all the rest of the flock, depends upon the owner's modes of

---

* It is understood, of course, that the above are merely imaginary cases to illustrate the mode of keeping a record. Such a sheep as No. 2, would hardly be found in any good breeding flock.

treatment, breeding, &c. In "Sheep Husbandry in the South," I recommended Von Thaer's elaborate system of permanently numbering lambs, by notches on the ear. By this, one notch over the left ear signifies 1; two notches over the same, 2; one notch under the same, 3; three notches under the left ear, 9; one notch over the right ear, 10; two over same, 20; a notch under the right ear, 30; three notches under right ear, 90; a notch in *end* of left ear, 100; in the *end* of right ear, 200; these added together, 300; the point of the left ear cut square off, 400; the point of the right ear cut square off, 500; the latter and the notch for 100 added, 600, and so on.

Von Thaer indicated the age by round holes in the ears. As there could not be a mistake of ten years in the age of a sheep, the holes are the same for every succeeding ten years. The absence of any hole indicates the beginning of each decade of years, as 1840, 1850, or 1860; one hole in left ear, 1861; two holes in left, 1862; one hole in right, 1863; one hole in right, and one in left, 1864; one hole in right and two in left, 1865; two in right, 1866; two in the right, and one in left, 1867; two in each, 1868; three in the right, 1869; none in either, 1870.[*]

I have again given this system of numbering because it has proved a highly satisfactory one to some pains-taking men; but I confess I long since got tired of and abandoned it. It requires considerable trouble; and if the holes and notches are not made large enough to mutilate the ear, they are liable to heal up or become obscure; and they therefore require watching while healing. Even when made as small as will answer, they still, in high numbers, cause a disagreeable mutilation.

There is another German system by which the different numerals are made by rows of sharp, steel points inserted in metallic types, as in the two upper figures on following page; and these types have dovetails which can be slid into corresponding grooves (*a a a a* in cut on next page) in the lower jaw of a pair of nippers constructed for the purpose, and thus will be made ready for use.

The inside of the ear is smeared with a thick paint made of vermillion, indigo, or gunpowder and whiskey. By means of the nippers, the steel points giving the proper numbers, are

---

[*] The proper instrument to use is a spring punch like those used by railroad conductors — cutting a hole a little less than one-fourth of an inch in diameter. James Martin, 20 Beaver Street, Albany, manufactures beautiful ones of any size, to order.

forced into the skin inside the ear as far as is practicable without causing bleeding, and when they are withdrawn the paint is rubbed into the punctures. Mr. Fleichmann — t

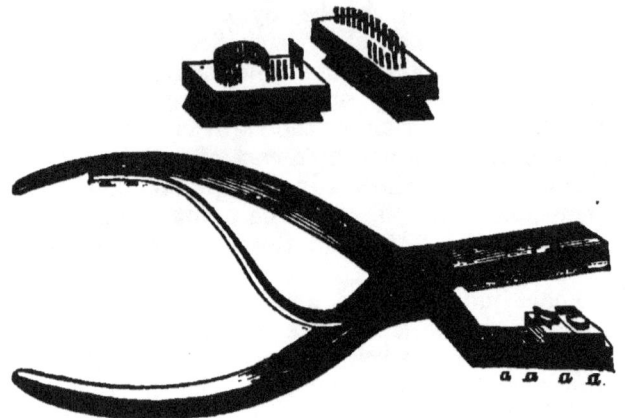

MARKING IMPLEMENTS.

whose Report on German Sheep* I am indebted for the illustrations of this process—declares, as the result of his own observation and experience, that it succeeds fully, and that the numbers remain visible "in old sheep which have been marked for several years."

I have seen imported sheep which had been perfectly tattooed in this way; and it constitutes a very beautiful mode of marking for those who have time and taste for manipula-

INSIDE EAR MARKS.

tions demanding so much care. They must be performed with great exactness to be successful. Mr. George Campbell, of West Westminster, Vt., writes me that "he likes the system very much when the figures can be made plain; that he has been using gunpowder, but does not get all the figures legible; that he is now experimenting with India ink."

A third mode of permanent marking is performed by punching a hole an eighth of an inch in diameter through the ear and inserting a lead rivet of the size and form of the ordinary No. 8 copper belt rivet, sold in hardware shops. Like the belt rivet, it has a bur on which the opposite end

---

* In United States Patent Office Report for 1847.

of the rivet is headed down, on the inner side of the ear. The head is about half an inch in diameter, and on this is stamped the number of the sheep. I have never tested it; but learn that it has given satisfaction to those who have done so. The copper belt rivet itself might be used.

A fourth mode of permanent marking was introduced to some extent among the breeders of New York in 1862. To a ring three-fourths of an inch in circumference, and formed of smallish No. 14 brass wire, was suspended a plate of copper of the form exhibited in the annexed cut, on which were stamped the initials of the owner's name, and the number of the sheep. The ring was inserted about the middle of the ear, so that the plate would remain visible outside the wool. It was found, however, that the ring sometimes cut down through the ear, and sometimes that it was itself cut through by the plate. The cutting of the ear might doubtless be prevented by making the holes with a punch, and allowing them to heal fully before inserting the rings,* and, if necessary, reducing the weight of the plate by making it no larger than in the cut, or even no larger than a five or three cent piece, and as thin as the last named coin. This reduction of weight would probably also prevent the ring from being cut through. Or a split steel ring, or a small T might take the place of the brass ring.† This is so neat and convenient a mode of permanent marking, that it ought to be brought to perfection.

METAL EARMARK.

If not permanently numbered, every large flock of any considerable value, from which sales of breeding sheep are to be made, or which is to be bred with particular reference to individual characteristics, should be annually numbered — for without this there can be no registration. It is performed by stamping figures about 2½ inches long, on the side or rump, with paint, by means of iron or wooden brands. The latter are cut like a type on the end of blocks of soft wood. It is convenient to have a box of brands (arranged and kept in their order,) with special marks for wethers, .cull or draft

---

* Brass is corrosive to a new wound, and by keeping the edges of the hole raw, works down through the ear more readily.
† The ring turning freely in a hole on sound healed up flesh, would be less likely to cut through. The split ring is inserted with considerable difficulty. The T, half an inch long, inserted through a hole already healed and lying across the upper side of the ear could not cut through. But if the plate is lightened, as suggested, (its upper edge might also be thickened and rounded,) I have little doubt the present brass ring would suffice.

sheep, those of particular crosses, etc., etc. It is a great convenience to have even permanently numbered sheep also receive this annual numbering on the body, so that they can be readily distinguished in the field, without catching, and at some distance. All marks should be put on near the spine to prevent rubbing before the paint is dry.

STORMS AFTER SHEARING.— It is remarkable how readily even hardy sheep perish if exposed to very cold storms soon after shearing. A cold rain-storm accompanied with a northwest wind, occurred in Central New York in 1860, during the height of shearing, a little after the middle of June. It came on a day which had opened pleasantly, and many farmers having made their preparations and having their sheep under cover, shut their doors and kept on shearing. Some, with singular thoughtlessness, turned the new-shorn sheep out as usual. Probably three hundred perished within a circle of a few miles. In one case within my knowledge, a wool buyer approaching a barn found a number of dead and dying sheep lying about. On entering the *closed* barn he found the farmer and his assistants shearing away in high glee and turning out new victims. They had not even thought to look out!

When death is not directly produced by such exposure, the sheep are apt to contract obstinate catarrhs, and exhibit other symptoms of unthriftiness for a considerable period afterwards — a very bad way of commencing the summer, particularly for ewes having lambs. Sheep should be housed on cold nights and during cold storms for a few days after shearing; and in default of conveniences for this, they should be driven into dense forests and to situations most sheltered from cold winds.

Very early shearing should be considered out of the question in climates like those of the Northern States, without a sufficient supply of barns and sheds to shelter every sheep on the farm in case of necessity. But, in truth, the early shorn sheep do not appear to suffer as much, in proportion, from cold. The change to them is not so great or sudden as when cold storms follow shearing after they have been sweltering in their fleeces in hot weather. New-shorn sheep rapidly become inured to much colder weather than they could endure at first, and this long before their wool has grown enough to offer them any additional protection.

SUN-SCALD.— This is very rare now, but was not so when Saxon sheep abounded in the country. It was the fashion to shear them very close, and their skins were so thin and delicate, that they not unfrequently blistered, and became sore under the scorching sun. Some greased these sores — others gave the sheep shade and paid no further attention to them.

TICKS.— A very ticky flock of lambs can not be kept in good order, and when they become poor and weak, toward spring, these destructive parasites rapidly reduce them lower and render it extremely difficult to save their lives. Ticks are found on all sheep in neglected flocks, but the heat and cold, and the rubbing and biting to which they are exposed on new shorn sheep, drive them to take shelter in the long wool of the lambs. Here they are so readily exterminated, that it is as much of a disgrace as a loss to the flock-master to suffer them to remain in a breeding flock. About a fortnight after shearing, every lamb should be dipped in a decoction of

DIPPING BOX.

tobacco strong enough to kill the ticks. The last point can be readily settled by an experiment on a few of these insects.* The decoction is poured into a narrow, deep box, which has an inclined shelf on one side, covered with a grate, as shown in the cut. One man holds the lamb by the fore-legs with one

---

* The rule used to be to boil 5 lbs. of plug tobacco (after chopping it fine) or 10 lbs. of stems for a hundred late Saxon lambs. The larger, earlier and longer fleeced lambs of the present day require more — say 6½ lbs. or 7 lbs. The decoction is used cold or blood-warm. Care must be taken not to dilute it so that it will fail to kill both the tick and its eggs.

hand, and with the other clasps the nose so as to prevent any of the fluid from entering the nostrils or mouth; another holds the lamb by the hind-legs, and they then entirely immerse it in the fluid. It is immediately taken out, placed on the grate, and every part of its wool carefully squeezed. The grated shelf conducts the liquor back into the box. In default of a dipping box, two tubs may be used. After dipping the lamb in one it is set on its feet in the empty one, its wool squeezed out, and the liquor returned to the dipping tub as often as is necessary.

Mr. Thorne informs me that he mixes whale oil with the tobacco water, until the latter is considerably thickened by it; and he thinks this renders the wash beneficial to the fleece.

A solution of arsenic has long been used for the same purpose in Great Britain, and at the present time it is vastly more economical than tobacco. Three pounds of white arsenic, in powder, are dissolved in six gallons of boiling water, and forty gallons of cold water are added. The whole is well stirred with a stick, and the lamb is then immersed precisely in the same way as in the tobacco water. The remaining liquor, containing this deadly poison, should be poured where no animal can get to it; and the dipping box, after being well rinsed, should be put in a safe place and used for no other purpose. Arsenic is not poisonous to the hands, if they are sound; and even if the skin should be a little broken, a couple of hours exposure to the above described solution would be attended with no danger. If large surfaces of the hands were denuded of skin, an injurious absorption of the arsenic might take place. The old sheep are frequently dipped at the same time with lambs, in arsenic water, in England.

If the lambs of a breeding flock are properly dipped, but very few ticks will be found either on the old sheep or lambs at the next shearing. If killed in the same way on the succeeding years' lambs, they will generally be wholly exterminated from the flock; and if no ticky sheep are subsequently introduced into it, and it is kept in good order, two or three or more years may elapse before another tick will be found in it.

When lambs have been suffered to go until winter without dipping, and are covered with ticks, arsenic boiled in water, an ounce to a gallon, is poured on them; but the Mountain Shepherd's Manual, which recommends this, adds:—"In this method, however, several of the ticks escape by crawling to the extremities of the filaments." The common mercurial ointment of the shops, mixed with seven parts of lard, is an effectual remedy. It is rubbed on the skin in furrows made by opening the wool, and should be most freely applied to the parts which are especially frequented by the insects, viz., the neck and brisket. Half an ounce of it may thus be used with entire safety on a common sized Merino lamb, having the ordinary access to shelter, in any but exceedingly tempestuous or changeable weather; and this would be more than sufficient for the purpose. In England, where mercurial ointment is frequently used, it is believed to have a generally salutary effect on the skin and on the growth of the wool. Indeed, it is often applied for this express purpose, about the first of October, to lambs which were dipped at shearing, and which, therefore, have no vermin on them. It is also applied to grown sheep for the same purposes, at the close of the coupling season — 2 lbs. to twenty head — or $1\frac{3}{5}$ oz. per head. An ounce would be sufficient on a grown Merino.

SHORTENING HORNS, ETC.—Every horn in the flock should be examined at marking time. When those of the ram press upon the side of the head or neck, a longitudinal section should be sawed from the inside of each, so as to relieve the parts of their contact — and the edges should be rasped smooth. Ewes' horns sometimes grow into the eyes or sides of the face. They should be sawed off, and it will save the trouble of repeating the operation often if they be taken off near the head. By far the best saw I have ever used for these different purposes is a butcher's bow saw.

MAGGOTS.— New-shorn rams do not recognize each other at once after shearing; and those often fight which have previously run kindly together. If the skin of the head becomes broken, and especially if blood oozes from the wound to a part where a horn presses on the flesh, or where the shearer has left a mass of wool between the flesh and horn, maggots are promptly generated, and they soon burrow in the flesh and produce death under the most distressing form. Where they have entered the flesh deeply it is difficult to exterminate them by one application of the proper substances—and they should be carefully re-examined at intervals of a day or two, according to appearances. Spirit of turpentine will kill the maggots it comes in contact with, and prevent the fly from again attacking the parts until its effects are dissipated. It is common also to daub tar over the wound. Having always found these applications sufficient, I have not experimented with others. Spirit of tar is said to be more effective than turpentine. A flock-master who is an excellent practical shepherd writes me that he has found that "two ounces of corrosive sublimate in a quart of any spirits that will dissolve it" is a sure remedy in such cases; and that the flies will not return to a wound to which it has been applied.*

Prevention here, as in most other cases, is much the best remedy. There is no excuse for leaving a horn pressing on the head, or wool under the horns. Rams should be smeared back of and between the horns immediately after shearing, with tar and turpentine, or with fish oil, to repel the flies in case the skin becomes broken. A ram attacked by maggots will soon show it by his rapid emaciation and by his agonized movements, but the mischief has then proceeded to a serious extent. When rams fight, or when it is necessary to keep them in considerable flocks together, they should be frequently examined: and it would be labor well spent to renew the smearing of fish oil on their heads once a fortnight through the months of July and August.

Maggots are sometimes generated under adhering dung on the breech. They are to be removed and the same remedies applied. Maggots in the feet will be mentioned under the head of Hoof-Rot.

CONFINING RAMS.—It is not often that a properly trained ram gives much trouble by leaping *good* fences — particularly if he is allowed one or two companions. But it is not very safe to allow very valuable grown rams to run together, even if acquainted and ordinarily peaceable. Nobody can tell how soon a sudden and fatal battle between them will occur. A choice ram should only be mated with a weather or two, or after lamb-weaning with some ram lambs. I would sooner, if necessary, build a high board fence round a sufficient enclosure for stock rams, than hopple or clog them. Hoppling, when resorted to, is effected by fastening a leather strap around a fore and a hind leg, just above the pastern joints, leaving the legs about the natural distance apart. The ends should be broad enough not to cut into the flesh.

---

* My informant is Mr. Prosper Ellithorp, of Bridport, Vermont. He considers it much more effectual than turpentine in *continuing* to repel the attack of flies. It is soluble in two and a third parts of alcohol. It dissolves in about 20 parts of cold water, and in three of boiling water. But a boiling saturated solution deposits it again in crystals after cooling. Applied externally it is an active stimulant and caustic and has been much used with other substances in applications to ill-conditioned ulcers

Clogging is effected by fastening a billet of wood to one foreleg by a strap. It used to be quite customary to fasten two rams together by a long yoke having bows like an ox yoke. These and similar modes of confinement are injurious to the sheep, and they are at best insecure.

TRAINING RAMS.—Great pains should be taken to teach stock rams the most perfect docility. They should be so tame that their keeper can anywhere walk up and put his hands on them. They should be taught to lead by the halter and to stand confined by the halter as quietly as well broken horses. But a rope should never be put around their heads, as it rubs and tears off the wool. An iron ring about an inch and a half in diameter, should be attached by an eye to a small bolt passing through the thin part of the (left) horn, confined on the other side by a nut. The halter should be a strap of leather with an iron snap, so that it can be readily fastened to or detached from the ring. On the hornless English ram the strap must buckle around the neck.

From being teazed or petted—or from natural viciousness of temper—a ram sometimes acquires a habit of attacking strangers who enter its enclosure—and occasionally even its keeper. Another will strike only when some other sheep in the flock is caught. A cross ram that requires constant watching, is not only an annoyance but a serious danger—for the full blow of one might inflict material injury and even death. Unless of great value, such an animal should be castrated at once. If kept, he should have a blind put on him—that is, his face should be covered and his line of sight forward cut off by a flap of leather in front of his face, secured to the horns. If very quarrelsome, he may be entirely blinded by tying back the ends of the flap over his eyes.

A ram that is not seriously disposed to be vicious, is often made so by the cowardice of those who are in the habit of meeting him. If he finds his attendant is afraid of him, he will soon exert his mastery to the utmost. It is not expedient to court an issue, but as soon as it is discovered that a ram is determined to test the question of mastery, his first motion toward an attack should be followed by carrying the war into Africa. He should be punished until he is taught the complete and absolute superiority of his attendant.*

---

* He should be sprung in upon with a good tough whip—with two or three in the left hand to supply the place of broken ones—and such a storm of blows rained on

FENCES.—It does not require a fence of more than very ordinary height, if it is kept constantly in repair, for the Merino or for the improved English breeds of sheep. But if portions of it are suffered to get partly down, and the flock pass over these low portions a few times, some of the more restless ones learn to be constantly on the look-out for such opportunities to escape; and they will gradually leap higher and higher, until they are ready to scale any ordinary fence that lies in their way. Therefore, the fences of sheep pastures ought in all cases to be thoroughly repaired before turning out flocks in the spring; and they should be frequently examined through the season, particularly after heavy winds.

If sheep are to be driven through an opening in the fence, that opening should be extended to the ground—so as never unnecessarily to make them acquainted with the fact that they can even leap over two rails. One "breachy" sheep will rapidly teach its habits to the whole flock; and it ought to be considered a fraud to sell one, without giving notice of its vice. Such a sheep should not be tolerated in an "orderly" flock, for a single day.

Stone walls unless very high and smooth, or unless surmounted by rough coping stones, set up on edge, do not turn sheep as well as rail or board fences. Sloping sod fences are still worse. In new cleared countries, where inclosures are very imperfectly made with brush, logs, etc., poorly kept sheep sometimes acquire a habit, almost equal to that of swine, of crawling through every opening.

SALT.—Salt is admitted by all to be necessary for the health of sheep. It may be kept in the fields, under cover, where they can have constant access to it: or as much as they will eat may be fed to them once a week on the grass. It is common to throw it in handfulls on mossy knolls, on tufts of coarse grass not eaten down by sheep, on new sprouting bull thistles, or around the roots of Canada thistles, or other weeds—so that it shall call in the aid of the sheep to extirpate vegetable enemies, and so that, if any of it is left, it

---

his head that he stands confused, not daring to open his eyes. If he retreats he should be pursued, and if recently shorn, whipped over the back as he runs, until thoroughly cowed. If he makes his attack on a person not prepared with whips, a few rapid and hearty kicks in the face will generally settle the contest. If he charges, the assailed person should stand firm until he is close upon him and then he should spring suddenly aside, and as the ram rushes past dash in upon him and so punish him that he will have no desire to renew the onset. If after one sound beating he is not quelled permanently, or for a considerable period, resort should at once be had to the knife or the blind.

shall aid in the same particular, and in preparing the soil for better products. I prefer weekly salting, because it is just as well for the health of the sheep; because it keeps them tame and ready to come at the call; and because it compels the owner or shepherd to see them once a week, and consequently to observe whether anything is amiss among them. He should make it an invariable rule to count them if practicable at salting.

TAR, SULPHUR, ALUM, ETC.—Some persons compel *healthy* sheep to eat these substances by mixing them with salt, on the supposition that like salt, they tend to preserve health and increase thrift. There is no proof of this; and we have every reason to believe that nature would prompt healthy sheep to eat these substances as it does salt, were they in like manner necessary to the animal economy. Tar is an impure turpentine, containing, however, some different principles, of which the principal medicinal one is creosote. Turpentine taken internally is stimulant, diuretic and in large doses laxative. The creosote, which adds greatly to the value of tar as an external application to old sores, has been used internally for various human maladies,* but it is one of the last things which would be administered in a state of perfect health. Sulphur is laxative, diaphoretic—i. e., it tends to produce a greater degree of perspiration than is natural, but less than in sweating—and resolvent, or in other words, possesses the power of repelling or dispersing tumors. Alum is astringent in moderate doses, purgative in large, and does not possess a property which gives it a place among the internal remedies of sheep, except as an astringent, and there it is inferior to other astringents† and is scarcely in use. Of what use can such a compound as this be to a *healthy* animal?

If there is a practice in sheep or any other animal husbandry, which more than all others lacks the shadow of an excuse, it is, in my opinion, that of cramming drugs or any substances which nature does not prompt them to eat, down the throats of *healthy* brutes, under the idea that these will, or can, make them *healthier;* or under the wholly mistaken idea that the medicines which are appropriate to particular diseases, are therefore preventives of those diseases, or even exert a tendency in that direction. On the contrary, by dis-

---

* Diabetes, epilepsy, neuralgia, chronic catarrh, hysteria, etc.

† Both Youatt and Spooner concur in this opinion.

arranging the habitual and orderly action of the functions, they actually increase the tendency to disease; and if there is any prevailing malady at the time, they, as it were, open the door for its entrance. To what an innumerable number of domestic animals of all sorts would the epitaph of the Spaniard apply, with a slight change: "I was well; *my owner* wanted me to be better, and I am here."

Some extremely intelligent men, however, attach much virtue to the articles under consideration, in combination with salt, as a general remedy for certain obscure diseases. A. B. Allen, Esq., formerly editor of the American Agriculturist, writes me:—"My brother Lewis had a flock of about two hundred sheep which were dying off with what was supposed to be the rot. They were on Grand Island. He called on me in despair, said he had done everything he could think of, and asked if I could help him. I told him to get large scows, load them with sheep and send them to my farm, nearly opposite to him on the main land. I then took long troughs made of two narrow boards put together in the form of a V. Into these I poured tar about three inches deep; then I sprinkled sulphur profusely; then salt and pulverized alum sparingly. Then I took each sheep and examined its feet thoroughly. If in the least diseased, I washed the feet clean with soap suds and applied the above mixture to them. The sheep would come to these troughs many times per day, just lick a little and go away. I believe I also placed some boards before and behind the troughs (for they stood in an open position) smeared with the above, so that they would be obliged to tread in the mixture when they went to the troughs. The tar, etc., was renewed as often as was necessary, for several weeks. The result was that only three or four sheep died after this: all the rest were soon restored to health, and in six weeks or so, my brother had the pleasure of selling as fine and healthy a fat flock to the butchers as was seen in Buffalo that season. I presume change of pastures and air were beneficial to my brother's flock, but let me tell you that there is nothing like plenty of tar, sulphur, salt and a modicum of pulverized alum to keep sheep in good health, especially on heavy soils, low grounds, and when the water is not over pure and abundant."

WATER IN PASTURES.—Water is not indispensable in summer pastures, but it is unquestionably beneficial to all sheep, and highly important for ewes suckling lambs. It will do at any time in the summer to change sheep from a dry to a watered field or range; but the reverse of this I have always found injurious, particularly to nursing ewes and their lambs.

SHADE IN PASTURES.—The eagerness with which sheep seek shade from the full glare of the summer sun, is of itself a sufficient proof of its utility. Occasional trees or clumps of trees in each pasture afford the most natural shade. Where these and all others (except those made by open rail fences,) are lacking, I believe it would repay the flock-master to form artificial ones by the cheapest means within his reach; and planting at the same time young, rapidly growing shade trees, for the future, would be a judicious and economical measure.

HOUSING SHEEP IN SUMMER.—The comparatively small, choice, high-priced breeding flocks of Merinos are frequently, as has already been mentioned, housed from all summer rain-storms. They are put up nights when there is any prospect of rain, and some put them up nights habitually after the lapse of a few weeks after shearing. The object is to preserve the yolk in the wool, and thereby obtain *color and weight of fleece.*

Sheltering in warm weather is unnecessary, and in the case of the sheep, as in that of all other animals, it is the tendency of habitual non-exposure to beget an inability to withstand exposure. But the Merino is not only an exceedingly hardy animal, but one which possesses a remarkable power of adapting itself to different circumstances. I have repeatedly bought sheep out of these summer housed flocks, and found no difficulty whatever in accustoming them to ordinary treatment. Housing in summer is not, then, of itself of much consequence, if it and its effects are, as I now believe them to be, universally understood. This being the case it would be binding the sheep breeder by more stringent restrictions than we impose on other breeders, if public opinion refused to tolerate the practice.*

---

* I expressed different views in my Report on Fine-Wool Husbandry, 1862. While I stated that the leading breeders were guilty of no deception in this particular, because they avowed their treatment and their motives for it, I urged that it led to disappointments on the part of the buyer, and that it was a purely unnecessary waste of labor and capital. Further information has convinced me that the effect of summer housing sheep is about as generally understood among sheep men, as the effect of stabling and currying horses is understood among horsemen. And the animals subjected to it or not subjected to it can be as readily distinguished from each other, in the fall, when the selling of breeding sheep commences. It is a waste of time; but why shall not the sheep breeder be permitted to waste his time as well as the cattle

PAMPERING.— But when housing is connected with pampering, with a high and forced system of feeding, the case is different. To make show sheep, to make rams saleable, to stimulate an unnatural growth of wool and secretion of yolk, and thus produce what are termed "brag fleeces"— to cover up defects of carcass, to convey false impressions as to the natural size and substance of the animal — some persons feed their sheep a good portion of the summer and all winter, as much as they can safely get them to eat of the richest feed. This treatment is not often given to breeding ewes, at least in its full extent, for it materially interferes with their own safety in lambing, and the lambs are small, weak and difficult to raise. But to young ewes kept for sale and for show sheep, and to rams kept for sale, it is applied to the fullest extent. Thus a good sized Merino ram is made to produce three or four more pounds, and a good sized ewe one or two more pounds, of wool and yolk, than they would if only kept in good ordinary condition.

But he who buys such sheep (for other purposes than slaughtering) — particularly if they are descended from several generations of ancestors which have been pampered in the same way — buys a spent hot-bed. It never will produce again the monster fleece which tempted him to give a monster price for it. If its feed is kept up, it has little value for breeding purposes; if its feed is taken off, it runs down, becomes debilitated and incapable of withstanding ordinary hardships, is subject to every malady, and succumbs to the first one. This was the case with that tribe of monster French rams which first spread over this country, and died within a year like mushrooms — ruining the reputation of the breed. Some of them had been so thoroughly pampered, that they could not sustain themselves on good pasturage, and perished almost without disease, of mere debility. This mode of preparing breeding sheep for sale is not a legal fraud; but it is dishonest and dishonorable by whomsoever it may be practiced.

No one will deny that every man has a right to keep his sheep *well*, whether he proposes to sell them or not. Good keeping may be pronounced the custom of all breeders. I am not sure, indeed, that it is not necessary to certain

---

breeder, the horse breeder, and the breeder of every other description? The world has agreed to find fault with no class of producers for "putting the best side out," provided no deception is practiced and no injury done to the thing produced in thus fitting it for market.

improvements. For example, size cannot be increased, nor even kept up without abundant feed. The highest bred Short-Horn dwindles rapidly in size in each succeeding generation — however strong the individual and family tendency to size — if put on thin upland pasturage and fed only hay in winter. I do not suppose that Mr. Ellman could ever have raised the flat rib of the unimproved South Down to its present almost horizontal spring from the back-bone, had he suffered his sheep to remain ill-fed and empty — because, while it is true that the viscera adapt their size to the inclosing structures, it is equally true that the bony and muscular inclosing structures adapt their size and shape to the viscera. Whatever *we* may do, nature insists on and enforces harmony!

Good keep may be pronounced necessary to improvement in other particulars: but while the fire warms and cheers and strengthens, the conflagration destroys! Knaves are generally very much puzzled to ascertain, in all such cases, where the good agency ends and the bad one begins. Men of common sense, common experience, and *common honesty*, labor under no such difficulties. They can decide at once between good keep and destructive pampering.

CHAPTER XVIII.

FALL MANAGEMENT.

WEANING AND FALL FEEDING LAMBS — SHELTERING LAMBS IN FALL — FALL FEEDING AND SHELTERING BREEDING EWES — SELECTING EWES FOR THE RAM — COUPLING — PERIOD OF GESTATION — MANAGEMENT OF RAMS DURING COUPLING — DIVIDING FLOCKS FOR WINTER.

WEANING AND FALL FEEDING LAMBS.—Lambs of all breeds should be weaned at about four months old; and if drouth or other circumstances have occasioned a particular scarcity of pasturage for the lambs and their dams, and the former can be put on good feed by separating them, it would be advisable to take off the lambs three or even four weeks earlier. The somewhat prevalent idea that it is improper to wean them in "dog days," has not a particle of foundation. But whatever the period of weaning, sweet, tender pasturage is indispensable for them. New seeded stubbles and the rowen of meadows are usually reserved for them in this country. But many flock-masters prefer rested pastures — i. e., those which, after being fed close, are cleared of stock and allowed to spring up fresh. A few of our breeders of English sheep fold their ram lambs on rape.

The modes of weaning and fall feeding lambs now practiced in England may interest the breeders of English sheep in this country. The following directions are from the Royal Agricultural Society's prize essay on the Management of Sheep, written by Mr. Robert Smith, of Burley, 1847:

"Lambs should never be placed upon rested summer-eaten clover pastures, however tempting they may appear, as they invariably cause scouring, fever and other severe ailments. Old grass, clover, or grass-eddish [after-math] is preferable until the autumn quarter commences, which is considered an important one, as much depends upon the manner in which the lambs are started, or taught to eat their winter feed. In the middle of September the lambs are placed in moderate lots upon grass or seeds, as, from the domestic habits peculiar to the race, they are fond of picking their food at this season of the year, cabbages being thrown to them upon the pastures, or cut for them in troughs: after a short time a few white turnips are mixed with them as a preparation for the winter. As October advances they are placed upon the common or white turnips. Some breeders mix a little cole seed in the first sowing, which is an excellent plan. After a short time the wether lambs are given ¼ lb. of oil cake, or corn to that value, each per day; at Christmas they are placed upon the Swedes which are cut for them, as also the white ones upon bad layer."

In the "commended essay"[*] of Mr. T. E. Pawlett, on the same subject, 1847, occur the following statements:—"I have found lambs to thrive much better on old keeping — as red clover, sanfoin, or grass — than upon what are termed eddishes; yet I must state that old white clover, or trefoil stubbles, are, when they are seeded and have become dry, the very worst of all kinds of food for young lambs. If, however, proper food cannot be provided for them, they should often have their pastures changed to keep them healthy, when a little oil cake or a few split peas or beans (one pint a day among four lambs,) would do them no harm. Having proved by many experiments the advantages of putting young lambs, after weaning, upon old keeping — namely, pastures that have been stocked from the commencement of the spring — over eddishes or pastures that have been previously mown the same season, I will state one experiment as a sample of the rest. In the year 1834, I put a lot of lambs on some old sanfoin, having a few tares carried to them, and another lot of lambs were put on young sanfoin, or an eddish which had grown to a pasture; these, also, had some tares. Each lot was weighed at the commencement, and again at the end of the trial:

"Gain in weight on a lot of lambs fed on old sanfoin, from July 10 to August 10, each on the average,... 14½ lbs
Lambs fed on sanfoin eddish, gained each in the same time,.............. 8½ lbs

Difference,.. 6 lbs."

The moist, mild climate and constant rain, in England, affect pastures very differently from the scorching and often

---

[*] This is headed as follows:—"A Commended Essay, written in competition for the premium awarded to Mr. R. Smith, by the Royal Agricultural Society, 1847." Mr. Pawlett is known as a distinguished breeder of Leicesters.

very dry summers of the United States; and as a general thing I have found good fresh rowen or after-math on meadows, or the new seeded grass in grain stubbles, better feed for lambs than rested pastures, unless the latter have been seeded the same or the previous year, and the grass on them is tender and fresh.

Both of the above quotations, however, teach one valuable lesson to those who have not already learned it — the high importance of giving lambs generous keep from the time of weaning until winter in order that they may continue growing rapidly during that entire period. If by poor keep or any other cause, their growth is seriously arrested, and instead of the rounded plumpness of thrifty lambs, they put on the dried-up appearance of "little old sheep" — the poorer ones are likely to perish outright before the close of winter; and by no amount of care or feed can the others be brought to the next spring equal with lambs which receive only common feed in winter, but which were kept properly through the fall months.

Lambs, when separated from their dams for weaning, should, if the feed is good enough, be left for a few days in the field where the flock has been previously kept—their dams being taken away to a new one. The lambs are more contented and make fewer efforts to escape when thus familiar with the place. The two fields should be so far apart that they cannot hear each others' bleating. If this is impracticable, the fence should be carefully stopped, for if a few lambs crawl through and again reach their dams, they will not give up renewing their efforts to escape and communicating their own restlessness to the others, for twice the usual weaning period. Two or three escapes establish a habit which it is difficult to overcome.

It is a great advantage to put two or three very tame old crones which have not lambs of their own, or a lead wether, among the lambs, to teach them to come at the call; and to lead them up to, and set them the example of eating salt, trough-feed, etc.

The dams should be put on the dryest feed on the farm for a fortnight after separation, to stop their flow of milk. The udders of some of them may require to be milked out once or twice, and if these exhibit much redness and warmth, they should be bathed as recommended at page 158. Smearing the udders with a thick, pasty mixture of soap and water, after a previous washing in cold water, is sometimes resorted to. I have already sufficiently adverted to the high importance of preserving the udders of breeding ewes in a perfectly normal condition. When entirely dried off, they should be put on good feed to get into condition for winter.

As soon as the fall frosts have touched the grass, it is highly beneficial—nay, it is indispensable in good sheep farming—to give lambs some kind of artificial feed. Turnips are (I am sorry to say,) but little raised among the great mass of our sheep farmers, and rape and cabbage are nearly unknown as field crops. Any of these would be vastly cheaper than grain feed; but in default of them, grain feed should be given. At first a little sprinkling of oats, shorts, bran or the like should be put once a day in troughs, in their pasture. By keeping them from salt on other occasions and salting their trough feed very slightly, they, led up by the crones, will first nibble at and then eat it; and when even a few do this, the rest will rapidly follow their example. A spoonful of oats a head is more than enough to begin with; and when they get well to eating, this may be gradually increased to half a gill per head—and before winter to a gill, or to its equvalent in shorts, bran, or other grain. Bran and shorts, or shorts and oats, mixed half-and-half, are proverbially good feed for lambs. An addition of turnips to these would leave nothing to desire. Indian corn, in despite of the fears entertained of it by some persons, for that object, is also an excellent lamb feed; but it must be given more sparingly. A bushel of it is equivalent to its *weight* in oats.*

SHELTERING LAMBS IN FALL.—Sheltering lambs from the heavy, cold rain-storms which fall for a month or a month and a half before the setting in of winter, in our northern latitudes, is now beginning to be practiced by all the best flock-masters; and when the ground becomes wet and cold, and frequently freezes, toward the close of autumn they should also be regularly housed every night. It is well to have racks of hay ready for them in their stables; and it is very easy to learn them to eat grain, etc., there. If it is regularly placed in the troughs over night, with a very light dusting of salt, as before mentioned, but two or three days will elapse before it will be regularly and entirely consumed. Getting

---

* A bushel of corn weighs 58 lbs., a bushel oats 32 lbs., by the rule established in New York.

the lambs accustomed to the stables before winter, is in itself no inconsiderable advantage.

FALL FEEDING AND SHELTERING BREEDING EWES.—It is a common and very truthful saying among observing flock-masters, that "a sheep well summered is half wintered." Breeding ewes should be brought into good condition by the time the first killing frosts occur. After that, they should not be suffered to fall off, but be kept rather improving by feeding them, if the condition of the pastures render it necessary, with pumpkins, turnip-tops, and any other perishable green feed on the farm — and after these are exhausted, with turnips. If some of the oldest and youngest ewes remain thin, they should be separated from the others and fed rather better — grain not being withheld, if it is necessary to bring them into plump condition before winter. Shelter from late, cold storms, though not as important as in the case of lambs, is very desirable, and there can be no doubt that with persons possessing convenient and commodious sheep stables, it will well pay for the trouble to put up breeding ewes nights whenever the weather is raw and the ground wet and cold.* In default of artificial green feed, hay or corn stalks should be regularly fed to sheep — once or twice a day, according to circumstances — as the pasturage becomes insufficient for their full support.

A singular idea prevails among a class of our farmers, in regard to fall feeding sheep, which has been handed down from those days when the two dozen gaunt, "native" sheep which belonged to a farm and which roamed nearly as unrestrained as wild deer through field and forest, did not "come in to the barn" before the ground was covered with snow. In coppices, on briars, and in swamps where the water kept the snow dissolved — and by digging in the fields — they even found subsistence until the snow became deep and so packed and crusted by sun and wind as to prevent their reaching the ground. They then retreated to the barn-yard, usually lank enough! But every farmer knows the immense difference whether in the fields in summer, or in the

---

* My own flocks have generally been too large and spread over too much surface, to render housing from storms practicable until the sheep are brought into their winter quarters; and if well kept, they certainly do well enough without it. But I housed a flock of lambs last fall, and I thought the benefit was very obvious. I have repeatedly observed the same thing in other men's flocks — particularly in Vermont. In that State, fall housing is almost as common, and is regarded as almost as indispensable, as winter housing. This is probably somewhat a question of climate.

stable or barn-yard in winter, between recruiting up and getting into condition two dozen, or two hundred lean, reduced sheep. The little handful of "natives" choosing every morsel of their food over one or two hundred acres of land, through the summer, had high condition to fall back on, in the pinch of the early winter; and when put into the barn-yards with the cattle and young horses, they still chose all the best morsels of the hay — robbing the latter animals — so that they not only made a shift to live, but usually got round to the next spring in tolerable order. True, when let out to grass again, their condition began to change so rapidly that they frequently shed off nearly all their wool — so that many of them had not half a pound a piece at shearing; and those which escaped this were very likely to have their fleeces half ruined by cotting. But what of all this? This was the way things were done in those days!

Brought up under such traditions, many of our older farmers who consider it highly essential as well as profitable to give their cows, horses and other animals, artificial and extra feed a month before the winter sets in, consider every pound of fodder bestowed on sheep at that time, so much taken from the profits which these animals *are bound*, under all circumstances, to yield to their owners — a total loss! A more absurd and pernicious notion could not prevail. If sheep could withstand the effects of such treatment with as little danger to life as the horse or cow, it would still occasion a much greater proportionable loss in their products.* But they can not. The former are capable of being raised at any period of the year, from the lowest condition of leanness, without danger. The muscular and vascular systems of the sheep are so much weaker, that if they become reduced below a certain point in winter — and if they herded together in considerable numbers — their restoration to good condition is always difficult and doubtful, and, in unfavorable winters, impracticable. Their progress thenceforth is frequently about as follows: If fed liberally with grain, their appetites become poor and capricious, or if they eat freely it is followed by

---

* I urge no "petting" or enervating system of treatment. I have not five times within thirty years fed hay or grain, or brought in the body of my store sheep from their summer pastures, before the fall of snow — which generally occurs in this climate not far from the first of December. But I should have done it in all cases, if they had not sufficient feed in their pastures. In this respect I would put them on precisely the same footing with cows and horses. And I would sooner limit the feed of either of them in the winter, than during the month preceding winter. Unless the fall feed was unusually abundant and good, I have always fed my lambs and crones pumpkins, turnip tops, grain, etc., and a little hay as soon as they would eat it.

obstinate and enfeebling diarrheas. Low, obscure forms of disease seem to attack them and become chronic. The strength of the lambs and of the very old sheep, rapidly fails. They scarcely move about. The skin around the eyes becomes bloodless. The eyes lose their bright, alert look, and yellow, waxy matter collects about and under them. A discharge frequently commences from the nose — perhaps the result of a cold, but how or when taken it is frequently difficult to say. The viscid mucus dries about the nostrils so that they cannot breathe freely without its removal. The evacuations become dark colored, viscid, and have an offensive odor. The strength fails more rapidly; the sheep becomes unable to rise without assistance; and it falls when jostled to the least degree by its associates. It will taste a few morsels of choice hay, but generally the appetite is nearly gone. Some, however, will eat grain pretty freely to the last. Finally, it becomes unable to stand, and after reaching this stage, it usually lingers along from two or three days to a week, and then, emaciated, covered with filth behind, and emitting a disgusting fetor, it perishes miserably.

Post mortem examination shows that this is not the rot of Europe. Some American flock-masters term it the "hunger rot." If to this could be added something to express the fact that the hunger which engenders it, usually occurs in the fall, before the setting in of winter, it would be an admirably descriptive name!* It is true, that entering the winter poor does not prove equally destructive in all instances. Its effects doubtless may be materially enhanced or diminished by the regularity and excellence of the winter management, the nice condition of the feed, etc., or the reverse of these conditions. And the character of the winter itself exerts a very marked influence. Sheep thrive best when the temperature is comparatively steady — no matter how cold. A cold, blustering, stormy winter is preferable to one of greatly milder temperature, if its fluctuations are frequent and great — storm and thaw, rapidly succeeding to each other. There comes occasionally what farmers term a "dying winter," when almost any adverse conditions become fatal — and when almost every disorder assumes an epizootic, malignant and fatal type.

Certain specific diseases, like cold, catarrh, pulmonary affections, diarrhea, dysentery, etc. — the most common ones

---

* It might not inappropriately be termed the "fall-hunger rot."

which are of a dangerous description — are far more liable to attack sheep when in low condition. And it is surprising with what destructive effect ticks will work on very poor sheep and lambs. The latter are sometimes literally depleted and irritated to death by their blood sucking.

I have specially and strenuously urged the point of bringing sheep into the winter in good condition, because it admits of no doubt that this, far more than any other one item of management, constitutes the sheet anchor of all successful sheep farming.

There is a point of importance which I have overlooked in the preceding statements. A flock of ewes which are in inferior condition, and especially if they are at the time running down, will not take the ram as readily as a fleshy, thriving flock. It will take six or seven weeks to get the bulk of them served, and then a number of them will "miss," especially if the weather is very cold. A high-conditioned flock is often served in about thirty days. The saving of time and trouble at lambing, and the superior evenness and value of a flock of lambs which is obtained by having them all yeaned within a few days of each other, is well known to all sheep farmers. Many flock-masters give their ewes extra feed during the coupling season, to promote this object. A little sharp exercise, like an occasional run across a field, is thought by many to excite ewes to heat — but I have never tried the experiment.

SELECTING EWES FOR THE RAM.—Where there is an opportunity to choose between several valuable rams, the selection of the ewes to breed to each, requires judgment and careful study. The flock of ewes should be examined, the individual excellencies and faults of each, and her hereditary predispositions and actual habits of breeding, so far as can be ascertained, fully taken into account; and then she should be marked for the ram, which, in himself, and by his previous get, appears, on the whole, best calculated to produce improvement in their united progeny. Many of the Vermont farmers thus divide their small flocks of ewes into parcels of ten or twenty each, and take them to rams owned by a number of different breeders: for, by a prevailing custom, the liberality of which cannot be too highly commended, all the most distinguished breeders of that State allow other persons to send ewes to their best stock rams for a merely nominal compensation, considering the advantages which are often thus secured.* This enables the owners of flocks who can not afford to incur the serious cost and risk of keeping a number of high-priced stock rams, to obtain, notwithstanding, the services of those which are best adapted to breeding with each class of their ewes. And the young or less skillful breeder can thus, too, obtain the immense advantage of using the most perfect sire rams in the country — those which are too costly for his purchase † — and those which will improve his flock more in the first generation than he could possibly otherwise improve it in five generations.

COUPLING.—Very few flock-masters now feel that they can afford to bestow the whole annual use of a choice, high-priced ram on the seventy-five, or at the very utmost, on the one hundred ewes he can serve, if he is permitted to run at large with them; and to accomplish this, he must be a very strong animal, and must be taken out of the flocks nights and fed by himself. And no even tolerably good manager turns two or more valuable rams at the same time into the same flock to waste their strength, ‡ excite, worry, fight, and perhaps kill each other. Even the ewes are frequently injured by the blows inflicted by a ram while another ram is covering her.

There are several different modes of putting ewes singly. Some keep "teasers" in the flock so "aproned"§ that they can not serve a ewe, and daubed with lard and Venetian red under the brisket, so that when a ewe will stand for them she is marked red on the rump. The flock is driven several times a day into a small inclosure (usually a sheep barn,) in apartments of which the stock rams are kept, the "redded" ewes are drawn out and each is taken to the ram for which she is marked. After being served *once* she is turned into the flock of served ewes.

---

* The customary price has been from $1 to $2 per ewe — but I am informed that some leading breeders will feel themselves under the necessity of raising the price of service.

† Some of the more celebrated stock rams whose services are thus let, would sell for more than the entire flocks of many of those who hire their services!

‡ The question is sometimes asked whether the cohabitation of two males with the same female, occasions superfetation, or conception after prior conception. When there are two or more progeny at the same birth, facts have occasionally occurred which appeared to show quite conclusively that they were begotten by different males, but such cases are exceptional; and when there is but one progeny, no facts ever go to show that it is the combined progeny of two male parents.

§ The apron is a piece of coarse, open sacking, which covers the belly from the fore to the hind-legs, and extends half way up each side. Careful persons tie or buckle it over the back at both ends and in the middle, and then fasten it from slipping back by a strap round the breast, and from slipping forward by a strap around the breech. Though allowed to bag a little in the middle, the urine soon renders it a very dirty affair. When I last used teasers, I kept the same one in a flock only every third day.

Another mode is to use no teasers, but to drive in the flock selected for a particular ram twice a day, and let him loose in it; and as soon as a ewe is served to draw her out. After three or four are served, the ram is returned to his quarters, and the remainder of the flock to the field. A very vigorous ram may be allowed to serve from eight to ten ewes a day. This last mode is now generally preferred. It takes up but little more time than the other. It saves the expense and trouble of keeping teazers, which must be frequently changed; for after making their fruitless efforts for two or three days, they generally almost cease to mark ewes. Lambs and yearlings are nearly useless as teazers. Good stock rams ought not to be put on this service, for it rapidly reduces them in condition.

Any mode of effecting the object in view — one on the correct management of which the success of breeding so much depends — must be conducted with rigid accuracy, so that the mark on the ewe shall in all cases indicate the ram actually used. An erroneous record is vastly worse than none. It misleads the owner, and cheats the purchaser who buys with reference to its showings.

The served ewes should be returned to the ram after the thirteenth day. If they come in heat again, it is usually from the fourteenth to the seventeenth day; but the number is ordinarily quite small if the ram is a good one, and is well managed.*

PERIOD OF GESTATION.— The time during which ewes go with young frequently varies upwards of a week — in some unusual cases, nearly two weeks. They usually go longer with ram than with ewe lambs. The average period of gestation does not usually vary much from one hundred and fifty-two days.

MANAGEMENT OF RAMS DURING COUPLING.— Whatever system of coupling is adopted, the ram demands extra care and feed during the season of it. Whether taken from the flock only at night, or kept from it entirely except when

---

* A ram which has been ill, or overworked, may not get lambs one year and may prove a sure lamb-getter the next. Sometimes rams fail in this respect in the opening of the season, but not subsequently — or *vice versa*. Occasionally a Merino ram is hung so low in the sheath that he cannot serve a ewe. If he is valuable, some persons give him the advantage of a platform, raised three or four inches. Others buckle a broad strap tight enough around his body to elevate the point of the sheath sufficiently. With some rams confinement to dry feed a few days is all that is necessary.

covering, his separate inclosure should of course be dry, clean and comfortable — properly ventilated and lighted: and it is better that it entirely seclude him from seeing or hearing the ewes, except when he is admitted to them. It should also be strong enough to defy his utmost efforts to escape.* He should have fresh water in a clean bucket (no sheep freely drinks dirty water, or out of a dirty bucket,) at least three times a day — the choicest of hay — and be fed on grain morning and evening. That mixture of oats and peas which is produced by sowing three bushels of the former to one of the latter — with one-quarter part of wheat added, constitutes an admirable grain feed, when the ram's powers are severely taxed. A quart of this mixture daily, and sometimes even more, is often fed to a good-sized, mature animal, which has been used to hard service and high feed. It would, however, cloy the appetite, if the feeding was not commenced two or three weeks in advance of the coupling season and gradually raised to that point. This should be done not only to prevent that result, but to give the ram a degree of preparation for his work. He ought, by no means, however, to be shut up in his stall without exercise during this preparatory period.

It is not to be understood that the precise mixture of feed above recommended, is indispensable. But all the articles named contain a very large proportion of those nitrogenized matters which produce muscle, or lean meat, and consequently strength, energy and activity,—while Indian corn, oil meal, etc., contain an excess of carbon which tends to the production of fat. The ram demands the former, and is only encumbered by any excess of the latter.

One rule is to be kept steadily in view in feeding a ram during the coupling season. He should not be fed more at a meal than he will consume briskly and cleanly. If he leaves any part of his allowance, it should be removed from his manger; and if this is found to be habitual, the allowance should be reduced.

I regard it as highly inexpedient to keep two rams in the same inclosure or room at this period, however well one may seem to be subjected to the other. Jealousy often provokes even the weaker one to make battle: and an animal of great value may be sacrificed by a chance blow.

The modes of putting ewes and managing rams I have

---

* Powerful Merino rams which have acquired the habit of breaking inclosures, will often dash through the side of a barn, or knock a stable door from its hinges, at the second or third blow. They are "battering-rams," indeed!

recommended demand some expenditure of time and labor. It would probably consume all the time of an active shepherd properly to take care of four hundred and fifty or five hundred ewes and the number of rams required to serve them, during the ordinary coupling period of thirty-five or forty days: and if he had but two hundred and fifty or three hundred to take care of, it would still consume all his time. But the labor of one or two men for that period, would be a very trifling matter compared with the benefits thus secured. These directions are not, of course, intended for the owners of cheap, common flocks who are aiming at no important improvements, and who would regard $25 an enormous price to pay for a ram, and who oftener do not pay more than $5.* But for the last ten if not twenty years preceding the late rise in the price of sheep, those Merino and English rams which *breeders* regard as first class ones, have sold for at least $100 a piece—frequently for twice or three times that amount, and, as already remarked, no property is more precarious.

When the period fixed on for coupling is over, it is generally decidedly best to separate the rams from the flock and keep them separated until that period again recurs. If rams are allowed to run with the ewes either in winter or summer, there is always a chance of having lambs come at very unseasonable times. Eating at the same rack or trough in winter with horned rams, is dangerous to breeding ewes. If the former are cross the danger is great; but even if not, the ram, in making his way to the rack through a crowd of ewes, is liable to inflict unintentional injury on those in advanced stages of pregnancy.

DIVIDING FLOCKS FOR WINTER.—In latitudes where sheep are fed dry feed, and are kept confined to stables and small yards in winter, even Merinos will not bear herding together in large numbers. They should be divided into separate lots before, and preparatory to, going into winter quarters. It is better that these lots be made as small as convenience permits, and not exceed 100 each. The sheep in each should be as nearly uniform in size and strength as practicable, or otherwise the stronger will rob the weaker, both at the rack and trough, and will jostle them about

---

* I could illustrate the curious kind of economy sometimes exhibited in regard to rams, by *naming* an individual residing on the borders of this (Cortland) county, who has within the last five years allowed 60 good ewes owned by him to go without the ram one year, rather than pay $10 for a decent one, which was offered him at that price!

whenever they come in contact. Breeding ewes, wethers and weaned lambs, should always be kept in separate parcels from each other, in well regulated flocks.

Sheep which are old and feeble, late born lambs, etc., had better be sold at any price or given to a poor neighbor who has time to nurse and take care of them. But if kept by the flock-master, they should be put by themselves in a particularly sheltered and comfortable place where they can receive extra feed and attention. This is usually called "the hospital."

English sheep should be divided into still smaller parcels, and with the same regard to age, condition and sex.

CHAPTER XIX.

WINTER MANAGEMENT.

WINTER SHELTER — TEMPORARY SHEDS — HAY BARNS WITH OPEN SHEDS — SHEEP BARNS OR STABLES — CLEANING OUT STABLES IN WINTER — YARDS — LITTERING YARDS — CONFINING SHEEP IN YARDS AND TO DRY FEED.

WINTER SHELTER.— It has already been assumed that a degree of winter shelter is requisite for the most profitable management of sheep in all parts of the United States. The Merino can withstand far greater exposures to extremes and to rapid fluctuations of weather, than any other improved or really valuable breed. In Spain it was unsheltered. In Western Texas — in that magnificent sheep-growing region which lies immediately north of San Antonio — it has been claimed that it requires no shelter; but facts which I shall allude to hereafter incontestibly prove the contrary.

SHED OF POLES.

TEMPORARY SHEDS.— Adequate shelter in warm regions like Western Texas, demands no arrangements which would be at all expensive in a well-wooded region, or where sawed timber could be obtained at moderate prices — for the cheapest form of open shed (i. e., open on one side,) would answer the purpose. Or, excellent sheds might be constructed with logs or poles. The pole shed is made as shown above.

This is covered with straw, reeds, sods, brush, clay, or anything else which will prevent the wind and rain from driving through it. It is decidedly improved by raising the lower ends of the poles two feet by means of a log, stone-wall, or a bank of earth or sods.

CLUMPS OF TREES AND STELLS.—If one generation would be persuaded to make arrangements for another generation, good sheep shelters could be cheaply formed, and on the most comprehensive scale, by planting clumps or belts of woodland, for that purpose, on the vast timberless plains of the Southwest. Evergreen trees would be far preferable, if they could be obtained, and would flourish in the situations where they are required. With stone walls or hedges on the west and north, even a small clump of such trees would form a far better stell than many of those which are used on the bleak and storm-swept highlands of Scotland,—which consist of walls alone. Larger clumps would answer without the walls; but they should be sufficient to protect sheep from the fury of the wind, which renders cold vastly less endurable by them — particularly when it follows a rain which has penetrated to their skins. For this object, and indeed for all objects, naked stells composed merely of high stone walls, board fences, or double lines of poles with straw, sods or earth filled in between them, are far better than no protection.

HAY BARNS WITH OPEN SHEDS.—In all the States lying south of 40 deg., open sheds are sufficient winter protection for Merino sheep, and probably so for the English mutton varieties,—though perhaps the high-bred New Leicester would, in many situations, find more protection profitable at some periods of the year.

Hay barns and sheep sheds like those on the following page, or of some analagous construction, were much in vogue in the Northern and Eastern States, a few years since.

But there were many difficulties about them, in the climates of those States. Snow often blew under the sheds when the wind was in front; and in severe gales, even when the wind was in their rear, it drifted over from behind — piling up large banks immediately in front, which gradually encroached on the sheltered space, and filled its bottom with water whenever there was a thaw.

If a cold storm, or a very freezing temperature occurred at lambing time, these open sheds did not sufficiently exclude the cold; and they did not prevent the ewes going out of them to lamb, or from leading their new-born lambs out at very unseasonable times, to follow the movements of the flock.

SHEEP BARN.

No female animal is more attached to her young than the ewe, but none exhibits less providence in protecting it from any danger, except by setting it an example of running from those which terrify and demand flight.* If the ewe needed

---

* Even then, if seriously frightened, she generally runs directly away from the danger without stopping for her lamb if it cannot keep up. She has not the remotest idea of sheltering it from cold by the warmth of her own person, or any apparent consciousness that anywhere, or under any circumstances, it is weaker or tenderer or more exposed to danger than herself. We read anecdotes of a very contrary tenor among sentimental writers, and naturalists who wish to enliven their narrations, or sustain some favorite theory. These anecdotes are very pretty—sometimes affecting; but unfortunately in ninety-nine cases out of a hundred, untrue! Jessie, for example, expatiates on the fact that the ewe with twins does not allow one of them to suck until the other is ready to share in the meal. Now every practical sheep farmer has been a thousand times provoked by seeing a ewe, followed by one strong, fat twin lamb which she allowed to fill itself at pleasure, moving restlessly about, without waiting for, or seeming to have any care for, its mate, which was born weaker and less able to follow—and which is being starved to death in consequence of its weakness. Even Mr. Youatt talks of special attachments between particular sheep, and of their "alternately sheltering each other from the biting blast and the suffocating drift." He quotes from the Shepherd's Calender the following statement:—"When a sheep becomes blind it is rarely abandoned to itself in this hapless and helpless state: some one of the flock attaches himself to it and, by bleating, calls it back from the precipice, and the lake, and the pool, and every kind of danger." (Youatt on Sheep, p. 375.) I have no doubt that the half wild breeds in the mountains of Scotland, and in other regions where they are left almost in a state of nature to obtain their food and take care of themselves, retain far more of their natural instincts than the more thoroughly domesticated sheep. They will band together to fight an enemy, and it is said the ewe will fight a fox or small dog in defence of her lamb. I never saw an instance of either, among the Merinos. I never saw one sheep render another any direct or intentional assistance of any kind unless the following are instances of it. There are a few rams which will not permit a stranger to catch out one of their ewes when they are together in the winter yard. I own such a ram now, and even his attendant has to act with great caution under such circumstances. Whether the precise object of the ram is to *protect* its associates, I am unable to say. The Merino, removed to mountains or great plains, and removed from the constant control and supervision of man, may acquire, or *resume* habits more necessary in such situations.

assistance in lambing, or if the lamb required to be helped to the teat, it was difficult to catch her conveniently in an open shed.

SHEEP BARNS OR STABLES. — For all the preceding reasons, barns or stables for the winter shelter of sheep, now receive universal preference in the Northern and Eastern States. These are generally constructed — and always should be — so that they can be closed as tightly as ordinary horse or cow-barns. But they require doors sufficient for ventilation and exposure to the sun in fine weather, and for the ingress of a farm wagon to haul out manure. And by means of movable windows, or slides covering apertures in the walls, they should be capable of being thoroughly ventilated at any time, with the doors closed.

When these close sheep barns first came into use, each was generally made large enough for seventy-five or one hundred sheep; and they were scattered about the farm so as to be contiguous to the meadows from which they were to be filled with hay, and so the manure made in and about them would only require hauling a short distance. There was another argument in their favor. If a contagious or infectious disease broke out in one of the divisions of the flock, it did not necessarily extend to all; and, theoretically speaking at least, the fewer the sheep which inhale the same local atmosphere the freer from impurities it must remain.

But serious inconveniences were found to attend this system. It required almost a double outlay of materials and expense to build separate barns and prepare separate yards, arrangements for watering, etc., for each flock. These scattered barns required the farmer or his shepherd to wade wearily two or three times a day, mounted or on foot, for long distances through sheets of snow which the winds generally rendered pathless; and oftentimes, and even for days together, to do this amidst blinding snow-storms or the most terrible extremes of cold. Much shoveling was constantly necessary to give the sheep access to water, etc. If the supply of hay happened to fail at one of these distant barns, it was often more trouble to get it there, than it would have been to cart all the hay consumed in the barn to a central one near the farm-house, and haul all the manure made from it back. These barns were inconvenient at lambing time, because the constant attention which one man could give to all the breeding ewes at once, if in the same or contiguous buildings, was necessarily divided up between the several scattered parcels of them, leaving but little time, comparatively, for each. And, finally, the farmer was not so apt, under such circumstances, to see all his sheep daily *with his own eyes ;* nor was either he or his shepherd half so prone to turn out *in the night* to take care of the sheep or the lambs, provided a change of weather, the rising of a gale, or any other circumstance rendered it expedient.*

It is now usual to construct the sheep, like the horse and cow barns, near the farm-house. When the farm flock does not exceed about three hundred, it is often wintered in a single barn which has separate apartments, holding from seventy-five to one hundred sheep each; and each apartment has a separate outside yard. The upper story of these barns is devoted to hay for the sheep: the under one is eight feet high, and floored on the bottom if it is necessary to insure perfect dryness.

It is common to take advantage of a slope in the ground, and by means of a small amount of excavation, so to place the sheep barn that while the doors of the basement story open on a lower level, those of the second story open upon a higher level, or on the surface of an ascent, on the opposite side — so that hay can be drawn on wagons into the upper story. This is something of a convenience, and was a great one before the invention of the horse pitch-fork. The side of the lower story which supports the bank of earth resting against it, is generally composed of stone-wall — this being necessary both for strength and durability. In various states of the atmosphere this wall exudes moisture, or, as it is termed, "sweats,"— diffusing dampness through the apartment. Unless that apartment is far higher, more spacious and better ventilated than would otherwise be necessary, this dampness is unquestionably prejudicial to the health of sheep. The better course would be, where such a barn is thought desirable, to build it entirely independent of the bank-wall and connect them with a short bridge.

The usual way of dividing the lower story of the sheep barn into apartments for different parcels of sheep, is simply

---

* For example, I remember some twenty or twenty-five years since to have had several hundred ewes with young lambs left out on a warm and beautiful night in early May, in four *adjoining* fields. A little after midnight I was wakened by the first howl of a north-easter, which was accompanied by a blinding snow-storm. This was a case to say *come* instead of *go*. In fifteen minutes three of us, with our lanterns, had started for the fields about half a mile off: and we worked on until 9 o'clock the next morning in getting in the sheep, and half frozen lambs, and in resuscitating the latter. We probably saved a hundred lambs which would have perished before morning. Had these sheep been out in the same number of parcels half a mile from each other — some of them a mile and a half from my house — what chance would there have been to save the great body of the younger lambs?

by placing feeding racks across them — so that in reality the sheep are all in one room. This mode is a material saving both of space and expense; and it is highly convenient, inasmuch as the partitions can be changed in a moment to adapt them to any change which it is desirable to make in the relative number of sheep in the different apartments. But it must be obvious that any considerable number of sheep when thus kept breathing the same indoor atmosphere, require that the means of ventilation be abundant and most thoroughly kept in operation. Indeed, I should prefer, as a matter of prudence, not to place more than one hundred and fifty sheep in the *same room*, though divided into smaller flocks on the floor. With different rooms, and with independent means of communicating with the external air, four hundred or six hundred could be kept, perhaps, just as safely, under the same roof, unless during the prevalence of infectious or epizootic diseases. But who can be certain that these will remain absent? On the whole, such large and close aggregations of sheep are inexpedient.

The room required for a given number of Merino breeding ewes in a barn is, for Paulars, about ten and two-thirds square feet of area on the floor each; in other words, an apartment twenty by forty feet in the clear will accommodate seventy-five, so that they can all eat at the same time at single or wall racks placed round the entire walls, except before the doors. A room forty feet square will accommodate one hundred and fifty, but it requires forty feet of double rack* to be placed in the area inside of the wall racks. Larger Merino, or English ewes, require more room in proportion to their size. Some of the last would probably require nearly twice as much room per head.

A sheep barn should open on the side least exposed to the prevailing winter winds; and its yards should be placed as much as practicable under its shelter. Some persons build these barns in the form of an L, to break off the winds from different quarters; others make a high stone wall or board fence a substitute for one of the limbs of the L. The yards are inconveniently narrow if restricted to the breadth of the inside apartments; and should, therefore, be widened according to circumstances.

The following ground plan is intentionally confined to a

---

* I here use the word single or wall-rack to signify one made to set against a wall, which can only be eaten from on one side — the word double rack, to signify one which can be eaten from on both sides, so that forty feet of one is equivalent to eighty feet of the other.

mere outline of a very simple and compact sheep barn, which is under a single roof, has no waste space, and makes the utmost use of all its materials. Three different modes of watering are presented, either of which is sufficient, and the choice between them should depend upon circumstances.

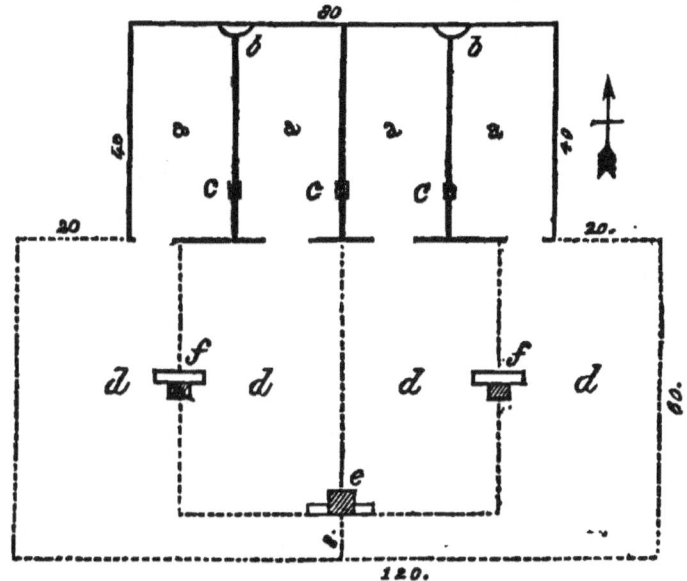

PLAN OF SHEEP BARN AND YARDS.

a, a, a, a, Apartments or stables in sheep barn, 20 by 40 feet. The central partition a close one, with single racks on each side. The other two partitions composed of double racks. Single racks round all the outside walls except at doors.

b, b, Watering tubs, when water is brought into barn in pipes.

c, c, c, A door in central partition and gates in the other two partitions.

d, d, d, d, Sheep yards, 30 feet wide; the two outside ones 60 feet long; the two inside ones 52 feet long; thus arranged to allow the four flocks of sheep to drink from the troughs of one pump-house at *e*.

e, Pump-house and troughs for four yards, if water is not carried into the barn at *b, b*.

f, f, Pump-houses and troughs, each accommodating two yards, provided neither of preceding plans of watering are available or desirable.

Sheep barns are often connected with other farm buildings, such as horse stables, wool rooms, ram stables, etc. The following is the plan of Mr. Hammond's sheep establishment.* His house, wood-sheds, etc., stand south of the barns, so that they principally break the force of the wind from that quarter.

----
* Except a slight change in respect to wool room, which stands detached from barn.

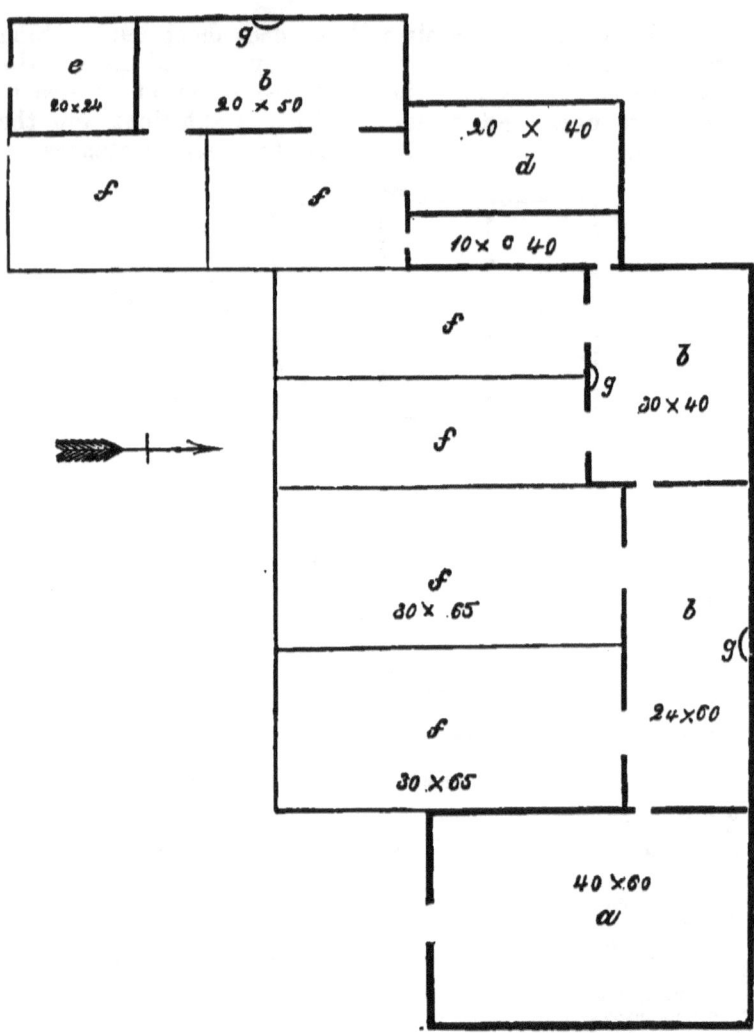

PLAN OF A SHEEP ESTABLISHMENT.

  a. Horse barn.
  b, b, b, Sheep stables, each divided into two apartments by racks across them in the middle.
  c, Ram stable, divided into two apartments.
  d, Hay barn.
  e, Wool room.
  f, f, f, f, f, f, Yards before each apartment of sheep stables.
  g, g, g, Watering places, each supplying two apartments.

Whatever plan is adopted for a sheep barn, certain things are indispensable. It should stand on and be surrounded by dry ground; occupy an elevated, airy position, but one as little exposed as possible to prevailing winter winds; be of easy access to water; possess ample capacity for the number of sheep to be kept in it; and have means of thorough ventilation in every state of the atmosphere. The hay floor above the sheep stables should be matched or battened, so as entirely to prevent dust, hay seeds, or chaff from sifting through on the sheep. It should have pens in the sheep stables to throw the hay in from above when feeding, so that it cannot fall on the backs of the sheep or be run over by them.* Every gate, door, fastening and fixture about it should be strong and secure.

CLEANING OUT STABLES IN WINTER.—It is rather the prevailing custom among Northern flock-masters not to clean out their sheep stables in winter, but merely to cover the manure occasionally with fresh litter. This is unquestionably bad practice, in two particulars. It certainly prevents making anything like the amount of manure which could be formed by mixing the dung and urine of the sheep with an amount of litter which would half fill the sheep stable, if suffered to accumulate there throughout the winter. And there can be no reasonable doubt that a deep bed of manure, which, except during severely cold weather, is constantly heating, evolving gases, and filling the apartment with a warm steam and the odor of fermenting dung, and which, after a decided thaw of a few days, positively produces an offensive stench, can not form a very healthy lair for sheep. It is rather the prevailing opinion now among the best flock-masters, that the increased practice among Merino sheep of pulling their own and each other's wool in the winter, is occasioned by an irritation of the skin caused by lying on these beds of heating manure. Unstabled flocks do not, so far as I have observed, thus become addicted to "wool-biting." Stables should be cleaned out three times during each winter, say in the early part of January, the latter part of February, and in April. And in the intermediate periods, it is an excellent practice always to strew the manure on the floor with plaster (gypsum,) prior to covering it with fresh straw. This absorbs the escaping gases, and thus not only preserves the purity of the atmosphere, but vastly enhances the value of the manure.

---

* Some, instead of this, shut the sheep out of doors when filling the racks. But the state of the weather, as, for instance, in a winter rain-storm, or the situation of the sheep—say when they are lambing—sometimes renders this highly improper.

YARDS.—I by no means wish to be understood to express the opinion that sheep yards should, for any purpose of utility, be restricted to the narrow dimensions of those given in the preceding ground plans. I rather consider those the least dimensions which can be regarded as proper; and if convenience equally admitted of it, I would prefer to have them much more spacious. They should be constructed on dry, firm, thoroughly drained ground; and a gravelly soil rapidly permeable by surface water, and which quickly dries, is much preferable to a clayey, tenacious soil, or a peaty or mucky one which retains moisture. All the yards ought to have separate access to water, and, if practicable, separate access to different fields. This last fact renders the plan of yards given with the first of the preceding ground plans objectionable, unless the two middle flocks can be let into different fields through doors in the opposite side of the barn. That plan merely saves the digging of one well; and I should much prefer to dig the two wells (at $f, f,$) and have the yards of equal length, and each possessed of separate and independent egress and ingress.*

LITTERING YARDS.—Strawing or otherwise littering sheep yards in winter in the most thorough manner, is a matter of prime importance. If sheep are compelled to stand or move about in mud or water whenever out of doors, the most liberal feeding and good management in every other particular, will hardly preserve them in the best condition. They should have a comparatively dry out-door bed to stand on in wet weather, and a warm one in cold weather. The sheep — or at least all the upland breeds of sheep — find one of the worst enemies of their health and thrift in habitual wetness under foot. Muddy yards prevent sheep from moving about out of doors and spending a portion of the time in the sun and fresh air, in pleasant winter weather; promote fouls; render hoof rot incurable; and cause lameness and annoyance to sheep which have sound feet, when a sudden freeze converts the small pellets of mud which adhere to the hairs in the forward part of the cleft of the foot, into pellets of stone. A little straw is excellent feed for sheep. If it is scattered over

---

* By gates opposite each other on the eight-feet passage — one of them opening entirely across it on the side of the outer yards—a separate passage could be obtained; but this would not be very convenient, and when the passage was thus closed, the sheep in the outside yards would not have access to the water trough at e.

the yard they will "pick it over," eating the best parts, and leaving enough to keep the littering constantly renewed.

CONFINING SHEEP IN YARDS AND TO DRY FEED.—A decided majority of Northern flock-masters prefer the strict confinement of sheep to their yards during the entire winter. They contend that the slightest taste of the pasture during thawing weather takes off the appetite from hay, and that sheep are equally healthy and even more thrifty under such confinement. I dissent from both conclusions.

If sheep, long kept from the grass by deep snows, are suddenly admitted to it in consequence of a winter thaw, and if they are allowed wholly to subsist on it for a number of days — as long as the thaw continues — they unquestionably lose condition and strength on herbage which has been rendered innutritious by age and by repeated freezings and thawings. Thin breeding ewes and young sheep sometimes suffer materially in this way, particularly in the critical month of March. When returned to their confinement and to dry feed, they have no vigorous appetite for it, and consequently do not recover from their debility. In certain unfavorable seasons they pine, and eventually perish, if not solely from this cause, yet with the fatal termination accelerated and rendered more inevitable by it. Stronger sheep recover from its effects — but of course any check in the thrift of a flock results in a proportionable loss in some of its products.

Having habitually and regularly fed turnips daily to breeding ewes, young ewes, rams, and wethers, (when I have kept the latter,) for the last fifteen or twenty winters, I am enabled to affirm, of my own positive knowledge, that green feed, administered in proper quantities, does not in the least diminish the appetite for dry feed; and that proper green feed, so far from weakening, adds to the condition and strength of sheep, besides producing other good effects which will be adverted to when I speak of the relative value and influence of winter feeds. The experience of the great body of English farmers fully sustains these conclusions. The practice of wintering sheep exclusively on dry feed — say on meadow hay and straw, with or without grain or pulse — is substantially unknown in the arable districts of England. For sheep of every class not to receive green feed daily would there be an exception; and *fattening* sheep receive it in abundant quantities.

The winter grass in our own Northern States, though

comparatively innutritious, is, in the absence of better green feed, a healthful change in the diet of pregnant ewes. It keeps down the tendency to costiveness, habitual to females in that situation, and in conjunction with that exercise which is required to obtain it, renders the system less subject to the plethora, which is also natural in pregnancy, but which is greatly fostered by rich food and inactivity. But to attain these objects, the sheep should be let out an hour a day, instead of the entire day, in warm winter weather. It should obtain a small portion of its feed, instead of the whole of it, from the fields.

Sheep, like other animals, spontaneously diminish their amount of exercise as they advance in pregnancy, and it thence may very properly be inferred that they require less of it than at other times to preserve a healthy condition. It is also undoubtedly true that excessive or fatiguing exercise is positively injurious at this period. But if we can trust to established physiological principles, or to the teachings of analogy, the sudden change produced in the habits of an active, roving animal, by rigid confinement from the commencement to the close of gestation—accompanied by a complete alteration of diet—must be attended by baneful consequences. Are we told that pregnant sheep thrive and grow fat in this confinement—fatter than when they are let out on the fields? This is true, and it is one of the dangerous incidents of the system. Pregnancy of itself favors the taking on of flesh; and when this tendency is aided by concentrated and highly nutritious food, and by entire inactivity, the condition established is rather that of *plethora* —high condition attended by an unnatural excess of blood— than of the healthy fleshiness which comes with natural feed and exercise.

We know that the sow which is confined closely to the pen and fed to fatness on wholly artificial food never farrows in safety. We should esteem that farmer beside himself who confined his mares and cows to little dry yards and to dry feed during the whole term of pregnancy. The most celebrated practitioners of medicine allow no such changes of habit among their human subjects during this period. I can not do better than to quote the sensible remarks of Dr. Bedford on this subject. He says:

"Allow me here to remark that, *as a general principle*, if the pregnant female observe strictly the ordinances which nature has inculcated for her guidance; if, for example, she *take* *her regular exercise in the open air*, avoid, as far as may be, all causes of mental or physical excitement, employ herself in the ordinary duties of her household, partake of nutritious and digestible food, repudiate luxurious habits, * * * * if, I say, she will steadfastly adhere to these common sense rules, the reward she will receive at the hands of nature will be general good health during her gestation, and an auspicious delivery, resulting in what will most gladden and amply repay her for her discretion—the *birth of a healthy child*. * * But if in lieu of these observances, the pregnant woman pursue a life of luxury, 'eat, drink, and become merry,' *neglect to take her daily exercise*, and prefer her lounge — the case is entirely reversed, etc.*"

I might swell quotations of the same tenor to a volume: for such are the settled opinions of the whole medical profession.

Am I asked where the injurious effects of the close confinement of sheep to small yards and dry feed have manifested themselves? I *suspect* that they have manifested themselves in the prevailing and destructive loss of lambs which annually takes place in our flocks. Why is it that with better shelters and conveniences of every kind, and with greatly increased skill as shepherds, the body of American Merino flock-masters do not raise a larger per centage of lambs than they did twenty or thirty years ago? I have already expressed the opinion that eighty per cent. is still as high as the general average, taking a series of years together, though I know many small flocks in which 90, 95, and occasionally 100 per cent. are raised. The American Merino is a much larger and better formed animal than it was twenty years since, and though it has undoubtedly lost something of that locomotive power and energy which it possessed when it was compelled to make a journey of eight hundred miles each year in Spain, it remains a far hardier animal than the improved English sheep, and it is less subject to parturient difficulties and diseases.† Yet the English sheep rear from

---

* Principles and Practice of Obstetrics, by Gunning S. Bedford, etc., etc. New York, 1862, p. 131.

† Mr. Youatt enumerates among the defects of the Merino, "partly attributable to the breed, but more to the *improper mode of treatment to which they are occasionally subjected*," in Spain, "a tendency to abortion or to barrenness; a difficulty of yeaning; a paucity of milk, and a too frequent neglect of their young." (Youatt, p. 149.) The tendency to abortion is not greater in the American Merino than in the English ewe: the former does not so often experience difficulty in yeaning: and it is decidedly less subject to parturient fevers. It has, however, a greater "paucity of milk," a greater tendency to barrenness, in the sense in which I presume Mr. Youatt

thirty to fifty per cent. more lambs! Our English flocks, it is true, are usually small; and among the established natural characteristics of the ewes are those of bringing forth twin lambs and having a sufficient supply of milk to raise them properly. But they also, so far as my knowledge extends, *lose* fewer lambs. How is this to be accounted for? If the Merino is a hardier animal than the mutton sheep, its lamb, it would seem, ought also to be hardier. And so I have no doubt it is, if it is born in a perfectly well developed, normal condition, and if it gets anything like a corresponding supply of milk. It is not among such that the annual losses among our lambs occur. Those which perish are generally undersized and feeble, or else they do not obtain sufficient support from their dams. It is these causes and *failure to take the ram* which keeps the rate of increase so low in Merino flocks.

This comparative want of prolificacy is the weak point — now really the only one for the purposes for which they are grown — of our American Merino sheep. Yet no other point has received more of the care of those breeders who have been so successful in improving them in every other particular. Their comparative failure is occasioned by no obstacle inherent in the breed, as I could show from a variety of considerations and direct proofs, did space admit of it. If it can be shown that there is a radical error in our modes of management — that we habitually compel the pregnant ewe to violate "the ordinances which nature has inculcated for her guidance" — need we go further to find the causes of that failure? Can we wonder that lambs are born imperfectly developed when ewes are rigidly confined for five or five and a half months — through the entire term of pregnancy — in little yards; and even then fed almost invariably within doors — so that they have no inducements left to take the least degree of exercise — and so that more than four-fifths of the whole time they are inhaling the atmosphere of a stable, without going out into the fresh air and sunlight? Can we wonder that an animal which obtains its entire summer subsistence from green vegetation does not secrete milk abundantly, and can not be bred to secrete it abundantly, when, from the first to the last day of gestation, it is unnaturally restricted to exclusively dry food? And when young and not fully matured ewes, or old and decaying ones, or

---

here uses the word, i. e., it oftener fails to take the ram. Literal barrenness, or a want of the power of conception, is almost unknown in the Merino; and its failure to take the ram, generally, springs from incidental and not necessary causes.

poor ones of any age — the classes which furnish the principal portion of those which do not breed*—are suddenly subjected to the commencement of the preceding changes, about contemporaneously with that great fall of temperature which usually attends the setting in of winter, can we wonder that the depressing effects of all these combined causes should prevent cohabitation? It has already been stated as a well established fact, that not only low condition, but anything which, for the time being, lowers the condition, tends to produce that effect. Even ewes in the most suitable situation for coupling, viz., in good, plump, store order and *improving* in condition, *at the time*, often wholly cease to take the ram in severely cold weather. And as winter advances, the heats of the Merino ewe are less to be relied upon.

Many American Merino sheep breeders, on reading this, will say:—"I have used small yards, fed generally in the stables, fed nothing but dry feed in winter, for ten, fifteen, or twenty years, and I have always had good success in lamb raising." But what proportion of these breeders, whose breeding ewes count up even to one hundred and fifty, would be able to show from contemporaneous records, or would dare to affirm as a matter of positive recollection, that they had on the average, for any considerable term of years, raised either 100 per cent. of lambs, or any *very* close approximation to that number? Yet can lamb raising be considered *successfully* carried on, or a breed to have reached its highest attainable standard in this particular, when a selected flock of only one hundred and fifty breeding ewes can not be made annually to raise their own number of lambs?

There is a material difference in the prolificacy of the English and Merino sheep — first produced, in all probability, by the different modes of artificial treatment to which they were subjected †— but long since established as permanent and hereditary characteristics of the different breeds: but I do not entertain a shadow of doubt that were the most prolific English families of sheep subjected to the same winter treat-

* If there are "dry ewes" in the flock, i. e., those which raised no lambs the preceding year, and they are allowed to become very fat, they too, are very apt not to become, as the English Shepherds say, "inlambed."

† The Spanish sheep were subjected neither to confinement nor dry feed, in the winter, in Spain — but there being no object to increase their number they were not allowed to raise over 50 per cent. of lambs; and consequently prolificacy was not cultivated. While their constant migrations gave them extraordinary general vigor, they did not tend to develop their milking properties.

ment which we give to the bulk of our American Merinos, half a dozen generations would find them seriously degenerated in prolificacy.

Occasionally there comes a year when double, treble and even quadruple the usual number of our lambs perish. The causes and symptoms appear to be the usual ones, but aggravated and extended by an epizootic influence. I have (at page 154,) described the appearance of the lambs, and the singular degree of mortality which prevailed among them in the spring of 1862. An extraordinarily deep snow fell in the early part of winter, and it was replenished about as fast as it wasted away until the opening of spring. It was remarked that most of the breeding ewes clung very closely to their stables — doing little more than rising to eat and then lying down again. Those flocks most accustomed to close yarding in many instances did not tread down the snow a dozen yards from their stables during the winter. But the weather was steady and cold, so that they continued to eat well, and the hay of that season was generally of good quality. Thus their inactivity increased their fleshiness, and their fleshiness re-acted and increased their inactivity. They generally reached the spring in uncommonly high order. They appeared to be well — but yet there were unmistakable symptoms of a plethoric habit in the best fed flocks: and it was in the best fed flocks that the loss of lambs was, as a general thing, far most severe.

Putting all these facts together, I have been disposed to trace this mortality in lambs to the condition of the mothers — the unfavorable condition being aided by an epizootic influence.* Is it asked why a proportionable degree of mortality does not habitually attend all unusual confinement of breeding ewes, and why, in 1862, it did not extend its destructive ravages to Vermont, where the snow was equally deep and laid still longer on the ground? When it is explained why the directly exciting causes of various destructive diseases among human beings, lie comparatively dormant

---

* Having, from inability to fix upon any descriptive or definite name, termed this imperfect state of the lambs of 1862, which resulted in such wide spread death, "the lamb epizootic of 1862," (in some articles which I published on the subject in the Country Gentleman,) several writers appeared to think that I intended to characterise it as a *contagious*, or *infectious* disease. An epidemic, or epidemy, is defined in Dunglison's Medical Dictionary to be "a disease which attacks at the same time a number of individuals, and which is depending upon some particular *constitutio aeris*, or condition of the atmosphere, with which we are utterly ignorant." And he defines epizootia (epizootic) to be "a disease which reigns among animals — corresponding in the veterinary art to epidemy in medicine." This correction is made simply to prevent similar misconceptions in regard to the use of the word in this work.

for years in a particular region — producing only sporadic or separate cases; why, in other years, when all the proximate causes *appear* to be the same, some one of those diseases assumes an endemic or epidemic form, desolating neighborhoods or provinces; and, finally, why, at the height of its fury, it passes round and spares this household or that, or this neighborhood or that, and frequently leaves as well defined margins as the track of a tornado, although the population was as dense without as within its track;—when, I say, these anomalies are explained, we shall be able to explain the one under consideration. And let it be remembered that the same anomalous facts will continue to exist, to stand as much in the way of the true as of a false theory of explanation.

I am not tenacious for the acceptance of this explanation. I merely offer it as the most probable one within my knowledge. Better observed facts may hereafter throw more light on the subject.

I do not wish to be understood that restriction to *dry feed* is necessary to produce that condition of the ewe which I have assumed to be so prejudicial to the offspring. On the contrary, I think it would be produced, though hardly so readily or to so dangerous an extent, by an over-supply of good, green feed, attended with the same other unhealthy auxiliaries. It is the high condition, the excess of blood, the excited vascular system ready to assume or produce inflammatory action, which produce or co-operate with the morbid tendency to non-development in the fœtus. Indeed, high condition alone, may, to some extent, offer a mechanical obstruction to its development. The internal fat of the dam may so far obstruct the full distension of the womb that the fœtus can not grow to its full size anterior to birth.

I urge letting out breeding ewes on the fields for a limited time each day, because no animal more intensely craves a portion of green food in the winter; and I consider nature or instinct a first-rate judge of its own wants: because the small portion of green feed obtained from the fields can exert no injurious influence whatever in any direction, while it prevents the costiveness peculiarly incidental to pregnancy, and by keeping the bowels in an open and regular state, has a strong tendency to avert all unhealthy action or agencies; because traveling about and digging in the snow for green feed affords a most necessary and healthful exercise; and, finally, because a neglect "of these ordinances which nature has inculcated" for the guidance of the pregnant ewe, has been followed by wide-spread disaster, under circumstances which at least give much color to the hypothesis that they are connected together as cause and effect.

It by no means follows from anything which has been said, that sheep require a very extensive winter range on grass. I should decidedly object to their being allowed to feed down all the grass lands on the farm at this period of the year, and particularly the meadows.

A few moderate-sized old seeded pastures about the sheep barn, with a good amount of grass left on them, in the fall, would answer every purpose; for the sheep with its fluted teeth will not only take the grass but some portion of its very roots. It wants but little each day, and the harder it works to obtain it the better it is.

Those who raise turnips for the sheep must obtain exercise for them in some other way. A stack to feed from at noon in fine weather, a quarter of a mile from the sheep barn, is an excellent arrangement; and who does not recollect the old-fashioned, lively and merry scene of hauling out hay on an ox-sled far from the dirty farm yard — the great oxen hurrying forward as if satisfied some frolic was going on — the feeder tossing the fragrant flakes right and left — each succeeding flock pursuing with a Babel of cries — some of the young ones bounding and kicking their heels into the air as if greatly enjoying their fine run over the snow!

I made it a rule in entering upon the writing of this book, to look little after *authorities* where I believed the facts were established by my own observations; but the necessity of winter exercise for sheep seems to be a much controverted question in this country, and therefore I have largely consulted the best European writers on the subject. I have thus far been unable to find one who mentions the subject at all, without distinctly insisting on the necessity of exercise; and when the destructive lamb epizootic of 1862 was terminating its ravages, I addressed letters to a number of the oldest and soundest breeders in our country, describing the disease as I saw it, and asking their opinions as to it origin. To no one did I suggest my own theory of that origin. In every instance, I believe, the *want of exercise* was put forward as either the leading cause, or as a cause second to no other in its effects. Several also stated that they thought the sheep "had been kept too long from the ground."

CHAPTER XX.

WINTER MANAGEMENT — CONTINUED.

HAY RACKS — WATER FOR SHEEP IN WINTER — AMOUNT OF FOOD CONSUMED BY SHEEP IN WINTER — VALUE OF DIFFERENT FODDERS — NUTRITIVE EQUIVALENTS — MIXED FEEDS — FATTENING SHEEP IN WINTER — REGULARITY IN FEEDING.

HAY RACKS.—A great variety of racks for sheep have been introduced into use, but for double and portable ones for ordinary purposes, those of the form exhibited in the annexed cut are generally preferred. The corner posts are 2 by 2½ or 3 inches in size, and are 2 feet 8 or 10 inches long — sometimes 3 feet, where the racks are to be used as partitions. The side and end boards are an inch thick, the upper ones six and the lower ones nine inches wide. The perpendicular slats are three-fourths of an inch thick, seven inches wide and seven inches apart, fastened to their places by wrought and well clenched nails. Each slat requires four nails, instead of two as represented in cut. The slats are highly useful in keeping in hay, but their principal object is to prevent the sheep from crowding. They give every sheep fourteen inches at the rack while eating. This is a liberal allowance for the Merino; but the English sheep requires more room. The ordinary breadth of the rack is two and a half feet, and the length depends upon circumstances. Those intended to be moved often are usually made ten feet long. They should be so light that a man standing inside of one of them can readily carry it about.

SLATTED BOX RACK.

Single or wall racks to be used against the walls of stables and other places where the sheep can approach them but on one side, are often constructed like one side of the box rack and attached to the walls by stay-laths. Some arrange them so that they can be raised as the manure accumulates; but there is no need of this if they are made with the bottom boards a foot instead of nine inches wide, and if the manure is cleaned out as often as it should be.

But a far neater and more convenient wall rack, having troughs also connected with it, was invented by Mr. Virtulan Rich, of Richville, Vermont.* The following cut, from a drawing kindly furnished me by that gentleman, gives an easily understood general view of it:

WALL RACK AND TROUGH.

a, Plank 2 inches thick and 9 inches wide, placed 20 inches from wall (*e*,) to form bottom rail of outside rack.
b, Scantling 3 by 3 inches, forming top rail of outside rack.
c, Bottom of trough, being a board placed on floor, or if there is no floor, on scantling to raise it sufficiently from ground.
d, Board five inches wide, to support the board 4 inches wide, which forms bottom of the inside rack (*f*.) These would be better made of plank. Bottom of inside rack should be 6 inches above bottom of trough.
e, Outside wall of barn or stable.
f, Inside rack hung with hinges to bottom board. It is made by nailing slats 1½ inches wide, 3 inches apart, on upper and lower rails, which are about 1½ by 2 inches in diameter.
g, Slats to outside rack 7 inches wide and 7 inches apart.
h, Slanting board, from bottom of inside rack to bottom of trough and forming back side of trough.

The end-views of the same rack (on next page) render the details of its construction a little more apparent. The left hand cut shows the inside rack (*f*,) in its place as when filled with hay. In the right hand cut, it is turned up or thrown

---

* I have previously, in this volume, named the Messrs. Rich as of Shoreham. This is the name of the *town* in which they reside, and was until recently the name of their Post-Office. The latter is now Richville.

back on its hinges as when grain or roots are being put in the trough (c,) or the trough is being cleaned out.

The advantages of this rack are, 1, That it prevents crowding as well as the slatted box-rack; 2, That it prevents sheep from thrusting their heads and necks into the hay, as they can do to some extent in the slatted box-rack, thereby

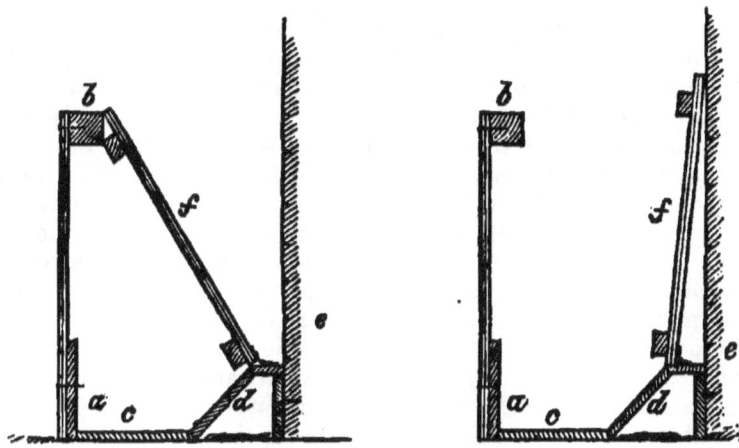

END VIEW OF WALL RACK.

getting dust, hay-seeds and chaff into their wool; 3, That it almost entirely prevents the hay which is pulled from the inside rack from being dropped under foot and wasted;* 4, That it combines the advantages of a good stationary feeding-trough with the rack; 5, That the trough, apart from its ordinary uses, is found very convenient to keep hay-seed out of the manure when it is is desirable to do so, and to catch and save hay-seed for use.

WATER FOR SHEEP IN WINTER. — Sheep, and particularly sheep fed with roots, will do very well in winter without water if they have a constant supply of clean snow; but that supply can never be relied on. And when watered at a pump or stream a portion of the time, they (particularly pregnant

---

* A considerable quantity is wasted from all slanting racks with small, close rounds (like the inside rack *f*, in the cut;) and some is thus wasted even from the slatted box rack. A sheep on being jostled by another, steps back from the rack frequently dragging out quite a lock of hay, which is immediately trodden under foot and hardly ever picked up.

ewes) suffer if again forced to depend exclusively on eating snow. Consequently, a regular supply of water throughout the winter should be regarded as indispensable. It becomes still more so, where sheep are housed and yarded. In winter climates cold enough frequently to congeal water, the most convenient arrangement, where it is practicable, is to bring it directly into the sheep barn, by means of underground pipes from a spring or dam of sufficient elevation to force it up into tubs. These should be placed in the middle partitions, (as seen in the two plans of sheep barns in the preceding Chapter,) so that each tub shall supply two flocks of sheep. If different tubs are supplied from the same spring, each must have a different pipe, or else the tubs must be at different elevations, so that a waste pipe from the higher one will go up into the bottom of and fill the lower one. When the surplus water is finally discharged into the ground, it should be by a waste-pipe emptying into a deep, well-made drain, which will never become clogged. An accumulation of ice in a sheep stable, or any overflow of water into the bedding, would be a nuisance far more than overbalancing all the conveniences of indoor watering. The tubs should rise but a few inches above the floor, and should, if they have much depth, have well secured but movable covers to prevent sheep and lambs from falling into them—the covers having holes cut through them barely large enough to enable the sheep to drink.*

Two plans for outdoor watering are given in the ground plan at page 217. As I have already stated, I decidedly prefer that which exhibits two wells and pump-houses (at f, f,) because free egress from all the yards, independently of each other, could thus be much more conveniently secured. Each well or cistern should be fitted with a pump of a construction which forces up water very rapidly, and which does not admit of its being frozen in the body of the pump, if some special precaution chances to be forgotten. Small pump houses, which can be shut tight and provided with proper conductors to the troughs, guard against numerous accidents to pumps, prevent ice accumulating inconveniently about them, and render it so comparatively comfortable to water sheep in very cold and blustering weather, that there is

---

* As the tubs are constantly forced full of water the sheep need not even put its head through the cover to drink; and elliptical holes through it 4½ by 5 or 5½ inches, for the mere insertion of the nose, are all that is required. If the tub waters two apartments it should have two holes on each side.

much greater probability of its being properly attended to. Some persons place sheds over the troughs also, to prevent snow from accumulating about them, and to offer greater inducements to the sheep to visit them in stormy weather. The troughs are placed lengthwise with and under the fence (as at *e*, in cut page 217,) or crosswise with the ends projecting (as at *f, f*, in same cut.) If the sheep are watered pretty early in the day, the water will generally be lowered so often by drinking that thick ice will not form over it, and the sheep will usually keep drinking holes open. But the shepherd should look to this; and in severely cold weather he should water the flock two or three times a day, (so that all will be likely to drink once,) and then by withdrawing a plug in the bottom of the trough, let off the water into a drain underneath.

A brook of sufficient volume and current not to freeze deeply, brought near to the sheep yards, is an admirable addition to a sheep farm, both in summer and winter; and when it can be had, no other mode of watering is necessary. The banks at the drinking places should be so sloped that there will be no difficulty in a number drinking at once, and no liability of a sheep being crowded off a high bank or into deep water; and the approach to and bottom of the drinking place should be thoroughly gravelled. I should, however, consider such a brook bought quite too dearly, if the sheep were compelled to wade through it whenever they entered or left their yards — even if the water did not usually exceed three or four inches in depth. Every approach to the yards, crossed by a stream, requires a bridge.

AMOUNT OF FOOD CONSUMED BY SHEEP IN WINTER.— It is now generally estimated that, taking the average of winter weather in our Northern States, American Merino and grade Merino sheep kept exclusively on hay, require about one pound of good hay, or its equivalent, per diem, for every 30 lbs. of their own live weight—to be kept in that plump condition somewhat short of fatness, which is usually regarded as the most desirable one for store sheep. Mr. Spooner adopts the same rule in regard to the consumption of English sheep.*

VALUE OF DIFFERENT FODDERS.—In most of the Eastern and Southern counties of New York, in similar regions of

---

* He says "sheep grown up take 3 1-3 per cent. of their weight in hay per day to keep in store condition." Spooner on Sheep p. 217.

Pennsylvania and throughout New England — the grazing region proper of the older-settled Northern States — the favorite meadow hay for sheep is produced by sowing about three parts of timothy (*Phleum pratense*) to one of red clover, (*Trifolium pratense*.) The first and second years, the clover is in excess, but after that it only appears in moderate quantities; and in the meantime many spontaneous clovers and grasses come in, such as June or spear grass, (*Poa pratensis*,) white clover, (*Trifolium repens*,) red-top or herds-grass (*Agrostis vulgaris*) in moist places, and various others in minor quantities and in special situations, such as the rough-stalked meadow grass, (*Poa trivialis*,) rye or ray grass, (*Lolium perenne*,) and several of the fescue grasses. For sheep, this collection of grasses and clover is cut down rather early and cured as bright as possible. Where meadows are not brought into a course of arable husbandry, and are only plowed at long intervals, no better hay could be obtained from the soil; and, indeed, better would hardly seem desirable. But those who have tested it, know that red clover cut early and cured bright is preferred by sheep, and will fatten them more. It is a prevailing impression, too, among clover growers, that it more specially conduces to the secretion of milk when fed to breeding ewes.

NUTRITIVE EQUIVALENTS.—But it is not economical in most situations, to winter sheep exclusively on any kind of hay. There are incidental products raised with other crops which are regarded as necessary in even that limited extent of mixed husbandry which is practiced on our sheep farms, such as corn-stalks, the straws of the different grains, pea-haulm, etc., which must be consumed in part by the sheep, or be wasted; and there are other crops which, like turnips and beets, are, so far as they can properly be fed, vastly cheaper than hay. Moreover, a well-selected *variety* in food is better, other things being equal, than uniformity: because the different products furnish more of all the different substances which go to form wool and meat. It is, therefore, incumbent on the intelligent sheep farmer carefully to study both in theory and practice, the effect of each of the kinds of available food, separately or in combination, to produce these results. Agricultural Chemistry has made new and important disclosures in this particular; and though its theoretical deductions cannot be implicitly relied on, owing to exceptional or incidental circumstances which have thus far eluded detection, still they usually approximate sufficiently near to the truth to be of great value to the farmer. Before offering any comments on them, I will proceed to lay some of these before the reader, in connection with a very valuable table of experimental deductions.

TABLE OF NUTRITIVE EQUIVALENTS.

	Theoretical values according to		Practical values, as estimated by direct feeding experiments, according to									
	Boussingault.	Fresenius.	Block.	Petri.	Meyer.	Thaer.	Pabst.	Schwerz.	Middleton.	Boussingault.	Schweitzer.	Rham.
Meadow hay,	100	100	100	100	100	100	100	100	100	100	100	100
Red Clover hay,	75	77	100	90	....	90	100	100	....	....	....	90*
Rye straw,	479	527	200	500	150	666	350	....	....	....	267	442
Oat straw,	383	445	200	200	150	190	200	400	....	....	200	195
Barley straw,	460	471	193	180	150	150	200	400	....	....	200	....
Wheat straw,	426	433	200	360	150	450	300	....	....	....	233	374
Pea straw,	64	....	165	200	150	130	150	....	....	....	....	153
Swedes,	676	....	....	300	....	300	250	200	....	....	....	308
Mangel wurzel,	....	391	366	400	250	460	250	333	....	....	366	339
Carrots,	382	542	366	250	225	300	250	270	338	380	300	....
Potatoes,	319	330	216	200	150	200	200	200	....	....	200	....
Beans,	23	34	30	54	50	73	40	....	....	....	30	45
Peas,	27	34	30	54	48	66	40	....	....	....	30	45
Indian corn,	70	....	....	52	....	....	....	....	59	....	....	....
Buckwheat,	55	93	....	64	....	....	....	....	....	....	....	....
Barley,	65	....	33	61	53	76	50	....	....	....	35	54
Oats,	60	58	39	71	....	86	60	....	....	....	37	59
Rye,	58	58	33	55	51	71	50	....	....	....	33	....
Wheat,	55	38	27	52	46	64	40	....	....	....	30	45
Bran,	50	....	....	96	....	....	....	....	....	....	....	105
Linseed cake,	22	....	....	42	108	....	....	....	....	....	....	....

* When blossom is completely developed.

To this Mr. Rham adds the following as equivalents of 100 pounds of "good hay:"—102 lbs. latter-math hay; 88 lbs. of clover hay made before the blossom expands; 98 lbs. of clover of second crop; 98 lbs. Lucerne hay; 89 lbs. sanfoin hay; 91 lbs. tare hay; 146 lbs. of clover after the seed; 410 lbs. of green clover; 457 lbs. of green vetches or tares; 541 lbs. of cow cabbage leaves; 504 lbs. turnips; 50 lbs. vetches; 167 lbs. of wheat, peas and oat chaff.*

No one will understand that because a certain weight of one product is a nutritive equivalent for a certain weight of another, that each will necessarily answer as a substitute for

---

* Rev. W. Rham's statements are not made from his own experiments, but Mr. Spooner (from whom I borrow this column of the above table,) says they were translated from the French by him, and are "the mean of the result of the experiments made by some of the most eminent agriculturists of Europe in the actual feeding of cattle."

the other in feeding. For example, taking the mean of the experimental results in the above table, 367¾ lbs. of rye straw contain as much nutriment as 100 lbs. of meadow hay. A Merino sheep weighing 90 pounds, daily consumes 3 pounds of hay: and to consume its equivalent in rye straw, it would have daily to masticate, digest, etc., a fraction over eleven pounds of it — a feat impracticable for a variety of reasons, and among others for the very obvious one that its stomachs could not be made to hold it, even though digestion should go on with twice its natural rapidity.

The experiments made in feeding Saxon sheep in Silesia, by Reaumur, show in what manner the nutritive parts of certain ordinary vegetable products enter into the composition of different animal products.

Kinds of Food.	Increased live weight of animal.	Produced wool. lbs. oz.	Produced tallow. lbs. oz.	Per cent. of nitrogen in such food
1,000 lbs. raw potatoes with salt	40½	6 8¾	12 5½	0.36
1,000 " " " without salt	44	6 8	10 14½	0.36
1,000 " raw mangel wurzel..	38	5 3½	6 5½	0.21
1,000 " peas	134	14 11	41 6	3.83
1,000 " wheat	155	13 13½	59 9	2.09
1,000 " rye with salt	90	13 14½	35 11½	2.00
1,000 " rye without salt	83	12 10½	33 8½	2.00
1,000 " oats	146	9 12	40 8	1.70
1,000 " barley	136	11 6½	60 1	1.90
1,000 " buckwheat	120	10 4½	33 8	2.10
1,000 " good hay	58	7 10½	12 14	1.15
1,000 " hay with straw, without other fodder	81	15 8	6 11	
1,000 " whisky still grains or wash	35	6 1	4 0	

At first view, there is a degree of incongruity between the theoretical and practical results exhibited in the first of the above tables, which, without due reflection, might materially tend to impair our confidence in the accuracy of the tests which are relied on in agricultural chemistry. But a further glance discloses the fact that these results do not differ more widely from each other than those obtained by practical experiments. How are we to explain these latter incongruities? If the results of actual experiments — experiments, too, conducted with care by men possessing unusual ability and means to do so understandingly and accurately — differ so widely, what then? Are we thence to conclude that experience is worth nothing, or that nature acts without any uniform laws? — that every agricultural result, whether successful or unsuccessful, depends upon chance — or that fatality which is expressed in the delusive and detestable word "luck?"

The explanation of such differences is, in truth, easy enough. The experiments were tried in different soils and seasons. Variations in the latter, every one knows, highly affect the comparative nutritiousness of vegetable products. And unfortunately, too, the standard taken, hay, is the subject of special variations. To say nothing of the natural difference in the nutritiousness of the various kinds of grasses, which, when cut and cured, are termed "meadow hay," we know that the same kinds grown in a wet or dry season — cut a week earlier or a week later — cured rapidly in the sun, slowly in the cock, or slower still and with difficulty during wet, cloudy weather — vary very essentially in quality and nutriment. Take, for a single example, the main meadow grass of the northern portions of the United States, viz., timothy, (*Phleum pratense*.) According to the Woburn experiments,* 64 drachms of it green give, when cut and cured in the flower, 2 dr. 2 gr.; in the seed, 5 dr. 3 gr.; latter-math, 2 dr. Thus, a difference of two weeks in the time of making timothy hay might cause a difference of more than 100 per cent. in the amount of nutriment it contains! †

While it is unfortunate that no unvarying standard can be obtained, or fixed set of conditions agreed upon and observed, in the trial of this class of agricultural experiments, still there is quite as much accord in their results as we are accustomed to find in the opinions of sound, intelligent, practical farmers in regard to any of the experimental facts of farming, which they have been familiar with all their lives. We do not disregard the opinions of such men because they differ. And if we find them all pointing towards the same conclusion, we accept that conclusion as one beyond reasonable doubt. This is the light in which the statements contained in the Table of Nutritive Equivalents, on page 235, should be regarded. When, for example, scientific theory declares that clover hay, pound for pound, contains more nutriment than meadow hay, and when out of six careful and intelligent practical experiments, three also find it more nutritious, and the other three equally so, we are bound, as reasonable men, unless we have

---

* Made some years since by Sinclair, on soils best adapted to each kind of grass, on the estate of the Duke of Bedford, at Woburn, England.

† But to prevent mistakes let me add, that it makes no such difference in the practical value of timothy as sheep fodder. In the seed it is a dry, tough, unpalatable feed for them — and no good sheep farmer intentionally cuts it in that state for his flocks. This, however, in no wise affects the particular fact under consideration. It is to be presumed that timothy composed no inconsiderable share of the meadow hay assumed as a standard by Block, Petri, Von Thaer, Bousseingault, etc. — but in neither instance are we informed whether it was cut in the flower or in the seed.

better proof to the contrary, to admit its equality and presume its superiority. When science and such an array of practice combine to pronounce peas and beans about equal with each other, and among the most nutritious of vegetable products, we ought to adopt that conclusion, if, indeed, we did not already know so notorious a fact. Accordingly, as few sheep farmers are able to make all these experiments for themselves in advance of trying them directly on the body of their flocks, all ought to see the expediency of a very careful study of such a table of Nutritive Equivalents as the preceding one.

Reaumur's experiments, given on page 236, are also especially valuable: and it is only to be wished that their accuracy had also been tested by numerous other experiments directed to the same specific objects of inquiry. Still, I have great general confidence in them. Some of the facts he arrives at are very striking, as, for instance, the superiority of peas over every other vegetable substance named in his list, in the specific production of wool, while barley and wheat considerably exceed it, and oats nearly equal it, in the production of tallow. And a still more striking fact is found in the increase of wool and diminution of tallow produced by adding straw to "good hay" as a habitual food. If there is no mistake in this showing, it is a high point of policy in the wool grower to feed straw, and in the mutton grower to avoid feeding it.

This brings me to another very important consideration, viz., the relative cost and general economy of the different kinds of feeds. According to Reaumur's Table, 1,000 pounds of peas produce 134 pounds live weight of carcass, 14 pounds 11 ounces of wool, and 41 pounds 6 ounces of tallow, while 1,000 pounds of mangel wurzel produce 38 pounds of live weight, 5 pounds 3½ ounces of wool, and 6 pounds 5½ ounces of tallow. Thus the latter produces between a third and a fourth as much live weight, a little more than a third as much wool, and nearly a seventh as much tallow. Peas weigh 60 lbs. to the bushel. If we assume that mangel wurzels weigh the same,* four bushels of them will produce more live weight and weight of wool than one bushel of peas. Not being personally familiar with the culture of mangel wurzel, I will, for the purposes of this illustration, substitute Swedish turnips

---

* This is the statutory weight of a bushel of potatoes in New York,— but no *weight* is prescribed for other roots. I have never raised or weighed a bushel of mangel wurzels — but there cannot be difference enough between their weight and that of potatoes to make any material difference for the purposes of the comparison instituted in the text.

for them—which, by the united testimony of the experimenters given in the table of Nutritive Equivalents, contain more nutriment. Which is most cheaply produced, one bushel of peas or four bushels of Swedes? An acre of ground is thought to do unusually well in the region where I reside, that produces, one year with another, 25 bushels of peas. That acre does very poorly that does not produce 500 bushels of Swedes* — 20 bushels for one of peas. The difference in the cost of preparing the ground, cultivating the crop and harvesting, is considerable; but it makes no approach to the difference in the product of nutriment. Oats compare equally unfavorably with turnips on the score of economy.

I wish to show by such facts as the above, that the sheep farmer in determining what crops he will grow for the winter keep of his sheep, is not merely to estimate the relative value of feeds per pound, but to ascertain how he can provide the most nutriment *suitable* for sheep, at a given cost. Knowing the adaptation of his farm to the different products, and the cost to himself of producing each, every intelligent farmer can, better than anybody else for him, institute comparisons like the above, between all the products named in the preceding tables.

The following records of experiments in feeding are from Mr. Robert Smith's essay "On the Management of Sheep," which received the prize of the Royal Agricultural Society of England, in 1847:

"Experiment No. 1.— On the 20th of December, 1842, eight lambs were weighed and placed upon the regular turnip land, (a red loam, with cold subsoil,) to consume the turnips where they grew, and were regularly supplied with what cut Swedes they would eat, which proved to be on an average of $23\frac{1}{2}$ pounds per day. They were again weighed on the 3d of April, 1843, being 15 weeks, and found to have gained, upon an average, during the time, $25\frac{1}{2}$ pounds each.

"No. 2.— On the same day eight lambs were placed in a grass paddock, under the same regulations, and found to have consumed, on an average, 19 lbs. of turnips per day, and gained, during the time, $26\frac{3}{4}$ lbs. each.

"No. 3.— On the same day, eight lambs were placed alongside the No. 2 lot in the grass paddock, and allowed to run in and out of an open shed during the day, but regularly shut up at night. They were allowed half a pound of mixed

---

* I think my own crops have averaged at least 700 or 800 bushels to the acre, for a period of 15 years or more; and one year they exceeded 1,100 bushels per acre.

oil cake and peas each per day, and consumed 20½ pounds of turnips per day, and gained 33¼ pounds each.

"No. 4.— On the same day, eight lambs were placed with the Nos. 2 and 3 lots in the grass paddock, under the same regulations as No. 3, but supplied with one pound of mixed corn* per day. They consumed 20 pounds of turnips per day, during the following ten weeks, being again weighed on the 28th of February, 1843, and gained, on an average, 26½ pounds each.

"No. 5.— Eight lambs were also placed in a warm paddock, with a shed to run under during the day, but were shut up at least 18 hours, and fed upon 1¼ lbs. of mixed corn per day, and consumed 18¼ lbs. of turnips per day. They were again weighed at the same time as No. 4, and found to have gained 33¼ pounds each during the ten weeks.

"No 6.— On the 5th of January, 1843, sixteen shearlings were equally divided, and eight placed upon a grass paddock, and allowed one pound of mixed corn each per day. They consumed 24 pounds of Swedish turnips each lot per day. They were again weighed on the 2d of March, being eight weeks, and were found to have gained 21½ pounds each.

"No. 7.— On the same day the other eight shearlings were placed alongside the No. 6 in the grass paddock, and allowed one pound of mixed corn each, and consumed 20½ pounds of turnips per day. They were allowed an open shed to run under during the day, and regularly shut in at nights— and again weighed at the same time as No. 6, and were found to have gained 24 pounds each during the eight weeks.

"No. 8.— On the third of April, the eight lambs (No. 3,) having been weighed, were placed upon young clover, and supplied with half a pound of mixed corn, as before. They consumed 12 lbs. of turnips per day during the following month. Being again weighed on the 1st of May, they were found to have gained 11¾ lbs. each. They had a shed to run under during the day, and were shut up at night.

"No. 9.— On the 29th of May, the eight lambs (No. 8,) were again weighed, having been allowed, as before, half a pound of mixed corn upon the clover, but no turnips, with a shed to run under at will. They were found to have gained 16 lbs. each during the month.

---

* Wherever the word "corn" occurs in this record of experiments, it is to be understood in its general sense of *grain;* and the mixed grain, referred to by Mr. Smith, did not even include Indian corn— that not being one of the grain crops of England.

"To prove the temperature of the animal body during the hot weather, I placed the two lots of shearlings, No. 6 and No. 7, upon moderate clover on the 1st of July, 1843.

"No. 10.— The eight shearlings, (No. 6,) were weighed, and allowed one pint of peas per day, and again weighed at the the end of 21 days, and were found to have gained 9¼ lbs. each.

"No. 11.— The eight shearlings, (No. 7,) were also weighed, and given one pint of old beans per day, and again weighed at the same time, and were found to have gained 6 lbs. each, the peas appearing most suitable to the animal temperature during the hot weather, and the beans far too hot. What is more important, those sheep fed upon beans were getting full of humors in this short space of time, while those fed upon peas were looking exceedingly healthy.

"In the autumn of 1843, after making the above experiments, I determined upon testing the qualities of the various vegetables open to our use at that season of the year. On the 2d of October, 1843, thirty lambs were equally divided into lots of ten each, and placed upon over-eaten seeds. They were all weighed, and the roots regularly given them by an experienced shepherd.

"No. 12.— Ten lambs, fed upon cut white turnips, were again weighed on the 13th of November, and were found to have gained, upon an average, 11 lbs. each.

"No. 13.— Ten lambs, fed upon cut Swedes, gained during the six weeks, upon an average, 11 lbs. each.

"No. 14. — Ten lambs fed upon cut cabbage, gained during the time, 16½ pounds each, showing, as I fully expected, a preference in favor of cabbage; but, to my equal surprise, a great difference in favor of the white turnip over the Swede. By subsequent experiments I found, as the cold weather advanced, the cabbage and white turnip became of less value, and that the Swede improved.

"In the autumn of 1844, having placed my ram lambs in their winter quarters, and observing that those placed upon cole-seed were going on apparently the best, I determined to weigh a part of them in comparison with those placed in pens upon grass land; consequently, on the 14th of October, 1844, the following lots were weighed, as in previous experiments, the ten upon the cole-seed being selected from 24 others, marked, and again placed with them:

"No. 15.— Ten lambs penned upon cole-seed,* with cut

---

* A species of cabbage.

clover chaff, were again weighed at the end of one month, and found to have gained 12½ pounds each.

"No. 16.— Ten lambs penned upon drum-head cabbage, with cut clover chaff, and weighed as above; they gained 10½ pounds each.

"No. 17.— Ten lambs placed upon grass and fed upon cut Swedes and cabbage, of equal quantities, with clover chaff, gained 9¾ lbs. each.

"No. 18.— Ten lambs placed upon grass and fed upon cut white turnips and cabbage, of equal quantities, with clover chaff, gained 11 lbs. each.

"Having frequently given my lambs carrots during the winter and spring months, and to no apparent advantage, when compared with other roots, I determined to test their qualities after the expiration of the above experiments, and the No. 16 lot were supplied with what Swedes they would eat, and the No. 17 lot with carrots.

"No. 19.— Ten lambs, fed upon cut Swedes and clover-chaff, having been weighed at the end of the other experiment, were again weighed on the 9th of December. They were found to have gained during the month 10 lbs. each, and consumed 22 lbs. of turnips per day.

"No. 20.— Ten lambs fed upon cut carrots and clover-chaff, were weighed as above on the 9th of December, and were found to have gained 9¼ lbs. each, and consumed 22½ lbs. of carrots per day.

"Thus proving that the carrot can not be given to sheep with equal profit, when compared with the Swede turnip, the carrot being more expensive and hazardous in its cultivation, and producing rather less animal food from a given weight at this season of the year."

I shall place a further list of English experiments in winter feeds in the appendix of this volume.[*]

Turnips are not adapted either to the soil or circumstances of all parts of our country where sheep are kept. I have been informed by many of the farmers in those regions of Vermont where the best sheep are raised, that this crop does not flourish on their farms.[†] And it would be folly to bring turnips into competition with Indian corn, as a habitual winter feed, in our Western States, where the latter crop can be raised for

---

[*] See APPENDIX C.

[†] I raised this question once in the presence of a number of the leading sheep breeders of Addison county — the first sheep breeding county in the State — and they without an exception concurred in the opinion stated in the text.

ten or fifteen cents a bushel. But I know of no cheaper feed, except the last; and that does not approach turnips in cheapness, on lands equally suited to their respective production in the Middle or Eastern States. In all the latter situations — even in those interior regions where the price of hay has hitherto averaged less than $8 a tun — it is more economical to feed turnips with hay and straw, than it is to feed hay alone. I have established that fact to my own satisfaction by the experience of many years.

The beet is not included in the above English experiments, and I have never used it as sheep feed myself. Mr. Chamberlain brought a variety of it with his sheep from Silesia, and is satisfied of the economy and high utility of the crop — but has not, so far as I am informed, tested it in comparison with turnips. My friend, Hon. George Geddes, of Fairmount, New York, has cultivated the same kind of beets, and also turnips, for sheep feed. On his soils (among the best in the State) he, thus far, gives preference to the beet. He has not instituted any comparisons between them by weighing respectively feed and product — but as a farmer who has no superior in our country in both the theory and practice of his occupation, his observations, although unaided by such tests, are entitled to very great weight. Carrots have failed as sheep feed in this country, for the same reasons assigned by Mr. Robert Smith for their failure in England. Rape is cultivated by a few of our growers of English sheep, and is thought highly of by them. Tares and cole seed are unknown to the great body of our sheep farmers, and I am not aware that common cabbage is cultivated by any of them as a field crop for sheep.

MIXED FEEDS.—In making up mixed feeds for sheep, composed of the different products which are found most available and economical, care should be taken to keep the proportion of nutriment to bulk such that a proper supply of the former can be taken into the stomach, without oppressing that organ. It has been seen that $3\frac{1}{4}$ per cent. of the live weight per diem in hay, about meets the demands of the animal economy; and it probably also about fills the stomach to a comfortable state of fullness. If then a sheep weighing 90 lbs. received half its nutriment in hay and half in the better kinds of straw (which contain half as much nutriment as hay,) it would be required to consume $1\frac{1}{2}$ lbs. of hay and 3 lbs. of straw daily — an aggregate of $4\frac{1}{2}$ lbs., which, I think, could not be daily taken into and digested in the stomach of a sheep of that size. Therefore, to put sheep on half straw feed, it is necessary that some other portion of their feed be more concentrated, or more nutritious in proportion to bulk than hay — as, for example, grain or roots — or else they will not get their proper supply of nutriment.

My own course, when feeding straw, has been to give a feed of hay at morning and evening, (intended to average about a pound per head each time,) all the straw the sheep will eat and about a pound of cut turnips each, at noon — the latter being a little increased if the hay and straw are not of prime quality. But I do not often give over two bushels, or 120 lbs. of turnips, to a hundred. Hay here does not average $8 a ton; and though I regard feeding turnips as economical, my major object in growing and feeding them is to promote the health and thrift of my breeding ewes, and the growth of my lambs.

Some excellent sheep farmers on grain and clover-seed farms lying a few miles north of me — where a contiguous city market raises the average price of hay about 50 per cent. higher than here — give their store sheep no hay until March, feeding them in lieu of it, bright, good straw in abundance, clover chaff,* and a daily feed of Indian corn ranging from one and a half to two gills per head, according to their size and to other circumstances. The straw and grain chaff are generally fed fresh from the thrashing floor half a dozen times a day, and the sheep are not required to eat it at all close. After the first of March a full supply of bright clover hay is given and the grain feed taken off. The sheep, as I have had repeated occasion to observe, winter well, and the breeding ewes raise good lambs.

I do not believe that breeding ewes or lambs could properly be fed enough straw and turnips — particularly if the straw was dry and ripe — to obtain the equivalent of a full supply of hay. If turnips are fed in excess, they render the evacuations too thin and active for severely cold weather. But a pound a head given to straw-fed sheep with a little diminution of the corn otherwise requisite, would, I think, constitute a better and cheaper feed than entire corn and straw.

The comparative nutriment of the different kinds of straw has been given in the table on page 235. Oat and barley

---

* That is, what is left of clover after thrashing or hulling — a black, unpromising looking mass.

straw cut quite green and cured bright, are highly relished by sheep. I had rather have them (particularly if thrashed with a flail so that a few small green kernels remain in the ends of the heads,) than hay in the situation in which it is frequently cured for use. Wheat straw ranks next, among the common varieties of straw. Sheep do not relish it, and will not eat it very well if they get any hay. But when confined to it and grain, they learn to eat it and thrive on it. They must not, however, be compelled to eat it as close as oat and barley straw. Ripe rye straw, unless cut fine and mixed with meal, is a dry, harsh, unprofitable and wholly unacceptable food to sheep. All straws are eaten much better by them when fresh thrashed and fed frequently in small quantities.

Corn-stalks are contained in neither of the preceding tables of nutrition. When cut and cured bright, before frost, no feed is better relished by sheep than the leaves and some finer portions of the stalks: and they thrive admirably on them.

Pea-haulm, if cut and cured green, is highly valuable and is highly relished by sheep; but when not harvested until dried up and dead — according to the more common mode — it is utterly worthless for them.

In seasons of great scarcity of hay and straw, sheep have been repeatedly and successfully wintered by feeding them almost exclusively on grain. Such a "hay-famine" occurred in the best sheep region of Vermont, in the winter of 1860–61, occasioned by a severe drouth the preceding summer. Flock-masters who were determined to keep well at all hazards, fed their sheep a pound (or quart) of oats per head, with such quantities of hay, straw, etc., as they could obtain. In better Indian corn growing regions, a pound of corn a day is given under like circumstances.

FATTENING SHEEP IN WINTER.—The present ordinary mode of fattening sheep in winter in New York, is thus described in a letter to me from John Johnston, Esq., of Geneva, New York, who is one of the oldest and most experienced feeders, as well as grain farmers in the United States:

"I generally buy my sheep in October. Then I have good pasture to put them on, and they gain a good deal before winter sets in. I have generally had to put them in the yards about the first of December. For the last 23 years I have fed straw the first two or two and a half months, a pound of oil cake, meal or grain to each sheep. When I commence feeding hay, if it is good, early cut clover, I generally reduce the quantity of meal or grain one-half; but that depends on the condition of the sheep. If they are not pretty fat, I continue the full feed of meal or grain with their clover, and on both they fatten wonderfully fast. This year (1862-3) I fed buckwheat, a pound to each per day, half in the morning and half at 4 o'clock P. M., with wheat and barley straw. I found the sheep gained a little over a pound each per week. It never was profitable for me to commence fattening lean sheep, or very fat ones. Sheep should be tolerably fair mutton when yarded. I keep their yards and sheds thoroughly littered with straw.

"Last year I only fed straw one month. The sheep were fed a pound of buckwheat each. From the 20th of October to the 1st of March, they gained nearly 1$\frac{1}{8}$ pounds each per week. They were full-blood Merinos — but not those with the large *cravats* around their necks. I have fed sheep for the eastern markets for more than 30 years, and I always made a profit on them except in 1841-2. I then fed at a loss. It was a tight squeeze in 1860-1 to get their dung for profit. Some years I have made largely. I did so this year, (1862-3,) and if I had held on two weeks longer I should have made much more. Taking all together it has been a good business for me."

Mr. Johnston by under-draining* and by the manure obtained by fattening sheep, has almost *created* one of the finest farms in New York. I think his land is not adapted to turnips.

REGULARITY IN FEEDING.—The utmost regularity should be observed in the *times* of feeding either store or fattening sheep, and in giving them just the requisite amount to last them until the next feeding. If permitted to waste hay, they rapidly acquire the habit of doing so — i. e., picking out the best and then waiting, even though quite hungry, for another feed. If the hay is coarse and was cut over-ripe, and especially if clover hay be thus circumstanced, it is not profitable to compel the sheep to eat all the orts or refuse; but even with such hay, sheep can soon be taught by over-feeding and carelessness, to make a most unnecessary degree of waste.

All experienced flock-masters concur in the opinion that

---

* He is the father of underground tile-draining in the United States.

sheep fed with perfect regularity as to time and amount (making proper allowance for the weather,) will do better on rather inferior keep, than on the best without that regularity. I prefer feeeding three times a day even in the shortest days of winter; but many good flock-masters feed but twice. If fed three times, it should be at sunrise, noon, and an hour before dark; if but twice, the last feeding should be an hour earlier. Sheep do not stand at their racks and eat well in the dark. It is not very important at what period of the day grain or roots are given provided the time is uniform.

SALT.—Salt is not perhaps quite as necessary to the health of sheep in winter as in summer, but still all good shepherds regard it as indispensable. It should be fed as often as once a week, in the feeding troughs, or by brining a quantity of hay or straw. The Vermont breeders almost universally keep it standing constantly before their sheep in boxes placed in the sheep-houses. My friend Gen. Otto F. Marshall, of Steuben County, New York, has an excellent and economical mode of feeding it. The orts when taken from the sheep racks are thrown into a box-rack wider and considerably higher than the common ones, and placed under a shed. The orts are sprinkled with brine, and the sheep when hungry for salt go to the ort rack and consume them. Thus all the hay is saved.

CHAPTER XXI.

PRAIRIE SHEEP HUSBANDRY.

PRAIRIE MANAGEMENT IN SUMMER — LAMBING — FOLDS AND DOGS — STABLES — HERDING — WASHING — SHEARING — STORING AND SELLING WOOL — TICKS — PRAIRIE DISEASES — SALT — WEANING LAMBS — PRAIRIE MANAGEMENT IN WINTER — WINTER FEED — SHEDS OR STABLES — WATER — LOCATION OF SHEEP ESTABLISHMENT.

The growing of sheep is rapidly increasing in nearly all the new States of the Union west of the Mississippi, and in those which lie on its east bank north of the Ohio.* In all these States are immense tracts of natural pasturage, usually lying in the form of level or rolling prairies — but occasionally in broken tracts containing hills of considerable elevation. The grasses which grow on them are invariably found to be well adapted to the support of domestic animals.

It has already been ascertained by direct experiment that flocks of sheep will obtain their support throughout the entire year, from these natural pastures, as far north as 33 deg. in Central and Western Texas. Ascending north on the banks of the Mississippi, the necessity for artificial winter feed gradually increases until in latitude 40 deg. — about the range of St. Joseph in Missouri, and Springfield in Illinois — it is required through six months of the year. But the domestic grasses will flourish a month longer there, so that the period of dry foddering is restricted to about five months.

Ascending north from Texas on the coast of the Pacific, the temperature decreases less rapidly. The variation of the isothermal line (the line of equal mean heat) on the shores of that ocean and of the Mississippi river, has been popularly claimed to equal ten degrees. While there are yet few settled data to enable us to draw definite general conclusions on the

---

* For Census of sheep and products of wool in all the States and Territories anterior to 1863, see APPENDIX D.

subject, the thermometrical observations already taken do not authorize the conclusion that the difference is so great. I have picked out the following examples of the annual mean heat at such points in Texas, on the Mississippi, and on the Pacific, as came nearest to the regions I wished to compare in this particular, from the multifarious tables contained in the Report of "The Results of Meteorological Observations made under the direction of the United States Patent Office and the Smithsonian Institution from 1854 to 1859 inclusive."*

	Latitude.	1854.	1855.	1856.	1857.	1858.	1859.
New Braunfels,† Texas,	29°.42′	64.61	68.85	—	—	—	70.07
Austin, Texas,	30.20	64.43	65.84	64.64	65.85	67.53	68.08
San Francisco, California,	38.00	55.28	—	—	57.43	56.23	—
Sacramento, California,	38.34	59.51	—	60.03	60.01	59.58	58.74
St. Louis, Missouri,	38.37	58.37	—	53.42	53.42	56.69	55.45
Ottawa, Illinois,	41.20	51.69	48.94	48.15	45.88	49.01	48.37

It will be observed that while the mean heat of St. Louis and Sacramento, in almost identically the same latitude, varies, on the average, 4.22 degrees, there is a much greater proportionable difference in the mean heat of Sacramento and Ottawa, which for six years averages 11.02 deg.‡ These facts render it obvious that the seasons of pasturage must be materially longer on our Pacific coast, than in corresponding latitudes on the Mississippi.

In all the newer States there are lands covered by natural pastures which are exceedingly cheap. In most of them it can be purchased in any quantities for $1.25 an acre. In the older prairie States, like Illinois, Missouri and Wisconsin, desirable tracts would cost considerably more — but still very greatly less than grazing lands of half their fertility in the old North and North-eastern States.

But, in reality, it is not necessary for the wool grower now, nor will it be for many years to come — in most of the above States—either to own or pay rent on a great proportion of the lands depastured by his sheep. We have no redundant population ready to take up with lands which are destitute of any of the essential requisites demanded by the settler. The comparative lack of wood and of running water in the interior of these vast western plains, prevents them from being

---

* Published by order of the Senate, 1861.

† New Braunfels is about twenty-five miles by a direct line north-east of San Antonio, and lies on the southern border of the sheep growing region proper, of Western Texas. It was rather the head-quarters of Mr. G. W. Kendall's different sheep establishments.

‡ To facilitate other comparisons I will here give the mean temperature of several of the points named in the table:—Austin, 66.39 mean of 6 years; Sacramento, 59.69 mean of 6 years; St. Louis, 55.47 mean of 5 years; Ottawa, 48.67 mean of 6 years.

settled, except on the edges and on water courses; and all the sheep farmer needs in such situations is sufficient land for his buildings, grain fields, and,— as his wealth and conveniences increase — for pastures of artificial grass for the early spring and late fall feed of his sheep. When the banks of the streams and the clumps of wood-land are occupied by settlers, they, in effect, have the permanent control of the interior pasturage, often many miles in extent. I have been informed of instances in Texas where an individual, or a small party of individuals, have bought a narrow strip on each bank of a river for a number of miles, and thus prevented the sale of and actually threw out of market hundreds of thousands of acres which were by this means cut off from all access to water, without traveling, perhaps for miles, to the next river bank. But, in truth, the vast extent of our Prairie lands defies all attempts at monopoly. Even in a State comparatively as old as Illinois — containing at the last census a population of over one million seven hundred thousand persons, and probably now containing 50,000 sheep * — immense tracts of land, owned in part by the Government, but principally by non-resident owners, ("speculators,") lie open and free to the use of all; and there is now actually a class of nomadic shepherds in that State who keep flocks of sheep, sometimes numbering upward of two thousand each, who, in the words of the dying Son of the Mist, "Take no hire — give no stipend — build no hut — inclose no pasture — sow no grain." These men are generally industrious Germans, who, after serving flock-masters as shepherds for a year or two, invested their earnings in enough sheep to commence flocks of their own. They follow their sheep by day over the prairies, herding them in little temporary inclosures at night to protect them from wolves and dogs. In the fall they buy a field of corn, drive their sheep to it for the winter, and in the spring resume their wanderings.

In all the new Western States, sheep have been found to acclimate without the least difficulty.† In Texas in the extreme South, in Minnesota in the extreme North, in California in the extreme West, and in every intermediate region where they have been introduced, sheep remain signally healthy, thrive to the highest degree, produce as much wool

---

* By the United States Census of 1860, there were then 33,822 sheep in Illinois, and they have increased much more rapidly than ever before, since that period.

† For a letter showing how sheep are got into the new States — how a sheep establishment is started — and how the first winter is got over, see APPENDIX E.

per head if as well fed, as in the old Eastern States, and the wool is not deteriorated in any apparent or real quality.

It can require no formal array of facts to show that the profits of sheep husbandry on the prairies must greatly exceed those obtained in States lying further east, where the land is no better and costs from five to fifty times as much. It seems now also to be a conceded fact that the profits of sheep production decidedly exceed those of horse, cattle, or swine production on the prairies.

The surplus wheat and Indian corn of the West finds its market on the eastern sea-board. It generally costs half of the crop of wheat, and from five-sixths to six-sevenths of the crop of corn to transport the remainder to New York by rail in the winter, from regions lying no further west than the east bank of the Mississippi. It costs less than two cents a pound to transport wool, which, at the average prices of wool for thirty-five years preceding the present war, is less than two forty-seconds of the value of the medium, and two thirty-fifths of the value of the coarse article. By the Mississippi, or by the northern river, lake and canal navigation which is available in summer, the transportation of the heavy, bulky Western products is considerably less. But when a pound of wool is worth on the farm about as much as four bushels of corn, and when that amount of corn is more than fifty times as bulky, and two hundred and twenty-four times as heavy* as a pound of wool, there must, under any circumstances, remain an insuperable obstacle to the comparative profitableness of corn as a marketable product—and indeed of all other bulky and heavy products.†

---

* In some of the States the weight of corn is established at 56 lbs., in others, 58 lbs. per bushel.

† Since the above was in the hands of the publisher, the articles on sheep, in the Report of the Commissioner of Agriculture, have fallen under my eye, and I find the following statements in an article on "Sheep on the Prairies," by Hon. J. B. Grinnell, of Grinnell, Iowa:—"At any point two hundred miles from Chicago this ratio of cost in freighting is well established; that to transport your products to the seaboard, on wheat you pay 80 per cent. of its value; on pork 30 per cent.; on beef 20 per cent.; gross *on wool* 4 *per cent.* This is not conjecture, but my own experience, that I give 80 per cent. of the value of my wheat which impoverishes my farm, to find a market; and 4 per cent. to find the best wool market, the production of which enriches my acres beyond computation."

The following statements occur in a paper entitled "Sheep Husbandy in the West," by Samuel Boardman, of Lincoln, Logan county, Illinois:—"With wheat worth sixty-five cents per bushel, it costs one bushel to send another from Central Illinois to market. With corn at ten cents per bushel, it takes over six bushels to carry the one to New York. It costs one cent and two-thirds of a cent to send a pound of wool to New York; less than two cents will carry fifty cents' worth of wool to market; to carry fifty cents' worth of corn costs about three dollars. In my own case, I could haul my wool to New York in less time than I could haul the corn I feed to my sheep in the winter six miles to the railroad, and I could also haul the wool to New York cheaper than I could ship the corn by rail. Even in this State, with its

Prairie sheep husbandry has the same general features everywhere, in the summer. In the winter there are essential differences in its operations in regions of perennial verdure, like Western Texas, and in those of six or seven months verdure, like Central Illinois, Northern Missouri and Kansas. I shall proceed briefly to describe the proper summer management in all these regions, and the different systems of winter management in the North and South. It will not be necessary to enter upon details, except when the management differs from that of the older regions already described in this work.

PRAIRIE MANAGEMENT IN SUMMER.—In latitude 40 deg., in the basin of the Mississippi — the latitude of Central Illinois and Northern Missouri — sheep can generally find subsistence on the prairies after about the middle of April. As soon as the new grass sprouts in the smallest degree, the immense range supplies them with food.

LAMBING.—Lambs in the last named regions, where they are, as it is termed, "raised on the range,"— i. e., where the ewes are kept on the open prairie during the lambing season — are not allowed to commence coming before the 1st of May, when the feed is expected to be abundant, and the danger of cold storms greatly over. Lambing on the range, however, is at best attended with great labor and care to the shepherd, and no little danger to the young of his charge. In a prairie flock, eight or ten hundred breeding ewes is a moderate number; and the same circumstances which compel their being turned out on the prairies to lamb — the want of suitable inclosures seeded to domestic grasses — also prevents any division of flocks. When from thirty to fifty lambs are dropped a day, it is a matter of difficulty to get the younger and weaker ones to the folds within the proper time at night, or on the appearance of a storm, without separating them from their dams. When such separation takes place, near nightfall, and twenty or thirty ewes are then running through the flock bleating distractedly for their young, it produces a scene of wild confusion; lambs are run over and trampled on; the ewes, in the increasing darkness, do not find their lambs;

---

more than three thousand miles of railroad, wool-growing is more profitable than wheat and corn, our great items of export. How much more, then, is it in the great portion of the North-west, which does not now, and may not for many years, possess the questionable advantages of railroads with which to market wheat or corn in the raw state?"

if new dropped and not well filled with milk, the latter are liable to perish before morning in cold weather; and when morning comes some of the ewes, particularly young ones, never again recognize their lambs. The small portable pens recommended at page 159, would not be available here, because they would not keep out the wolf. All folding pens on the prairies require to be five or six feet high for that purpose. I am not aware that it has been tried, but I am well satisfied that three or four, or half a dozen temporary pens, according to the size of the flock, put up on different parts of the range, each of which would conveniently hold half a dozen sheep, and into which the shepherd should be getting the youngest lambs and their dams some time before nightfall, would amply pay for themselves in one stormy lambing season — while they might be made to last through a man's life.*

FOLDS AND DOGS.—A permanent fold for the night, unless a good sheltered one, affords so few advantages and produces so many disadvantages, that it is highly desirable to dispense with it at all times, and particularly in lambing time, if any other way can be found to guard the flock from wolves and dogs. This is effectually done in other countries by means of suitable breeds of sheep dogs. The immense utility of introducing some of these varieties into our prairie States, and changing the system of folding, would seem to be obvious. Some information on this subject will be offered in the Chapter on Dogs.

STABLES.— But by far the best place for lambing, in northern prairie climates, is an inclosed field of domestic grass, immediately about sheltered close sheds or stables, which can be used as occasion requires. A large flock ought, for obvious reasons, if it is rendered practicable by the number of the fields, to be divided into smaller flocks — or

---

* It would be best to make them with materials prepared and kept for that express purpose. I should think it would be very convenient to construct them of four lengths, or panels of light, strong fence, capable of being put together without nails. Ten or twelve feet boards might be inserted in mortices or grooves in corner posts, the upper and lower boards being fastened in them by movable pins. The corner posts of these lengths might be fastened together by hooks and staples. Thus four lengths would form a pen of 10 or 12 feet square. This could be covered as far as desirable — a great improvement for inclement weather — by boards two feet longer than the side boards. This would form a pen which could be set up, or taken in pieces and loaded in a cart, by two men in less than half an hour, without any injury to the materials. The materials should of course be piled away under cover when not in use.

else the ewes having the older lambs ought to be frequently taken out and put by themselves. In other respects, the general management should substantially comport with that practiced in the Eastern States.

HERDING.—From the period of lambing to that of washing and shearing, there are no peculiarities in prairie management except in herding. The great art of doing this well, is to get out the sheep as soon as it is light in the morning; to conduct them to the best pasturage; to follow them about patiently, never losing sight of them, and allowing them to spread as far as is prudent over the face of the prairie; to avoid all unnecessary dogging; to avoid huddling them together with the dogs to enable the shepherd to take a siesta or attend to something else; to keep them out until there is barely enough time to fold them before dark; and, finally, to fold them at night carefully, gently and securely.

WASHING.—Some prairie flocks are necessarily driven from five to ten miles to reach running streams or "branches," as they are termed in the West, in which they can be conveniently washed; and owing to the level surfaces of most prairie regions, they generally have to be washed without any dams, and frequently in quite sluggish water. But washing is considered particularly necessary on account of the stained condition of the wool. The wild grasses on prairies grow up in separate stools or tufts, and do not sod over the ground like domestic grasses. Consequently the hoofs of the sheep detach the dirt in hot, dry weather, and it adheres to the wool as they lie down on it, or as it rises in clouds of dust under their feet. The sheep are usually washed at intervals, in parcels of 800 or 1,000 each, so they can all be sheared at about the same periods after washing, before the wool again becomes dirt-stained.

SHEARING.—Shearing is performed from a week to two weeks after washing. It is, or at least ought to be, conducted in the same general way as in the older States. The present practice is to pay hands five cents a head for shearing, and they shear from thirty to sixty sheep per day.

STORING AND SELLING WOOL.—Few prairie wool growers have yet constructed wool houses; and like growers everywhere else, most of them wish to obtain the avails of their wool as soon as practicable after shearing. The clip generally remains in the barn a few days, and if not sold, is sacked and sent to some eastern city market.

TICKS.— Prairie sheep generally suffer but little trouble from ticks, because they are kept in high condition the year round. But wherever these parasites obtain a foothold, they should be promptly exterminated.

PRAIRIE DISEASES.— Scab is by far the most formidable disease of sheep on the prairies, owing to its highly contagious character and to the labor it costs to eradicate it from large flocks running together. My attention has also been very frequently called by Texas correspondents to some minor forms of cutaneous disease, believed also to be infectious, which prevail in that State. Both of these maladies, and their proper treatment, will be considered in a subsequent portion of this work. Hoof-rot, the greatest scourge of the flocks of New England, New York, etc., does not yet appear to establish itself on the prairies. It is claimed, and no doubt is true, that flocks to some degree affected with this disease in the Eastern States, on being driven to the prairies lose all traces of it. That this is true in respect to sheep taken to the Southern and Southwestern States, I know from my own experience. The hoof-rot was introduced into my flocks about twelve years since, when I was receiving numerous orders for sheep from those States. Having got the disease subdued as far as practicable, for the time, I shipped several lots to Virginia, South Carolina and Georgia, apprising the purchasers of the facts, and making myself responsible for the consequences, by offering to refund the purchase money if the sheep should again exhibit the disease. I requested to be informed of their first lameness: and whether lame or not, to be informed of their condition after the lapse of a few months. Not one of the sheep again exhibited a trace of hoof-rot, or lameness of any kind; and their thriftiness was the occasion of especial remark. Before I exterminated the disease from my flocks, I, in like manner, sent colonies to nearly or quite every Southern State, except Florida, to all the Southwestern States, and the Indian Territory on the headwaters of the Arkansas, and always with the same result.

I am disposed to attribute this immunity from the disease in the South to the dry, sandy, permeable character of the soils, and to the *dust* which the sheep's foot constantly comes in contact with, in dry weather, between the stools of grass in the natural pastures. If the disease does not appear on the western prairies, I shall be disposed to attribute it to the same causes, where both exist, or entirely to the last. During active stages of the malady, dry dust might rather aggravate its symptoms than otherwise; but it has long been known that it will "dry up" and cure the old and partly subdued ulcers of the feet. Eastern farmers sometimes drive their sheep over dusty roads for this express purpose.

I am not aware that there are any other serious ovine maladies which are either peculiar to the prairies, or peculiar for not prevailing on them: although it is not at all improbable that further experience and closer observation may develop a number of this class.

SALT.—On the prairies, as elsewhere, salt is justly regarded as indispensable. It is usually fed once a week, about 40 pounds to a thousand sheep.

WEANING LAMBS.—The lambs are weaned about the first of September, when the prairie grasses in the North have become too tough and dry to put them into proper condition for winter. Accordingly the best prairie shepherds have a fresh field of domestic grass — generally blue-grass — to put their lambs on at weaning. Most of them have the corn-field, which is to subsist the sheep during the winter, next to the lamb pasture, and allow the lambs to run in each at will. This is done not only for the immediate benefit of the lambs, but to accustom them to eat corn before winter. Some sow the corn-field itself to winter rye, at the last plowing. This affords fall feed for the lambs, and good spring feed for the breeding ewes. And it is very common to turn the lambs on the stubbles to eat down the sprouts of the scattered grain.

Turnips, of suitable kinds, sown broadcast on the inverted prairie sward, would be likely to do extremely well on soils so rich and deep and so destitute of weeds; and they would furnish cheap and admirable fall feed for sheep of all ages.

PRAIRIE MANAGEMENT IN WINTER.—In regions where the pasturage is perennial, as in parts of Texas, and in latitudes much higher north on our Pacific coast, the winter management of sheep does not vary sufficiently from the summer management to require separate description. The sheep are daily driven out on the prairie in the same way, though they are necessarily driven further. They generally occupy the same folds or yards at nights — with no shelter whatever, overhead. The utility of some shelter and some artificial feed in winter, even in such climates, has been already urged.

For the purpose of giving a clearer view of the winter climate of the Prairie States, and particularly of Texas, I shall devote some pages to the subject in another place.*

Prairie management in regions as far north as Central Illinois, requires as much artificial preparation for winter as is required in New York or Pennsylvania. Should those preparations be the same?

He who embarks extensively in sheep husbandry in the older States must buy a large amount of comparatively high priced land, clear up the forest, fence his land carefully, sow pastures and meadows, build barns for winter storage and for shelter — or buy all these things already fitted to his hand — before he is ready to purchase a flock of sheep to commence his business. All this requires the outlay of much capital. The prairie sheep farmer can commence operations without buying anything but his sheep. Or, if he does not choose to be a pure nomad, he can buy acres for less than the annual interest of acres of the ordinary grazing lands of the old States. His principal necessary capital is a decent knowledge of his business, and enough energy to persevere in it.

Thus have started a large majority of the pioneer sheep farmers of the new States. The new settler builds a little log house, for himself and wife to sleep in — a rail pen covered with poles and prairie-grass, for his "team" and his cow, if he is so fortunate as to own such luxuries — a high yard for a fold, and then he is ready to commence wool growing! And in ten years he can count more sheep, and sometimes more dollars worth of property, than his eastern competitor, who commenced with everything prepared to his hand. The rail pen gives place to the stable, and the uncovered fold yard is succeeded by the fold yard and spacious sheds. Fine fields of domestic grass for spring and fall feed, and of luxuriant corn for winter feed, surround the comfortable farm house. Noble flocks of thousands are driven up nightly by his boys and by the "hired men," — who, in five years more, will be flock-masters themselves!

Are such men to be told that they ought not to *commence*

---

* See Appendix F.

sheep husbandry on the prairies until they have this or that special preparation for it? The sooner the prairie wool grower can surround himself with all the convenient appliances for his occupation, the better: but he acts entirely wisely in not *waiting* for them!

WINTER FEED.—Hay made from the domestic grasses — the "tame grasses" as they are called in the West—or clover, is but little known on the prairies. The wild grasses make sufficiently good hay, but like the preceding, it probably, in most situations, has a cheaper substitute in Indian corn. The remarkable adaptation of most of our prairie soils to this crop is well known. Eighty bushels of it to the acre would be regarded as a heavy crop anywhere — but an extraordinary one nowhere, on the first-class virgin soils. The stalks properly cut and secured, yield nearly double the feed per acre of the small varieties cultivated in the grazing regions of the Eastern States. Its cultivation, too, on the mellow, weedless, prairie soils can be performed vastly more easily and cheaply. With two-horse corn planters, and two-horse corn plows or cultivators, it is estimated that one man can properly take care of fifty acres of it. It should be cut up before the leaves are injured by frost, and placed in shocks, where it remains until it is drawn out to be fed to the sheep. It is drawn out twice a day and scattered on the ground. One active man, with a suitable wagon and team, and devoting his whole time to it, can feed about two thousand sheep. A firm, sodded field of domestic grass is very desirable to feed on, instead of one of wild grass, which soon becomes poached and muddy in wet weather. If the field is large enough to change the feeding places often, very little of the corn is wasted. Some farmers, in place of cutting up the corn and drawing it out in this way, leave it standing on the hill, and fold the sheep on it a couple of hours twice a day; but it is a wasteful mode for the frost-bitten fodder is much less valuable.

The sheep are generally wintered in the feeding fields without shelter, and even the farmers who have sheds do not put their flocks into them except in very stormy nights, and at lambing time. Those who have a sufficient number of feeding fields divide the sheep in the beginning of winter into three or four lots. When this is impracticable, the lambs are merely separated from the flock, and all the rest run together. This last is very objectionable management, as it leaves the weaker and smaller to be pushed about and driven from the choicer portions of the feed by the strong, heavy wethers. Most flock-masters aim, however, to draft occasionally from the flock any that become poor or feeble, and to make some separate arrangement for them.

The object of the prairie farmer is to have his sheep consume as much corn as practicable; for it is more profitable to convert it into animal products than to sell it at ten cents a bushel.* A good sized grade Merino fed exclusively on it will consume and waste from three to three and a half or four bushels during the winter, and the stalks on which it grew. If the corn is good, the proportion of ears to stalks is greater than it should be for the benefit of the sheep. Some farmers provide for this by making enough "tame hay" to give their sheep one feed a day; some make a quantity of prairie hay; and others, instead of burning their wheat straw, according to a prevalent, wasteful method, thrash and stack it in the feeding lot, so that the sheep can get to it at will, or so it can be conveniently fed to them when necessary. If the straw should be slightly brined when stacked, and the sheep be fed salt in no other way, it would prove an acceptable fodder for them, and would be sufficiently nutritious to meet their wants when accompanied with so much corn.

SHEDS OR STABLES.—As has been seen, these are also mostly for storms and for lambing time, because the Western farmer feels that at the high prices for lumber which prevail in almost all our prairie regions, and with the high price and actual scarcity of the labor necessary for housing winter feed, he can not afford to build regular sheep barns with room for in-door feeding for his great flocks, or to bestow the time necessary for housing his feed. Besides, his favorite corn feed would not bear housing in great masses without injury. Well shocked, it winters in the field without any serious loss.

Accordingly, the prairie sheep shed is but one story high, and generally not more than seven or eight feet between the ground and the eaves. It is made with a roof pitching both ways; is generally, at the best sheep establishments, closed up all round; and is long and comparatively narrow, so that by a proper arrangement of fences, portions of it can be made accessible to different fields. The stable room required by sheep has already been considered.

---

* Or to have it consumed for *fuel*, as has repeatedly been done, because it made the cheapest fuel attainable in badly wooded regions. Will this fact be credited in Europe!

WATER.—Snow on the surface of the ground is neither very regular nor very abundant in many of the prairie regions — and, as already. said, many of these regions are very deficient in running water. For a sheep fed exclusively on dry feed, water is indispensable; and one fed highly on dry corn would undoubtedly require it in extra quantities. On very many prairies there are frequent sloughs which are dry in summer, but which, by deep, broad ditches, can be made to supply abundant water in winter. It would be worse than folly to locate the headquarters of a sheep farm where surface water of no kind is available, and where it can not be obtained abundantly by wells; and even wells are a very poor resort, when, by going elsewhere and further, a running stream, or spring, or permanent surface water in any other form, can be obtained.*

LOCATION OF SHEEP ESTABLISHMENT.—The most desirable place for locating a prairie sheep establishment is on the banks of some permanent stream, where the land is high, rolling, and gravelly, the grass abundant and of a fine quality, small clumps of timber frequent, and a railroad to market near by! An undesirable one is a low, wet, level plain — or a dry one without water or timber — remote from all present or prospective avenues to market.

NOTE.—While these sheets are going through the press, as I have mentioned in a preceding note, valuable articles on Prairie Sheep Husbandry by Hon. I. B. Grinnell, of Iowa, and Mr. S. P. Boardman, of Illinois, have made their appearance in the Report of the Commissioner of Agriculture — and also a very discriminating and able paper of a more general character, abounding in the most valuable statistics, entitled "Condition and Prospects of Sheep Husbandry in the United States." I much regret that they did not appear in time to allow me to quote their confirmatory testimony on several subjects treated in this volume.

---

* Artesian wells may become available at some future day when the country is far more thickly settled and land far higher priced. From ordinary wells, water is sometimes raised at no great expense for stock, by means of pumps worked by small wind mills.

CHAPTER XXII.

ANATOMY AND DISEASES OF SHEEP.—THE HEAD.

COMPARATIVELY SMALL NUMBER OF AMERICAN SHEEP DISEASES — LOW TYPE OF AMERICAN SHEEP DISEASES — ANATOMY OF THE SHEEP — THE SKELETON — THE SKULL — THE HORNS AND THEIR DISEASES — THE TEETH — SWELLED HEAD — SORE FACE — SWELLED LIPS — INFLAMMATION OF THE EYE.

COMPARATIVELY SMALL NUMBER OF AMERICAN SHEEP DISEASES. — Many of the diseases of sheep which are described as comparatively common in Europe, are unknown in the United States; and this remark applies particularly to those which have proved most destructive in the former.

I have owned sheep the entire period of my life — a little over half a century — my flock numbering at alternating periods from hundreds to thousands. I have for considerably more than half of this period been constantly concerned in their practical management, and a deeply interested observer of them. For more than twenty years I have been engaged in a constant and extensive correspondence in respect to sheep and their diseases, with flock-masters in various portions of the United States, and have been in the frequent habit of inspecting flocks of every size and description, and I never yet have witnessed or had satisfactory proof brought home to me of the existence of a single case of hydatid, water on the brain, palsy, rot, small pox, malignant inflammatory fever, (*La Maladie de Sologne,*) blain or inflammation of the cellular tissue about the tongue, enteritis or inflammation of the coats of the intestines, acute dropsy or red-water, acute inflammation of the lungs, or of a whole host of other formidable maladies described by every European writer on the diseases of sheep. I do not aver that they never occur in the United States, but the above facts would seem to show their occurrence must at least be very rare, or confined to localities where they are not recognized.

To correct or confirm my own impressions on this subject, I addressed letters, a few months since, to a large number of highly intelligent and experienced flock-masters residing in various States, and in situations differing widely in respect to climate, soil, elevation, etc.—asking them what diseases sheep were subject to in their respective regions, and what remedies were most successfully employed for their cure. The spirit and substance of nearly all the replies are contained in the following extract from a letter of my off-hand friend, Mr. Theodore C. Peters, of Darien, New York:

"You ask me for our sheep diseases and for the remedies. After years of experience I discarded all medicines except those to cure hoof-rot and scab; and I finally cured those diseases cheapest by selling the sheep. An ounce of prevention is worth a pound of cure. If sheep are well kept summer and winter, not over-crowded in pastures, and kept under dry and well ventilated covers in winter, and housed when the cold, fall rains come on, there will be no necessity for remedies of any kind. If not so handled, all the remedies in the world won't help them, and the sooner a careless, shiftless man loses his sheep the better. They are out of their misery and are not spreading contagious diseases among the neighboring flocks."

When to the two maladies above named, (hoof-rot and scab,) are added the obscure one described at page 204, a very fatal but infrequent one in the spring, ordinarily termed grub-in-the-head, catarrh or cold, colic, parturient fever, (the last quite rare and mostly confined to English sheep,) and the few minor diseases of sheep and lambs mentioned under the heads of Spring, Summer, Fall and Winter Management—we have almost the entire list with which the American sheep farmer is familiar. All the diseases named do not, in my opinion, cut off annually two per cent. of well fed and really well managed *grown* sheep! Nothing is more common than for years to pass by in the small flocks of our careful breeders, with scarcely a solitary instance of disease in them. I have not space to offer any conjectures as to the causes of an immunity from disease so remarkable, in comparison with the condition of England, France and Germany, in the same particular.

LOW TYPE OF AMERICAN SHEEP DISEASES.— A discriminating English veterinary writer, Mr. Spooner, has remarked that owing to its greatly weaker muscular and vascular structure, the diseases of the sheep are much less likely to take an inflammatory type than those of the horse, (and he might have added the ox,) and that the character of its maladies is generally that of debility.* Mr. Spooner wrote with his eye on the mutton sheep of England — constantly forced forward by the most nutritious food, in order to attain early maturity and excessive fatness. Still more strongly, then, do his remarks apply to the ordinarily fed wool-producing sheep of the United States. I long ago remarked that the depletory treatment, by bleeding and cathartics, resorted to in so many of the diseases of sheep in England, is inapplicable and dangerous here. The American Sheep, which has been kept in the common way, sinks from the outset, or after a mere transient flash of inflammatory action; and in any stage of its maladies, active depletion is likely to lead to fatal prostration.

It is not purposed here to enter upon any explanation of the anatomy of the sheep, further than is necessary to give a general view of the principal internal structures which determine the form, discharge some of the principal animal functions, and become the seats or subjects of disease. And in treating of maladies, I shall aim to adapt both the language and the prescriptions to the degree of knowledge already possessed on the subject by ordinary practical men, instead of learned veterinarians.

On the next page is given an illustration and description of the skeleton of a sheep, and on the following page the skull of a hornless sheep is represented and described.

* Spooner on Sheep, pp. 269, 271.

STRUCTURE OF THE SHEEP.

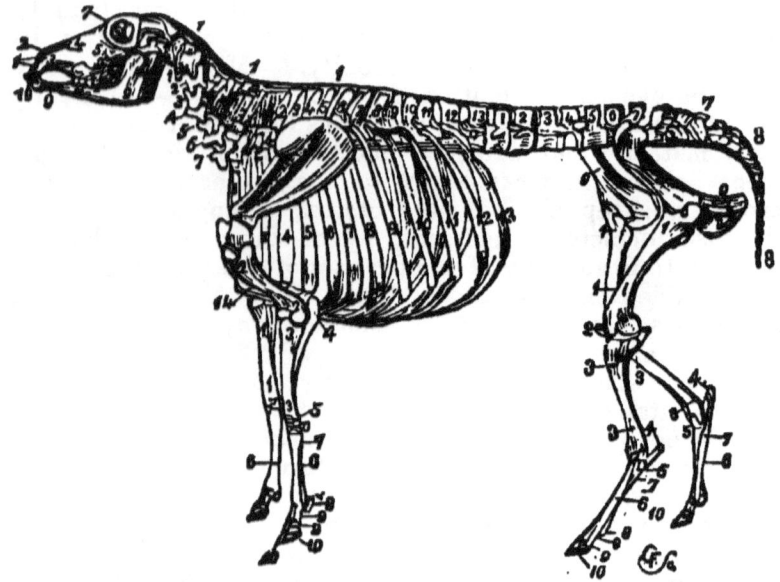

SKELETON OF THE SHEEP.

The Head.
1. The inter-maxillary-bone.
2. The nasal bones.
3. The upper jaw.
4. The union of the nasal and upper jaw bones.
5. The union of the malar and lachrymal bones.
6. The orbits of the eye.
7. The frontal bone.
9. The lower jaw.
10. The incisor teeth, or nippers.
11. The molars, or grinders.

The Trunk.
1. 1. The ligament of the neck, supporting the head.
1. 2. 3. 4. 5. 6. 7. The seven vertebræ, or bones of the neck.
1—13. The thirteen vertebræ, or bones of the back.
1—6. The six vertebræ of the loins.
7. The sacral bone.
8. The bones of the tail, varying in different breeds from 12 to 21.
9. The hannch and pelvis.
1—8. The eight true ribs with their cartilages.
9—13. The five false ribs, or those that are not attached to the breast-bone.
14. The breast-bone.

The Fore-Leg.
1. The scapula, or shoulder-blade.
2. The humerus, bone of the arm, or lower part of the shoulder.
3. The radius, or bone of the fore-arm.
4. The ulna, or elbow.
5. The knee, with its different bones.
6. The metacarpal, or shank-bones — the larger bones of the leg.
7. A rudiment of the smaller metacarpal.
8. One of the sessamoid bones.
9. The two first bones of the foot — the pasterns.
10. The proper bones of the foot.

The Hind-Leg.
1. The thigh-bone.
2. The stifle-joint and its bone — the patella.
3. The tibia, or bone of the upper part of the leg.
4. The point of the hock.
5. The other bones of the hock.
6. The metatarsal bone, or bone of the hind-leg.
7. Rudiment of the small metatarsal.
8. A sessamoid bone.
9. The two first bones of the foot — the pasterns.
10. The proper hone of the foot.

SKULL OF A HORNLESS SHEEP.

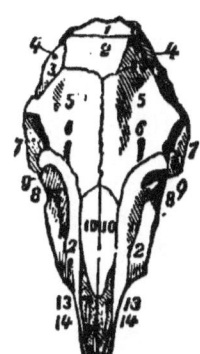

1. The Occipital bone.
2. The parietal bones, the suture having disappeared.
3. The squamous portions of the temporal bone.
4. The meatus auditorius, or bony opening into the ear.
5. The frontal bones.
6. The openings through which blood-vessels pass to supply the forehead.
7. The bony orbits of the eye.
8. The zygomatic or molar bones.
9. The lachrymal bones.
10. The bones of the nose.
11. The upper jaw bone.
12. The foramen, through which the nerve and blood-vessels pass to supply the lower part of the face.
13. The nasal processes of the intermaxillary bones.
14. The palatine processes.
15. The intermaxillary bone, supporting the cartilaginous pad, instead of containing teeth.

THE HORNS AND THEIR DISEASES.—Whether sheep should be bred to have horns or not depends upon the taste of the owner. In the abstract, they are, undoubtedly, a wholly useless appendage, render the lamb more difficult of parturition, and in their massive proportions on the head of the male Merino, cause him to be, however quiet his temper, a dangerous associate to breeding ewes in advanced stages of pregnancy. Yet I know no leading Merino breeder who would use a polled or hornless ram, any sooner than would a Down or Leicester breeder use a ram having horns! Each clings to the characteristics of his breed. Most Merino breeders, however, object to horns on ewes — though very small ones, having but one convolution, are not uncommon. I have never seen it remarked that the different families of Merinos in Spain exhibited any different characteristics in their horns — but the American Infantados and Paulars, as now modified, generally do so. In the former, the convolutions are nearer together, and the first one frequently passes down very close to the head and neck — in a few instances presses so closely on them that, in the case of valuable ram lambs, the horns are artificially spread apart by means of an iron brace placed between them (over the back side of the head) which can be lengthened by a screw as the horns give way to the pressure. In the Paular, the horns are usually quite divergent, and frequently of great size.*

---

* The fact that the Silesians, which are deep in Infantado blood, have also the close or convergent horn, would go to show that it is a family peculiarity. I owned a Paular ram two or three years since, which at two years old, measured three feet between the tips of his horns. He died before he was three years old; and a person sawed off his horns so as to take that portion of the skull covered by the base of each. He subsequently boiled them to detach them from the bones. They have lain dry two years. Weighed to-day, with the inside bones, they weigh 5 lbs.!

The proper mode of managing horns at shearing, was mentioned at page 189. I am not aware that they are subject to any diseases except those caused by fracture. They are sometimes broken in fighting; and I have seen an old ram which had one knocked clean from his head by the charge of a ram from behind, while another occupied his attention in front. The bleeding is very considerable in such cases, but a tarred rag securely bound over the part to keep away flies and irritating substances is all that is necessary.

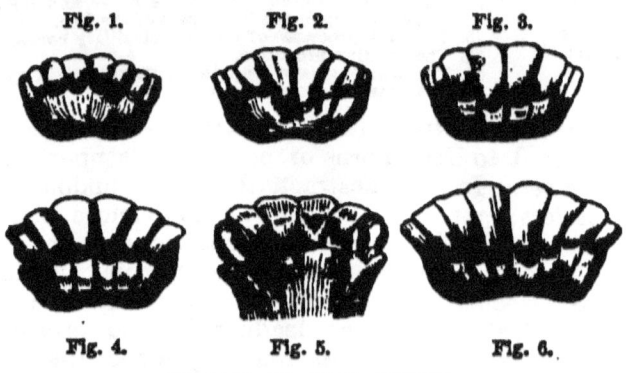

TEETH OF THE SHEEP.

THE TEETH.—The sheep has thirty-two teeth — eight incisors in front of lower jaw, and six molars on each side in the upper and lower jaw. The lamb at birth has two incisor teeth visible, or pressing through the gums. Usually before it is a month old it has eight comparatively short, narrow ones, as in Fig. 1. At about a year old, though sometimes not until the fourteenth or sixteenth month, the two central "lamb teeth" are shed and replaced by two "broad teeth," which gradually attain their full size. The sheep is then termed a yearling, or "yearling past." Two lamb teeth continue to be shed annually and replaced by broad teeth, until the sheep has eight incisors of second growth, when it is termed "full mouthed." Fig. 2 represents "the mouth" of a yearling past; Fig. 3 of a two-year-old past; Fig. 4 of a three-year-old past, and Fig. 6 of a four-year-old past.* Fig. 5 is a back or inside view of the teeth of a three-year-old, showing the narrow and dwindled appearance of the two last lamb

---

* The English, counting from the periods when each new pair of incisors become *fully developed*, usually speak of two broad teeth as indicating a two-year-old, four a three-year-old, six a four-year-old, and eight a five-year-old.

teeth, before they are shed; and they frequently, as in this cut, stand so far behind the third pair of incisors that they can not be seen, on looking into the mouth in front. Consequently, unless the broad incisors are *counted*, the sheep is often mistaken for a full-mouthed one.

The teeth afford the most decisive test there is of the age of a sheep, until it is four years old, though there is sometimes a variation of a number of months or even a year in their development. High kept and rapidly grown sheep acquire their second teeth earlier.

When perfect, the incisors are sharp, rounded on the edge, as in the cuts; a little concave without and convex within (or gouge-shaped;) and they project forward, so that with the firm, elastic. pad on the upper jaw with which they are brought into contact, they are capable of taking up the smallest body. They will not only crop the shortest grass, but scoop up its very roots. A sheep yarded on unpulled turnips usually scoops out the centers of them so far as they are in the ground, leaving little more than the mere skin of the sides and bottoms, remaining unbroken like cups in the soil.

At six years old the incisors of the Merino begin to diminish in breadth and lose their fan-like shape and position. At seven they become long and narrow, stand about perpendicular with respect to each other, and have lost their rounded, cutting edges. At eight they are still narrower, and their outer ends begin to converge considerably toward the middle. At nine the convergence is still greater, the teeth are not thicker than very small straws, and are very long, particularly the middle ones. At ten these appearances have increased and the teeth are becoming quite loose. At about this period of life the teeth begin to drop out, though frequently all are retained until twelve.* The sheep is then called "broken mouthed." In two or three years after beginning to lose them, all the incisors are usually gone but one or two. These should be pulled by a pair of nippers, as they prevent the sheep from cropping short grass.† The

---

* It is stated by Dillon, in his Travels in Spain, 1779, (quoted by Youatt,) that "the teeth of the Spanish ram do not fall out until the animal is eight years old; whereas the ewes, from the delicacy of their frame, or from other causes, lose theirs at five." These are undoubtedly the loose assertions of a misinformed traveler: at least, they do not approximate to accuracy in respect to the American Merino.

† Mr. Youatt is clearly mistaken, however, in saying "that if any of the teeth are loose they should be extracted," (*vide* p. 5.) *All* the incisors are frequently loose, to a considerable degree, a year or two before any of them drop out, and the

gum of the lower jaw hardens after their removal, so that it becomes, in a measure, a substitute for the lost incisors, in *separating* their food. The molars, though shortened and worn, are never shed, so that *mastication* continues complete. Old breeding ewes often live, thrive, and raise good lambs three or four years after ceasing to have any front teeth.

English sheep become broken-mouthed from three to four years earlier — the difference about corresponding with the difference in the longevity of the races. Sheep of all kinds differ not only as between individuals, but between flocks in the period of losing their teeth. If fed *uncut* and dirty roots, they lose them much earlier. The *prying* action of the incisors, as they are employed in scooping out a turnip, for example — particularly if it be partly frozen — or the obstruction of a bit of gravel (which often finds its way from the tap roots even among *cut* turnips) between an incisor and the pad above it, not unfrequently causes a loose one to be detached, or a comparatively firm one to snap off.

SWELLED HEAD.— The head of the sheep sometimes becomes swollen from causes which are not very well understood. I do not know of any special or characteristic disease among sheep which produces this effect.* It is occasionally heard of in this country — but I have never seen it, or heard its symptoms accurately described. According to Mr. Hogg, it appears in Scotland. An abscess is formed and breaks, and the sheep then speedily recovers unless too much reduced by the discharge. In England it is sometimes occasioned, Mr. Youatt thinks, from the sting of a venomous reptile or insect, in which case, he says, the wool should be cut off round the wound, the parts washed with warm water, olive oil well rubbed in, and small doses of hartshorn diluted with water, administered internally — "half a scruple of the hartshorn in an ounce of water every hour."

Mr. Youatt conjectures that the Scotch form of the disease may arise from eating poisonous plants, or from a species of catarrh or influenza.† To these causes, and to the last especially, I have been disposed to attribute such instances of

---

sheep remains capable not only of cropping grass, but of scooping out a turnip in the manner already mentioned. Nor should all be pulled when only one or two drop out. The judgment of the shepherd must be his guide in the matter; but as long as, say five incisors remain together or press together, it is not usually best to remove them.

* I should except blain, but this disease has not appeared in the United States.
† Youatt on Sheep, p. 371.

the disease as I have heard of in the United States. In this case, it should be treated like catarrh, (which see.)

SORE FACE.—The faces of sheep sometimes become so sore, in the summer, that the hair comes off. This is usually attributed either to coming in contact with, or eating St. John's-Wort, (*Hypericum perforatum.*)* Mr. Morrell states, in the American Shepherd, that the "irritation of the skin" will sometimes extend "over the whole body and legs of the sheep;" that "if *eaten* in too large quantities it produces violent inflammation of the bowels, and is frequently fatal to lambs, and sometimes to adults;" that "its effects, when inflammation is produced internally, are very singular;" that he "has witnessed the most fantastic capers of sheep in this situation, and once a lamb, while running, described a circle with all the precision of a circus horse," and that "this was continued until it fell from exhaustion." He recommends, if there are symptoms of internal inflammation, that tar be administered, but says that "simply hog's lard is used frequently with success." He recommends that the sheep should be removed to pastures free of the weed and salted freely; and remarks that "it is said that salt, if given often to sheep, is an effectual guard against the poisonous properties of the weed."

Mr. Morrell does not state how he traced these extensive, and especially these internal effects, to the consumption of St. John's-Wort. On consulting several works on Botany, and Dunglison's Medical Dictionary, lying before me, I do not, with the exception below, find it mentioned as a poisonous or noxious plant in any of them. Dr. Dunglison characterizes it as an aromatic and astringent, and states that an infusion of its flowers in olive oil is a vulnerary — or, is useful in curing wounds.

On the other hand, in Dr. John Torry's "Flora of the State of New York," (in the "Natural History of New York,") occurs the following remarks on the properties of this plant: — "This pernicious weed is generally believed, in this country, to be the most common cause of 'slabbers'† in horses and horned cattle; and likewise to cause sores on their skin, especially on animals whose noses and feet are white, and whose skin is thin and tender. Dr. Darlington remarks that the dew which collects on the plant appears to

---

* I gave this as the cause in Sheep Husbandry in the South, p. 271, but I did not suppose it was *eaten* by sheep.
† I had supposed that honor was more particularly assigned to lobelia (*L. inflata.*)

become acrid. He has seen the backs of white cows covered with sores wherever the bushy extremity of their tails has been applied, after draggling through the St. John's-Wort. Dr. J. M. Bigelow, of Ohio, states that he has known a high degree of inflammation of the mucous lining of the mouth and fauces produced by eating a few of the fresh leaves. It was formerly in considerable repute for its medicinal virtues, but was chiefly employed as a balsamic for wounds." *

What should induce sheep to *eat* a noxious plant which they are familiar with, and which is excessively acrid to the taste, it is difficult to conjecture. I doubt whether they do so. Indeed, I doubt whether it so often affects sheep in any way as I formerly believed, and as many persons continue to believe. It grows in most of my hill pastures; and having ceased to fear it and consequently to make special efforts for its extirpation, it being a hardy perennial-rooted plant, has increased so that it is readily found. Within a week of this writing† I have observed abundant plants of it in a field where I have kept one hundred and eighty ewes and lambs since they were first turned out in the spring, and not one of them has been in the least degree affected by it. I never saw a case where the sheep were affected beyond a soreness of the face: and I do not think I have seen even such a case within fifteen years. All my recollections of it go back to the days of the feeble little Saxon sheep, which were always *peeling* on some excuse or other!

I have some Short-Horn cows, too, with *white* noses, *white* spots on their backs, and long tails to draggle over the St. John's-Wort in their pasture, both when wet and dry; and none of them are affected by it. While I am not prepared to deny that it sometimes causes sores both on cattle and sheep, I am not disposed to concede much to mere popular belief on the subject without better proof than I have yet seen adduced.‡ Popular belief in France and Germany, says Loudon, cause the people "to gather it with great ceremony on St. John's Day, and hang it in their windows, as a charm against storms, thunder and evil spirits — mistaking the meaning of some medical writers who have fancifully given this plant the name of *Fuga Dæmonum*,§ from a supposition

---

* Flora of New York, Vol. 1, p. 87.

† August, 1863.

‡ I intended to make some experiments in regard to its effects, preparatory to writing this article, but have not had time to attend to it.

§ Flight of evil spirits or demons.

that it was good in maniacal and hypochondriacal disorders. In Scotland it was formerly carried about as a charm against witchcraft and enchantment!"

From whatever cause it arises, the sore face ascribed to the effects of St. John's-Wort is readily cured by sulphur ointment, composed of sulphur and hog's lard. If, as Mr. Morrell supposes, it produces "violent inflammation of the bowels," I should not like to trust either to tar or lard, but would resort to the treatment appropriate in the case of vegetable poisons. (See Poisons.)

SWELLED LIPS.—Sheep are sometimes quite suddenly affected with sore lips in the winter—and I think this oftenest occurs to the lambs of the preceding spring. The lips become swollen to several times their natural thickness, are hard, crack open, and are so stiff and sore that the animal eats with difficulty. This disease visited a flock owned by me five or six years since, and included nearly the entire number. It promptly disappeared on smearing their lips with tar rendered thin and soft by butter and slightly mixed with sulphur. A neighbor's sheep which were thus attacked, were simply, on my suggestion, smeared over the lips with pot-grease; and it likewise immediately relieved them. There appeared to be no observable constitutional disease in either case. Several other such attacks and cures have occurred within flocks which I am familiar with.

The causes of this affection are unknown. Some attribute it to St. John's-Wort, or other noxious weeds in the hay — but, in my own case, it can not possibly be explained in this way. A tun of the hay would have scarcely contained a handful of St. John's-Wort, and it contained no other weed even suspected of being noxious — while the malady was simultaneously exhibited by nearly every animal in the same flock. The hay, however, came from a new field, and contained an excessive quantity of bull-thistles. Whether the dry prickles of these had anything to do in producing the effect on the lips, I am unable to say. I have occasionally seen it stated in Agricultural papers that a disease of which swelled lips are one of the most prominent and characteristic symptoms, has resulted mortally. I think it must be a different malady from the one under consideration. I have never witnessed any instance of swelled lips which I think would have been likely to produce death without the application of any remedy.

INFLAMMATION OF THE EYE.—The eyes of sheep are subject to few diseases, in our country. The only serious one I have ever seen — and that is quite rare — is simple ophthalmia,* characterized by redness of the eye, and its appendages, with intolerance of light and a copious flow of tears. It is generally, however, attended with but moderate inflammation, and if neglected, its worst and that by no means the most common result is blindness, almost invariably confined to one eye. It might prove more serious among high fed mutton sheep. Mr. Grove, the best practical shepherd of his day, in our country, used to blow *red chalk* into the diseased eye. "Others squirt into it tobacco juice, from those ever ready reservoirs of this nauseous fluid, their mouths. Conceiving it a matter of humanity to do something, I have in some instances drawn blood from under the eye, bathed the eye in warm water, and occasionally with a weak solution of the sulphate of zinc combined with tincture of opium. These applications diminish the pain and accelerate the cure." †

---

* There is occasionally a case of cataract. Also see Art. Rabies in this volume.

† Sheep Husbandry in the South, p. 239.

CHAPTER XXIII.

ANATOMY AND DISEASES OF THE SHEEP'S HEAD, CONTINUED.

SECTION OF SHEEP'S HEAD — GRUB IN THE HEAD — HYDATID ON THE BRAIN — WATER ON THE BRAIN — APOPLEXY — INFLAMMATION OF THE BRAIN — TETANUS OR LOCKED JAW — EPILEPSY — PALSY — RABIES.

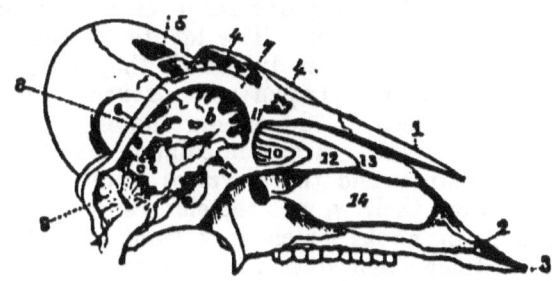

SECTION OF SHEEP'S HEAD.

1. The nasal bone.
2. The upper jaw bone.
3. The intermaxillary bone, the fore part of which supports the pad, against which the incisor teeth shut.
4. 4. The frontal sinuses, or cavities.
5. The sinus of the horn, communicating with frontal sinus, disclosed by removing a section of the base and bone of the horn.
6. The parietal bone.
7. The frontal bone.
8. A vertical section of the brain.
9. A vertical section of the cerebellum.
   a. The cineritious portion of the brain.
   b. The medullary portion of the brain.
10. The ethmoid bone, with its cells.
11. The cribriform or perforated plate of the ethmoid bone, pierced with numerous holes for the passage of the olfactory nerve.
12. The development of the lower cell of ethmoid bone.
13. The superior turbinated bone.
14. The inferior turbinated bone.
17. The sphenoid bone.

GRUB IN THE HEAD.— In the months of July and August sheep are often seen gathered in dense clumps with their heads turned inward and their noses held down to the ground. If driven away, they run without raising their heads, or rapidly thrust them down again, as if they had some very urgent motive for retaining them in that position.

Occasionally they stamp or strike violently with their forefeet near their noses as if an enemy, invisible to the spectator, were assailing them at that point. It is the Œstrus ovis, or gad-fly of the sheep, attempting to deposit its eggs within their nostrils. "The head and corslet" of this insect, says Mr. Youatt, "taken together, are as long as the body; and that is composed of five rings, tiger-colored on the back, with some small points, and larger patches of deep, brown color. The belly is of nearly the same color, but has only one large circular spot on the center of each of the rings. The length of the wings is nearly equal to that of the body, which they almost entirely cover. They are prettily striped and marked."*

SHEEP GAD-FLY.

If the fly succeeds in depositing its eggs within the nostrils of a sheep, they are immediately hatched by the warmth and moisture, and the larvæ or young grubs, crawl up the nose finding their way to the sinuses, where, by means of the tentacula or hooks which grow from the sides of their mouths, they attach themselves to the membrane lining those cavities, and there remain feeding on its mucus until the following year. As the minute worm ascends the nose, the sheep appears to be distracted with apprehension. It dashes wildly about the field, stamping, snorting and tossing its head.

Fig. 1 in the annexed cut, exhibits the larva or grub about half grown. It is then white, except two brown spots near the tail. Fig. 2 represents it of full size. The rings, and particularly those nearest the tail, are now dark brown.

  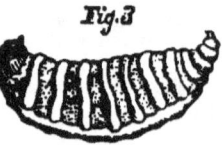

THE "GRUB" OR LARVA OF THE GAD-FLY.

Each ring has darker spots, and below them are others, as seen in Fig. 2. Fig. 3 exhibits a full grown larva on its back—the minute dots between the belly rings representing small red spines, the points of which turn backwards.

The larva, after remaining in the sinuses through the

---

* Those who wish a further description of this insect, will find more on the subject in Mr. Youatt's work on sheep,—and especially in Mr. Bracy Clark's monograph of the œstrus ovis.

winter and early part of the ensuing spring, abandons them as the warm weather advances. It crawls down the nose, again producing great irritation and excitement; drops on the ground; rapidly burrows into it; assumes the form of a chrysalis; and finally again hatches forth a perfect fly.

Many French and English writers consider these larvæ, while in the heads of sheep, the causes of most serious evils and of frequent death. On the other hand, Mr. Bracy Clark and Mr. Youatt are not only disposed to doubt this, but they even suggest that these parasites may be placed where they are for the benefit of the sheep, particularly those in high condition — to save them from determination of blood to the head by establishing counter irritation!

This is as far-fetched as a conclusion, as is the reasoning on which it is founded. Mr. Youatt declares:—" It is incompatible with that wisdom and goodness that are more and more evident in proportion as the phenomena of nature are closely examined, that the destined residence of the œstrus ovis should be productive of continued inconvenience or disease."* Had Mr. Youatt forgotten that the "destined residence" of the scab acarus, of the tick, of the common maggot, etc., are all productive of inconvenience, disease and death to the sheep?

If a sheep dies in the spring of the year, fat or poor, suddenly or lingeringly, with one or another set of symptoms, the popular belief generally traces the malady to "grub in the head." It is the convenient name which covers all the unknown fatal maladies of that season of the year. This probably arises from the fact that on making what may be termed the farmer's autopsy—viz., on *splitting open* the body and head of the dead sheep with an *axe* — the most striking deviation, if not the only one, from what is supposed to be the natural situation of things, which is discovered in the five minutes scrutiny, is a quantity of large, fat, ill-looking worms in the cavities of the head: and our rapid practitioner at once decides that these are cause enough for any disease! His theory is that the "grub" bores through the walls which separate the nasal cavities or the sinuses from the brain, and that they produce death by attacking the latter organ. I have been triumphantly shown the cribriform plate of the ethmoid bone (see 11, in Fig. on page 273,) with its natural perforations for the passage of the olfactory nerve, in proof

---

* Youatt on Sheep, p. 368.

that the "grubs" had already "got small holes opened to the brain," when their further operations were, it was supposed, suspended by the death of the subject!

I have had a singularly limited experience with any diseases which could reasonably be attributed to the presence of these parasites, and therefore do not feel myself at all well qualified to judge of their actual effects on the sheep. That want of experience is a strong proof of itself, that resulting maladies are not as frequent by any means as is popularly supposed. And knowing, as I do, that other and wholly dissimilar diseases are habitually termed "grub in the head," I can entertain no doubt that the extent of the injuries thus inflicted is enormously exaggerated.

Influenced by these latter considerations, and by the strong counter testimony of such really able veterinarians as Messrs. Clark and Youatt, and the silence on the subject of Mr. Spooner and some other modern writers, I was formerly led to doubt whether the larvæ of the œstrus ovis ever did more in the sheep's head than effect a degree of temporary irritation of the lining membranes, which might produce serious inconvenience when acting in concert with the inflammation already established by catarrhal or other cerebral affections, but which never caused death. Again reminding the reader that I speak from a very limited personal knowledge of the disease, I feel it due to frankness to say that my opinions have undergone some change. The testimony of intelligent men has satisfied me that the irritation and ultimate inflammation of the mucous lining of the head produced by the tentacula of the worm and by its constant feeding on the secretions if not even on the substance of the membrane itself, in certain stages of the disease,* are sufficient in some cases to cause death. I should not expect a sheep in high condition and apparent health to die suddenly from this cause without previous symptoms of disease, and under circumstances resembling those of apoplexy. I should not expect the powerful nervous disturbances of epilepsy. But if the sheep began to fall off rapidly in condition a little before the opening of spring, without any other traceable cause — if it wandered round with irregular movements, twisting about its head occasionally as if it was suffering pain — and especially if the mucus discharged from the nose was tinged with blood — I

---

* This may be more doubtful.

should *suspect* "grub in the head," and administer remedies or antidotes on that hypothesis. And after the death of patients, I should as carefully as practicable examine not only the sinuses of the head, but also the entire nasal cavities, to ascertain whether there were any traces of the supposed destructive action of the larvæ.

Some farmers protect their sheep measurably from the attacks of the œstrus ovis, by plowing a furrow or two in different portions of their pastures. The sheep thrust their noses into this on the approach of the fly. Others smear their noses with tar, or cause them to smear them themselves, by sprinkling their salt over tar. Those fish oils which repel the attacks of flies might be resorted to. Blacklock suggested the dislodgement of the larvæ from the head by blowing tobacco smoke up the nostrils,— as it is said to be effectual. It is blown from the tail of a pipe, the bowl being covered with cloth. Tobacco water is sometimes injected with a syringe for the same purpose. The last should be prevented from entering the throat in any considerable quantity.

I trust that scientific and impartial investigation will henceforth be more directed to a determination of the actual existence and proper treatment of this real or supposed malady.

HYDATID ON THE BRAIN. — This disease, known as turnsick, sturdy, staggers, etc., is spoken of by Chancellor Livingston, and other writers of reputation, as having occurred in this country within their own observation. I have never seen a case of it, and shall be obliged, therefore, to make use of the descriptions of others. Mr. Spooner says:

"The symptoms are a dull, moping appearance, the sheep separating from the flock, a wandering and *blue* appearance to the eye, and sometimes partial or total blindness; the sheep appears unsteady in its walk, will sometimes stop suddenly and fall down, at others gallop across the field, and after the disease has existed for some time will almost constantly move round in a circle — there seems, indeed, to be an aberration of the intellect of the animal. These symptoms, though rarely all present in the same subject, are yet sufficiently marked to prevent the disease being mistaken for any other. On examining the brain of sturdied sheep, we find what appears to be a watery bladder, termed a hydatid, which may be either small or of the size of a hen's egg. This hydatid, one of the class of entozoons, has been termed by naturalists the *hydatis polycephalus cerebralis*, which signifies the many-headed hydatid of the brain; these heads being irregularly distributed on the surface of the bladder, and on the front part of each head, there is a mouth surrounded by minute, sharp hooks within a ring of sucking disks. These disks serve as the means of attachment by forming a vacuum, and bring the mouth in contact with the surface, and thus by the aid of the hooks the parasite is nourished. The coats of the hydatid are disposed in several layers, one of which appears to possess a muscular power. These facts are developed by the microscope, which also discovers numerous little bodies adhering to the internal membrane. The fluid in the bladder is usually clear, but occasionally turbid, and then it has been found to contain a number of minute worms."

According to Mr. Youatt, this disease attacks many of the weakly lambs in the English flocks. It usually appears, he remarks, "during the first year of the animal's life, and when he is about or under six months old." It succeeds "a severe winter and a cold, wet spring." He says:

"If there is only one parasite inhabiting the brain of a sturdied sheep, its situation is very uncertain. It is mostly found beneath the pia-mater, lying upon the brain, and in or upon the scissure between the two hemispheres. If it is within the brain, it is generally in one of the ventricles, but occasionally in the substance of the brain, and, in a few instances, in that of the cerebellum. * * * This is a singular disease; but it is a sadly prevalent and fatal one in wet and moorish districts. * * * It is much more fatal in France than in Great Britain. It is supposed that nearly a million of sheep are destroyed in France every year by this pest of the ovine race. * * * The means of cure are exceedingly limited. They are confined to the removal or destruction of the vesicle. Medicine is altogether out of the question here."

Many barbarous methods have been adopted to rupture the hydatid. Mr. James Hogg thrust a wire up the nostrils of the sheep, and through the plate of the ethmoid bone *into the brain*, and thus, as he assures us, punctured the hydatid and "cured many a sheep!"* This practice, which I can not characterize otherwise than as atrocious, is justly condemned by Mr. Youatt. Mr. Parkinson "pulled the ears very hard for some time," and then cut them off close to the head!†

---

* Hogg on Sheep, p. 59.
† Parkinson on Sheep, Vol. 1, p. 412.

Where the hydatid is not imbedded in the brain, its constant pressure, singularly enough, causes a portion of the cranium to be absorbed, and finally the part immediately over the hydatid becomes thin and soft enough to yield under the pressure of the finger. When such a spot is discovered, the English veterinarians usually dissect back the muscular integuments, remove a portion of the bone, carefully divide the investing membranes of the brain, and then, if possible, remove the hydatid whole — or, failing to do this, remove its fluid contents. The membranes and integuments are then restored to their position, and an adhesive plaster placed over the whole. The French veterinarians usually simply puncture the cranium and the cist with a trochar, and laying the sheep on its back, permit the fluid to run out through the orifice thus made. A common awl would answer every purpose for such a puncture. The puncture would be the preferable method for the unskilled practitioner. But when we take into consideration the hazard and cruelty attending the operation at best, and the conceded liability of a return of the malady — the growth of new hydatids — it becomes apparent that, in this country, it would not be worth while, unless in the case of uncommonly valuable sheep, to resort to any other remedy than depriving the miserable animal of life.*

WATER ON THE BRAIN.—I have never seen this disease. It was first described by Mr. Youatt as an effusion of serous fluid, or water, without being confined in any sack or bladder, within the cavity occupied by the brain, or between its investing membranes. It is peculiar to young lambs, and sometimes occupies the head before birth, giving it unusual size, and rendering parturition difficult. The skull is a little enlarged; the bones of it are generally thin; but sometimes they are thickened. The appetite occasionally fails, but oftener is increased; the bowels are usually constipated — though sometimes they are relaxed; the lamb appears more or less stupid; is disinclined to move; staggers slightly; pines away "almost to a skeleton," and dies before it is two months old. Mr. Youatt, after pronouncing the disease generally incurable, advises the administration " of purgatives and tonics combined — the epsom salts with ginger and gentian and small doses of mercurial medicine, the blue pill, in doses

---

* I take the above remarks and quotations on the subject of this disease from my Sheep Husbandry in the South, having learned nothing new in relation to it since that work was written.

of four or five grains," with plenty of good milk, exercise and air. Mr. Spooner says:—"Nothing can be done in the way of treatment,"—and he soundly adds:—"But it will be prudent not to breed again from the ewe; and if there are many such cases, the ram, too, may be changed with advantage, for it is evident the disease is owing to some constitutional fault in the parents, or mismanagement during utero gestation."

APOPLEXY.— Apoplexy is frequent among the improved mutton breeds of sheep in England—which, from their birth to the time of their being butchered, are steadily forced forward into the utmost attainable growth by rich and stimulating food. During their whole lives, they are in a condition of over-fatness and plethora — and apoplexy is a natural result.

This disease is very rare among American sheep, and I have never personally seen an instance of it. Yet when fleshy sheep are first turned out to grass in the spring, and the sun beats down with that burning heat occasionally characteristic of our spring weather, one of the fattest sheep in the flock is, suddenly, without a premonition of disease, found lying dead on the ground. Sometimes one is seen to leap suddenly and frantically into the air, act as if unconscious, stagger, fall and die within five, ten, or fifteen minutes. The farmer has a ready-made name for the malady. The sheep has "grub in the head," and the grubs have just "bored through and penetrated the brain!" (See Grub in the Head.) If the perishing sheep was examined closely, it is probable that the eye would be found staring — the pupils dilated — the sight nearly gone. If additionally the membranes of the nose and eyes were found of a deep red or violet color, as if engorged with blood, I should not doubt the presence of apoplexy.

The treatment, when all treatment is not too late, is immediate bleeding from the jugular vein, until the animal shows signs of weakness. Mr. Youatt and Mr. Spooner speak of a pound as about the appropriate quantity of blood to be taken, but I am confident this would be found too much for the classes of American sheep — the Merino and its grades — which are kept for wool growing purposes. Mr. Youatt says four ounces of epsom salts should be administered as soon as possible after the bleeding, and an additional ounce every six hours until the bowels are opened. Mr. Spooner says that two or three ounces of salt should be administered, and to lambs half that dose. I should prefer Mr. Spooner's prescription — but for the Merino sheep, would be inclined to reduce it to two ounces of the salts, followed by an ounce in six hours, until a copious evacuation took place.

INFLAMMATION OF THE BRAIN. — This is a secondary effect of the causes which produce apoplexy, by which the substance of the brain, its membranes, or both, become the subject of inflammation. The symptoms are much more violent than the preceding. After a degree of dullness and inactivity, accompanied by redness and protrusion of the eyes, the animal becomes delirious, rushes about the field "with its tail cocked," attacks men and trees; and, says Mr. Spooner, "in lambs their motions are quite ridiculous, and have in consequence, among the ignorant, given origin to the idea of their being bewitched." The disease is treated in the same way with apoplexy.

TETANUS OR LOCKED JAW. — In the spring of 1861, I had about eighty rams — the "culls"* which had been accumulating for two or three years in a breeding flock then numbering three thousand. They were castrated — a portion of them by slitting the scrotum and tying the spermatic cords with waxed thread in the usual way that old rams are altered — a portion of them (mostly yearlings and those older ones which had small spermatic cords,) by cutting off the end of the scrotum and removing the testicles precisely as is done in the case of lambs — i. e., by pulling them out! This last novel mode I permitted as an experiment, on the urgent solicitation of the operator, a person of great experience and practical skill in such matters, and who had heard of its being practiced with success. And I am bound to state that not one, or not more than one, of the rams castrated in this unusual way was among the victims I am about to describe.†

Owing to preceding bad weather and other hindrances, the sheep were castrated rather late, and the flies caused much trouble. After the lapse of about ten days, when the animals appeared to be doing well enough, three or four of them were suddenly found entirely rigid and unable to walk, or only retaining some command over the muscles of the

---

* They were those which promised too well to be castrated when lambs, but which did not develop themselves satisfactorily as they grew older.

† This may have been because they were younger, or had smaller spermatic cords — but it at least shows that the mode is as safe as any other with such sheep.

fore-legs, while the hind parts were as immovable as if already stiffened by death. Their jaws were set. The parts of the abdomen near the scrotum were considerably swollen and very hard. They generally stood with their legs a little farther apart than usual, but their postures were so natural that at a few rods distance their situation, or that anything unusual was the matter with them, would not have been suspected by anybody. Some six or eight others were speedily attacked and the symptoms were the same. There was not, in a single instance, any peculiar protrusion or retraction of the head or any other member; and though I watched them for hours, I did not discover the least approach to a convulsion, or even a spasm involving a single muscle. They gave no peculiar evidences of pain — breathed without difficulty — and I think that they all died within about twenty-four hours from the time their situation was discovered.* As their jaws were immovably fixed, no internal remedies could be administered, and I thought that the administration of external ones under such circumstances would be labor thrown away.

The malady is very rare in the United States, but as it is liable to recur I will mention that the foreign veterinarians recommend prompt bleeding from the jugular vein, and aperient medicines, followed by opiates — also warmth and quiet. Mr. Spooner omits bleeding from his recommendations.

EPILEPSY. — Mr. Youatt remarks that "tetanus and epilepsy may be regarded as kindred diseases in all animals; but that in none do they assimilate to each other as in the sheep."

Epilepsy appears to be extremely prevalent in England and on the continent of Europe, but is unusual in this country. The sheep when laboring under its attack, suddenly ceases to feed, stares about stupidly, runs round with a staggering gait, falls to the ground, lays there struggling for a few moments, and then gets up and remains for some period in a semi-conscious state. These attacks recur, and a severer one ends in death. It is thought to result from high condition and the nature of the pasturage — aided by certain not very well understood incidental causes. In England it is commonest early in spring and late in autumn. It is so prevalent in

---

* I am ashamed to say that being peculiarly hurried at the time, I made no contemporaneous written record of the facts, — and therefore am compelled partly to *guess at them*, as do all persons who rely on their recollections for *minute and exact facts* in such cases. But the general course of symptoms I have described are distinctly remembered by me.

certain districts of France, that the people have given up sheep husbandry. Tessier ascribes it to the pasturage. Gasparin states that it is most destructive in Germany in spring and summer, but sometimes in the winter. He says the shepherds of that country attribute it to the sheep's eating the sproutings of the pine in spring, and some species of dock and garlic in the winter.* It would seem that in regions where it particularly prevails, flocks acquire a predisposition to this malady; and the farmers of Beauce, in France, either get rid of the whole flock in which it appears, or they kill every sheep in any degree affected by it.†

PALSY.— I never have seen an instance of this malady. It consists in a suspension of the nervous influence on the muscles — the opposite of tetanus and epilepsy, by which they are excited to unnatural action. The sheep sometimes becomes powerless in every limb and unable to move; sometimes the palsy extends only to the loins or hind-quarters. It is produced by cold and improper exposure,— and sometimes, it is thought, by improper feed. Young lambs when yeaned in very cold weather, and lambs soon after weaning when they receive too plentiful and stimulating food, are most subject to its attacks — though grown sheep are not exempt from them, and particularly, says Mr. Spooner, "the ewe that has aborted or produced her lamb with difficulty and after a tedious labor in cold weather."

The treatment of the disease consists, in the case of a chilled lamb, in the restoration of warmth, and the administration of warm gruel with a little ginger — and if activity is not soon restored, with the addition also of a small quantity of ale. If diarrhea ensues, the "sheep's cordial" is given. In the case of older sheep Mr. Youatt recommends removal to a more comfortable situation, and a purgative consisting of epsom salts and ginger, followed by a dose or two of the cordial.

RABIES.— On Christmas eve, 1862, some sheep belonging to my son, Henry P. Randall, were bitten by a dog. I saw them next morning. The flock consisted of about one hundred ewes, three years old last spring, and in lamb. I thought a dozen or more were wounded; but as their hurts did not appear dangerous, I did not go to the trouble of

---

* Quoted by Youatt, p. 398.   † Ibid.

ascertaining the precise number. All were bitten, so far as I could discover, only about the head, and principally about the nose and ears. The ears of some of them were torn into shreds, and their noses and lips covered with tooth marks, showing that the attack on them had been long persisted in. This was evidently the work of an animal which was unable to kill the sheep outright.

On Christmas morning, a small dog, belonging to a neighbor, was found attacking some sheep owned by the Messrs. Freer, kept about three-quarters of a mile from the preceding. He had wounded two of them in the same way, but more severely, when he was discovered and driven away. He returned to the attack not long afterwards, was again detected, followed home, and killed the same day. The idea of his being rabid did not then occur to any one, though the facts I have since learned lead to the impression that his disease would have been apparent to a person familiar with its symptoms.

The wounds on H. P. Randall's sheep were found to heal rapidly, and nothing was done for them. On the 12th of January, 1863, he informed me that he had found one of the bitten sheep on the ground unable to rise; that, on his helping it up, it moved about with difficulty. It had frothy saliva about its mouth. The next day it died. He had observed some ewes riding each other about, prior to the 12th, but did not know whether the dead one was one of these.

On the 14th of January, he informed me that two or three of the wounded sheep were riding and fighting each other; that one of them had suddenly butted him from behind; that on his turning and offering to kick it, it would not retreat. He confined it in the barn.

I saw the flock in the afternoon. It was in fine condition. The wounds of the bitten sheep were mostly healed; and, with two exceptions, they looked as healthy and full as any in the flock. Two of the sheep were obviously laboring under an attack of rabies. I continued to visit these and the succeeding cases daily, and generally twice a day, until the 29th of January, and until all the earlier cases observed by me (seven) terminated in death.* I usually remained from

---

* As each sheep was attacked it was immediately caught out of the flock. The two first cases were put first in a barn and afterwards in a small pen together, sheltered on the north by a stack. The third one was put in a pen about twenty by forty feet, partly sheltered on the north by a barn and on the west by an overhanging straw stack, and the other four were placed also in this larger pen as fast as attacked.

three-quarters of an hour to an hour at each visit, carefully noting the appearance and actions of the sheep, and keeping a separate and continuous record of each case, as I was able to do without the least danger of mistaking one animal for another — as every one exhibited its number clearly printed on its side.

The history of these cases is published fully in the annual volume of Transactions of the New York State Agricultural Society for 1862, and is quite too long for re-publication here. The recapitulation appended by me to that history is as follows:

SUMMARY.

The cases I have described present variations in the minor developments of rabies, owing perhaps to individual peculiarities of the different animals; but, as a whole, there has been a remarkable identity in the general symptoms.

Assuming that the rabid sheep, which I have designated as No. 3, was seen by me on the first day of the attack of the disease — a fact of which I entertain no doubt after comparing the subsequent symptoms with those of the later ones — and estimating the two first numbered cases to have had the average duration of the other five, the period of "incubation" in the whole seven, (that is, the period between the sheep's being bitten and the appearance of rabies,) ranged from fifteen to twenty-six days, and averaged about twenty-one days.

The first observed symptom, in every case which was seen at or near its commencement, was the same, viz., ungovernable apparent salacity, (lust,) manifested not according to the sex of the patients — all of which were ewes, and supposed to be in lamb — but in the manner in which the ram exhibits sexual heat. This resemblance extended to the minutest particulars in movements, postures, and in that characteristic note with which the male animal expresses desire as he approaches and importunes the female. In no instance did the rabid ewe show any of the usual indications of rutting. She incessantly attempted to ride her companions, but uniformly manifested rage, and turned and fought the one attempting to ride her. This propensity remained active until the sheep became too weak to exercise it, and never entirely ceased.

In all the cases, rumination was totally suspended from the first visible attack of the disease until death; and throughout the same period, all the patients, with perhaps one exception,* were not seen to consume an ounce of natural food, though the choicest was repeatedly offered to them — in some instances, where they had been purposely deprived of it for twenty-four hours. They, however, manifested a depraved appetite. All of them frequently ate wool from each other, and gnawed the rails of their pen. One was seen to eat dung balls from the breech of another — another, snow which had just been saturated with sheep's urine — and two eagerly to lick the mucus and saliva from the nose and mouth of a dead one, and afterwards the post-mortem discharges from the same parts. They preyed upon every substance within their reach which was unnatural as food, except the flesh of their dead companions. Their eating, as I have termed it, was attended, so far as could be observed, with no regular mastication. When they gnawed the rails of their pen, they held their heads down and extended, so that it could not be seen whether they masticated or not. They did not pause and raise their heads to do so, but continued intently gnawing. The only evidence I had of their swallowing the wood was, that considerable quantities of it were bitten from all parts of the pen and none of it could be found on the snow underneath; and as some of the wood gnawed was of a red, and much of it of a dark color, it would have been readily visible there. When they ate wool, dung balls and the like, they generally snatched them, as if in haste, and in all cases swallowed them after two or three rapid movements of the jaws, which were apparently only made to place the substance in a situation to be forced into the esophagus.

No exhibition of thirst was observed in any case, and, on the other hand, no dread of water, when it was placed in a pail before them. One played in the water with her nose, as a horse is often seen to do, and drank a little without apparent difficulty. One or two were seen to nibble a little ice or snow on two or three occasions.

The evacuation of both dung and urine was very slight. The feces appeared natural in color and consistency.

I came to the conclusion, after considerable hesitation, that the disease, in its earlier stages, and perhaps throughout, was accompanied by a slight unnatural expression of the

---

* No. 7 was seen for an instant attempting to ride another sheep the afternoon before the disease, apparently, was fully developed. She resumed eating hay while I stood looking on. I observed her eating for perhaps five minutes. When I next saw her she was rabid.

eyes, which, for the want of a more expressive term, I have called glistening. But I do not think any one could safely undertake to select a rabid sheep from a flock, even if one was known to be there, by this indication alone. Yet obscure as is this symptom, it is the only one which distinguishes the rabid sheep, in appearance, from one in perfect health, until emaciation and the other later effects of the malady exhibit themselves. The animal is as gregarious as ever; eats its food and ruminates as placidly as usual; looks as plump, bright and healthy as any sheep in the flock; half an hour later, with looks entirely unchanged, unless in the trifling particular named, it is moving round restlessly and incessantly among its companions, struck by a malady which has transformed the habits of its sex*—which no human power can arrest or even palliate—and which will know no respite until terminated in a miserable death.

The subsequent occurrence and progress of the symptoms, in the cases observed by me, were about as follows:—The rabid sheep both exhibited and provoked extreme rage when they were first put in a pen with other rabid sheep; they fought or pursued each other fiercely; but this mood soon subsided in the new comers, and for the next twenty-four hours they remained comparatively peaceable, at least unaggressive, but they were ever ready to fight on being ridden. On the second day the depraved appetite manifested itself, and they began to rub their heads against fences, walls, etc., and to scratch them with their own hind-feet, leading to the inference that they were suffering some cerebral pain. The part of the head invariably rubbed was that over the parietal bones. On the second or third day the scars left by the dog's teeth looked red and inflamed. The sheep were more restless and irritable; they frequently assailed their companions without any provocation; they fiercely butted, and two of them actually bit at a stick, as often as it was pushed against or towards them. On the third or fourth day they rushed at a man if he entered their pen—bounded forward and dashed against the fence which separated them from him, on his thrusting a stick at them. Three of them thus charged the fence, if only a hat or handkerchief was shaken towards them. Two were so ungovernably fierce at times that they sprung at a bystander if he uttered a sound or merely approached their pen. They bounded forward when they made these assaults,

---

* At least so in the case of ewes.

most of them emitting that loud, snuffing sound (caused by a violent expulsion of air through the nostrils,) by which rams, bulls, etc., often express their rage at the approach of some strange object. Two of them opened their mouths, gnashing and threatening to bite, whenever they attacked a man or a stick, but I did not see them offer to bite when fighting their companions. On the fourth or fifth day the wounds of a portion of them, more or less, re-opened. On the fifth or sixth day they began to exhibit considerable weakness, and most of them displayed less ferocity. No. 1, however, remained indomitably savage to the last; No. 3 remained so until near death; and No. 6, after a temporary lull, became more deeply re-excited and ferocious, and remained so until death. These three last named sheep would rush at a man, a stick, or another sheep, when they were so weak as frequently to fall before reaching their object, and as soon as they could rise they would renew the attack. They and others frequently fought each other when in this condition, constantly falling, and some of them uttering short, bleating sounds, or groaning piteously when they were hurt. Their voices on such occasions were more shrill and plaintive than the notes of the healthy sheep; but the only one I heard utter the usual prolonged bleat, with which sheep call to each other, or to their keeper, uttered it in the natural key; and this was on the sixth day of the disease.*

On the sixth day, one of the sheep began to rub her breech, often and hard, against the fence, and she continued this, more or less, until death. From the appearance of the parts, I inferred this was occasioned by an irritation of the vagina.

Those which exhibited the greatest decrease of aggressiveness, as their strength failed, never resumed the usual timid habits of their nature. They retreated from nothing; and to the last if a man entered their pen and threatened them with a stick, they instantly attacked him.

The prostration of strength progressed with different degrees of rapidity, owing probably to their different degrees of constitutional vigor; but all showed much and rapidly increasing debility by the close of the sixth day. Their respiration was labored and sometimes irregular. The pulse

---

* Their notes were in no case very "much altered" from the usual ones which indicate rage, pain, &c., and the "howl of the dog," said by Mr. Youatt to be "characteristic of the disease," was entirely wanting. I do not suppose, however, Mr. Youatt meant to be understood literally, but merely that the key of their voices was changed, and rendered high and plaintive, as in the case of the rabid dog.

of the one counted rose to one hundred and forty a minute. One became blind in one eye, one in both, and a third partly blind in one eye. The cornea, in each instance, became opaque and white; but this happened only where wounds of the dog's teeth could be found on the lids or close to the affected eye. At this stage the scabs of nearly all of them dried up, and their wounds appeared to be rapidly healing again. When standing quiet, their heads sunk down low and they trembled slightly all over, as an animal often does after drinking cold water. Froth exuded in rather small quantities from the front part of the mouths of two or three of them, and ropy saliva fell from the lips of one to the ground.

The last day or two of their lives they staggered in their gait, fell over their dead companions, and rose with difficulty. Finally they became unable to rise. The respiration was more labored and irregular, and, in one instance, stertorous. Their debility was extreme. Even at this stage, and until actually dying, they did not manifest that degree of "stupor" and "insensibility to all that is going forward," mentioned by Mr. Youatt. They looked up when a loud or unusual noise was made, and those which were not blind evidently took notice of objects of sight; and not one of them to the last showed the least indications of becoming paralytic, as the same distinguished author states that rabid sheep usually do in England. Neither the appearance of the ground, nor their postures, indicated convulsions or struggling at the time of their death. I saw none of them die.

The five cases which were seen throughout, extended respectively through nine, seven, eight, ten and six days, giving eight days as the average duration of the disease.

While the preceding statement of the symptoms of rabies accords in its leading features with that given by Mr. Youatt, there are even more discrepancies between them in detail than I have called attention to. I think it probable that these differences are due in some measure to local or incidental circumstances, such as the peculiar breed, constitution and habits of the animals, their previous keep, etc. In all these respects the American Merino differs widely from the English breeds. The season of the year when the cases were noted, may also have had an influence. And, finally, owing to climate or other undetected causes, the malady may not assume precisely the same form in different countries. But be all this as it may, I at least know that I carefully noticed, and instantly, and, so far as I could, faithfully, recorded the facts seen by my own eyes.

No remedies were administered to any of the sheep, under the impression that it would be utterly useless, and attended with disagreeable if not dangerous consequences.

Professor Hyde, of the Geneva Medical College, kindly promised to assist me in making post mortem examinations of the several patients — but necessary absence from home prevented it from being attended to until it was too late. I regretted this less, because it is well known that in all such cases, the post mortem appearances are irregular, unsatisfactory, and not characteristic of the special disease.

Two later cases occurred in the same flock, from the bites inflicted on the night of the 24th of December — in the last of February or first of March, according to my present recollection; but I can not speak with certainty, having given my memoranda to an Agricultural Editor. The general course of the disease was the same. The last animal exhibited peculiar violence, fighting a stick thrust toward her with a ferocity resembling that of an enraged dog; and, unlike its predecessors, it constantly uttered short, angry bleats when making its attacks. It remained equally furious after it was unable to rise.

CHAPTER XXIV.

DISEASES OF THE DIGESTIVE ORGANS.

BLAIN — OBSTRUCTIONS OF THE GULLET — THE STOMACHS AND THEIR DISEASES — EXTERNAL AND INTERNAL APPEARANCE OF THE STOMACHS — THE MODE OF ADMINISTERING MEDICINES INTO THE STOMACHS OF SHEEP — HOOVE — POISONS — INFLAMMATION OF THE RUMEN, OR PAUNCH — OBSTRUCTION OF THE MANIPLUS — ACUTE DROPSY, OR RED-WATER — ENTERITIS, OR INFLAMMATION OF THE COATS OF THE INTESTINES — DIARRHEA — DYSENTERY — CONSTIPATION — COLIC, OR STRETCHES — BRAXY, OR INFLAMMATION OF THE BOWELS — WORMS — PINING.

BLAIN.— This malady, as has been remarked, is unknown in the United States. The following is Mr. Youatt's description of its symptoms and treatment:

"Sheep are liable, although not so much as cattle, to that inflammation of the tongue, or rather of the cellular tissue on the side of and under the tongue, to which the above singular names are given. A few sheep in the flock are occasionally attacked by it, or it appears under the form of an epidemic. A discharge of saliva runs from the mouth; at first colorless and devoid of smell, but soon becoming bloody, purulent and stinking. The head and neck begin to swell, and the animal breathes with difficulty, and is sometimes suffocated. A succession of vesicles have risen along the side of the tongue — they have rapidly grown — they have broken — they have become gangrenous — they have formed deep ulcers, or deeper abscesses that occasionally break outwardly. When this is the case it is probably the "Greathead" of Mr. Hogg. The cause is some unknown atmospheric influence; but the sheep have been predisposed to be affected by it, either by previous unhealthy weather, by feeding on unwholesome herbage, or by unnecessary exposure to cold and wet.

"Whatever may be the case with regard to cattle, there is no doubt that the blain is often infectious among sheep. The diseased sheep should immediately be removed from the rest, and placed in a separate and somewhat distant pasture.

"The malady must first be attacked locally. If there are any vesicles in the mouth they must be freely lanced. If any tumors appear on the neck or face, and that evidently contain a fluid, they must be opened. The ulcers must be bathed with warm water at first, and until the matter is almost evacuated — then lotions of cold water, in each pint of which one drachm of the chloride of lime has been dissolved, must be diligently used. Aperients must be administered very cautiously, and not at all, unless there is considerable constipation. The strength of the animal must be supported by any farinaceous food that it can be induced to take — linseed mashes — bran mashes with oatmeal — and the best succulent vegetables, as carrots and mangel wurzel; plenty of good, thick gruel, if necessary, being horned down, and two drachms of powdered gentian root and one of ginger, with four grains of powdered cantharides, being given morning, noon and night. Bleeding will be very proper in this disease before the vesicles have broken, or the external tumors begun to soften, and there is an evident and considerable degree of fever; but after the purulent, fetid matter has begun to appear, it will only hasten the death of the animal."

OBSTRUCTIONS OF THE GULLET.—Sheep are much less liable to become "choked" than cattle, but it occasionally occurs when they are fed cut roots. The obstructing substance which is lodged in the esophagus or gullet, can sometimes be felt from the outside, and moved upward or downward by the pressure of the fingers. If this can not be done the sheep should be placed on its rump between a man's legs and held firmly with the head extended upward in a line with the neck. Some oil should then be poured into the throat, and a flexible probang very carefully inserted and pressed down with sufficient force to carry the obstruction before it into the stomach. I trust gutta-percha probangs for this purpose will soon be prepared for sale. The best implement now attainable on most farms is a strong, flexible, elastic rod of hickory or elm, made perfectly smooth, and not far from five-sixteenths of an inch in diameter. A little bag of flax seed is firmly secured to the lower end, and on dipping the rod into hot water to limber it for use, the bag becomes perfectly soft and slippery. Some wind the end of the probang with tow and dip it in oil.

There is usually no great difficulty in removing the obstruction, but the sheep is often injured so that it subsequently dies, in consequence of the lacerations inflicted on the parts by the haste or carelessness of the operator. Too much care and gentleness can not be manifested in every part of the process.

Where the obstruction can not be thus removed, veterinary practitioners cut down upon it from the outside, and having removed it, the edges of the esophagus are carefully brought together with two or three stitches, and the threads left long enough to project from the external wound. The skin is also stitched together, and a bandage placed without much pressure round the neck. If the sheep is fleshy a moderate cathartic should be administered, and it should be kept on mashes or gruel until the wound is closed. I would not, however, recommend this process to persons unfamiliar with surgical operations.

THE STOMACHS AND THEIR DISEASES.—I shall describe the stomachs to some extent for the better understanding of their diseases; and for this purpose I quote the following from my "Sheep Husbandry in the South":

"On opening the abdomen the *omentum* or caul is found covering the intestines. It is a thin, and, in a normal state, colorless and transparent structure, formed of two membranes, between which extend streaks of fat in the form of a net.

"The external appearance of the stomachs is given in the following cut (see next page) of those of a young sheep which died of disease. Their arrangement is slightly different in the animal.

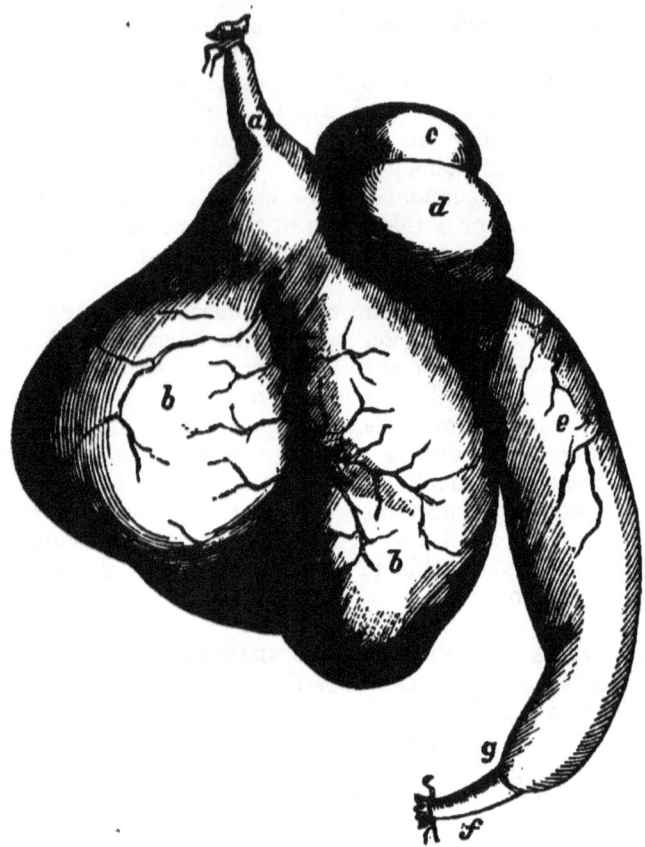

THE STOMACHS.

a, The esophagus or gullet, entering the rumen or paunch.
b, b. The rumen, or paunch, occupying three-fourths of the abdomen.
c, The reticulum, or honey-comb — the 2d stomach.
d, The maniplus, or many folds — the 3d stomach.
e, The abomasum, or 4th stomach.
f, The commencement of the duodenum or first intestine.
g, The place of the pylorus, a valve which separates the contents of the abomasum and duodenum.

"The walls of the rumen or paunch consist of four coats or tunics — 1st, The peritoneal or outer coat; 2d, The muscular; 3d, The mucous, covered with papillæ, or little protuberances, from which (or glands under which) is secreted a peculiar fluid to soften and prepare the food for re-mastication; and, 4th, The inner or cuticular coat, a thin, entirely insensible membrane, which defends the mucous coat from abrasion or erosion."

The following cut which I borrow from Mr. Youatt's work on sheep, exhibits the

INTERNAL APPEARANCE OF STOMACHS.

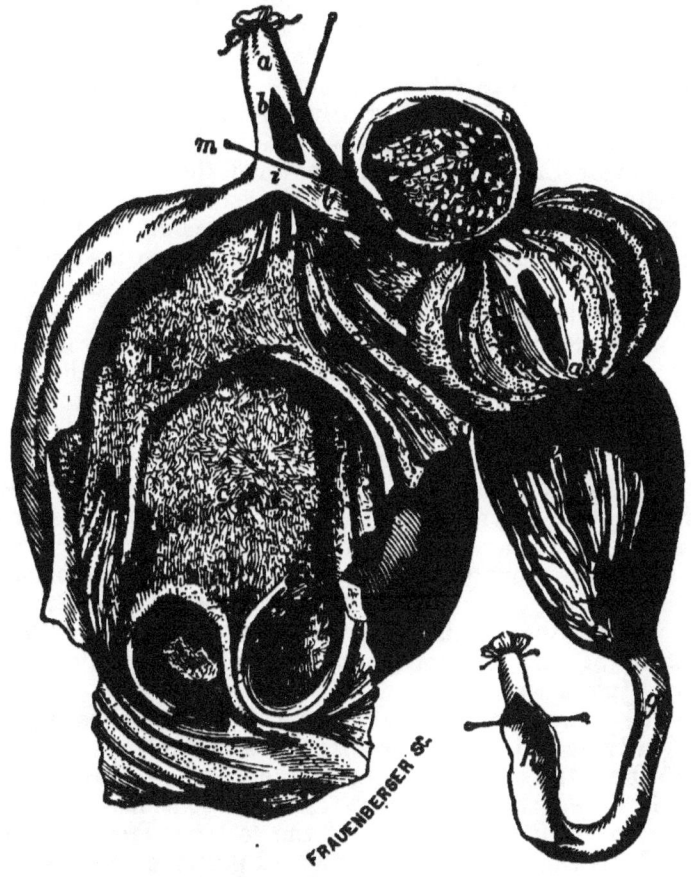

a, The esophagus or gullet.
b, The commencement of the esophagean canal, slit open, with muscular pillars underneath.
c, c, c, The rumen, paunch or first stomach, slit open.
d, The reticulum or honey-comb — slit open.
e, The maniplus or many-folds — slit open.
f, The abomasum — slit open.
g, The commencement of the duodenum or first intestine.
h, The duodenum slit open.
i, m, l, Wands, showing course of esophagean canal, opening of stomachs, etc.

"The reticulum or honey-comb is composed of the same number of coats as the rumen, fulfilling similar functions.

But the mucous coat, in addition to minute papillæ, is covered with elevations arranged in pentagons and sexagons of different sizes, somewhat resembling a honey-comb, except that the cells are larger and shallower.

"The maniplus has the same four coats. Its *floor* is a continuation of the esophageal canal. From its *roof* depend many parallel folds of the cuticular coat — here thicker and stronger than in the other stomachs — reaching nearly to its floor. The cuticle is covered toward the edges of the folds, with hard, bony processes, shaped like fangs, or cones bent in a curvelinear form, and pointing toward the entrance of the stomach. The interior of each fold or leaf contains muscles which impart to it the power of a peculiar and forcible motion. There are forty-two of these folds in the maniplus of the sheep — occasionally forty-eight. They do not all equally nearly approach the esophageal canal, but are disposed in groups of six — one of the central ones of each nearly reaching the canal or floor of the stomach — the others on each side growing shorter and shorter, so as to form a series of irregular re-entering angles.

"The abomasum is the digesting stomach, where the gastric juices are secreted, and where the pultaceous food is converted into chyme. It is funnel-shaped, and its lower extremity connects with the intestines as shown in the cut. The cuticular lining of the three preceding stomachs is wanting in this. The mucous coat is disposed in the form of *rugæ* or shallow folds, arranged longitudinally with the direction of the stomach, and from this membrane the gastric juices are secreted.

"The comparative size of the four stomachs will be sufficiently seen in the preceding illustration.

"Where the esophagus enters the rumen, it terminates in what is called the esophageal canal, a continuation of the former constituting the roof of the latter. The bottom or floor of this canal is formed of divided portions or folds of the upper parts of the rumen and reticulum — muscular "pillars" or "lips," as they are sometimes denominated — which may remain closed so that the food will pass over them into the third and fourth stomachs — or they may open, permitting the food to fall between them, as through a trap-door, into the first and second stomachs. It is probable that the opening of these lips, as food passes over them, depends somewhat upon a mechanical effect, and somewhat upon the will of the animal. Fluid and soft pultaceous food fit for immediate digestion glide over them. But most of the food of the sheep, like that of other ruminating animals, is swallowed with little preparatory mastication; and these untriturated solids drop down through the first opening above described into the rumen. It is certain, however, that the animal can, at will, also cause water to pass through the opening into the first stomach. This would be necessary in the animal economy, and the water is always found there.

"When the food has entered the rumen, the muscular action of that viscus compels it to make the circuit of its different compartments, and, in time, the food later swallowed forces it on and up to near the opening where it originally entered. In its passage it is macerated by a solvent alkaline fluid secreted by the mucous coat. The papillæ of that coat are supposed to influence the mechanical action of the contents of the stomach, and, perhaps, to a certain extent, to aid in triturating them. The food performs the circuit of the stomach, and is ready for re-mastication, according to Spallanzani, in from sixteen to eighteen hours. By a muscular effort of the stomach, a portion of it is then thrown over the membraneous valve or fold which guards the opening from this into the second stomach. The reticulum contracts upon it, forming it into a suitable pellet to be returned to the mouth, and also covers it with a mucus secreted in this stomach. By a spasmodic effort (always perceptible externally when the sheep or cow commences rumination) the pellet is forced through the roof of the reticulum, by the opening before described, and returned to the mouth by contractions of the spiral muscle of the œsophagus or gullet, for mastication.

"This explanation of the functions of the second stomach is not accepted by all the physiologists who have examined this subject. Some contend that all the solider portions of the food are returned directly from the rumen for re-mastication; that when raised to the floor of the esophagean canal, the hard parts are carried up to the mouth — the more pultaceous ones (but still not sufficiently pultaceous for the fourth stomach) passing into the reticulum, where they are again macerated — the fluid squeezed out of them by a contraction of the stomach and allowed to pass on to the fourth stomach — and then the drier parts raised, like those from the paunch, for re-mastication. More solid and indigestible substances 'may be submitted two or more times to the process of rumination.'"

"Let us now observe the course pursued by the food, and the process to which it is submitted, after rumination. It glides over the trap-doors which open into the first and second stomachs. As it passes over the floor of the third, or the maniplus, the pendant leaves of this viscus, armed with their beak-like protuberances, seize the advancing mass, and squeezing out the fluid and the more finely comminuted portions of the food which escape with it, commence triturating the bulkier fibrous portions between their folds. Their bony papillæ give to these folds something of the mechanical action of rasps, in grinding down the vegetable fiber. The food being now reduced to an entirely pultaceous state, passes into the fourth stomach, or abomasum, where it is acted upon by the gastric juice, and converted into chyme. The amount of food found between the folds of the maniplus, after death, depends upon the time that has elapsed since rumination. It is dry and hard, compared with the contents of the other stomachs.

"The entrance to the fourth stomach—the cardiac opening—is closed against regurgitation or vomiting, by a sort of valve, composed of a portion of one of the rugæ, before alluded to, which line the interior of this stomach. The pylorus is also closed by a valve, which prevents a premature passage of the contents of the stomach into the intestines.

"Before the duodenum enters into (or changes its name to) the jejunum, and about 18 inches from the pylorus, it is perforated by the biliary duct — *ductus choledochus* — which brings the bile eliminated by the liver, from the gall-bladder, and also the fluid which is secreted by the pancreas, or sweetbread, which last is introduced into the biliary duct two inches from its entrance into the duodenum, by another duct or small tube. The compound fluid thus introduced into the duodenum exercises various important offices in the digestive and assimilating processes. The bile is supposed to aid in the separation of the chyme into chyle and fecal matter — or the nutritive parts of the food which are assimilated into blood, from the innutritious parts which are discharged as excrement. It also prevents a putrid decomposition of the vegetable contents of the intestines, and serves various other useful purposes.

"The chyle — a white albuminous fluid, with a composition differing but little from that of blood — is taken from the intestines by a multitude of minute ducts called lacteals, which traverse the mesentary, constantly uniting as they advance, so as to form larger ducts. These enter the mesenteric glands — small glandular bodies attached to the mesentary — after the passage of which the chyle begins to change its color. The lacteals still continue to unite and enlarge, and finally terminate in the *thoracic duct.* In this the chyle is mingled with the *lymph* secreted from a portion of the *lymphatics* — another exceedingly minute system of absorbent ducts, which open on the internal and external surfaces of the whole system. From the thoracic duct, the chyle is conveyed to the heart, and enters into circulation as blood."

MODE OF INTRODUCING MEDICINES INTO THE STOMACHS OF SHEEP.— Owing to the peculiar arrangement and action of the stomachs above described, solids, and even fluids if forced down the throat rapidly, fall on the pillars or lips of the esophagean canal with enough momentum to cause them to open, so that the swallowed substance falls into the paunch: and the comparatively insensible walls or coatings of this stomach are scarcely acted upon, to any sensible degree, by medicines, when they are administered in the proper and usual quantities. Consequently, let him who administers medicines in "balls," or in thick, heavy forms, or pours down fluid ones with haste and violence through the usual horn, remember that he is, in most cases, substantially throwing away his medicine, or putting it where its effects will not be felt in time to be of any service in acute cases. The reader is requested to keep these facts distinctly in recollection whenever the administration of remedies is spoken of in this volume; for there can be no possible use of constantly repeating the caution.

HOOVE.— When sheep are suddenly turned from poor pastures on fresh clover, turnips, or other unusually succulent food, and allowed to fill themselves to excess, its fermentation in the first stomach or paunch causes an elimination of gas, which sometimes distends that organ almost to bursting. It presses against the diaphragm (midriff) so that the lungs can not be filled with air, and thereby directly produces suffocation; or the blood no longer circulates through the paunch, and is determined to the head, producing stupor and death.

It is most egregious folly in all cases, to make any such sudden change in feed. If dried-off ewes, for example, are to be put on rank clover, they should, at first, be admitted to it for only two or three hours a day—and driven in at mid-day, when their hunger is already, to a good extent, satisfied. This continued two or three days entirely prevents the danger of hoove.

When sheep are discovered to be "hooven," they should be driven gently about for an hour. If swollen to a very dangerous extent, and the distress and oppression are evidently increasing, they must be relieved by mechanical means. Those provided with such instruments either pass a flexible tube with a rounded perforated end* down the throat into the stomach, through which the pent-up gas escapes, or they plunge a trocar (a sharp stylet or puncturing instrument covered with a canula or sheath,) into the paunch through the left flank. The trocar is withdrawn, leaving its sheath in the wound, which keeps the openings in the side and paunch opposite to each other, thus allowing a freer exit to the gas, and preventing the other matters forced along with it from being left within the cavity of the abdomen or belly. Any solid or semi-solid matter deposited there leads to inflammation, and ultimately, if in any considerable quantity, to death. If a pocket knife is used instead of a trocar, the above dangers are incurred; but it is often the only available instrument at hand, and generally proves a safe one. The place for inserting it is in the left flank, half way between the haunch and ribs, and well up toward the back bone.

It is considered safest always to administer a purgative—usually one or two ounces of epsom salts with a drachm of powdered ginger—after a severe attack of hoove. Mr. Spooner prescribes:—"Sulphate of magnesia two oz., ginger one dr., gentian two dr., chloride of lime half dr., to be dissolved in a pint of warm water or gruel."

If gas continues to be developed, Mr. Youatt recommends the introduction into the stomach of chloride of lime—a drachm dissolved in a gill of water—either by means of a horn or through the canula of the trocar. This would also be an admirable remedy to administer (down the throat) in earlier stages of the disease when the case was not urgent, or the opening of the paunch yet called for. Once in the paunch it would produce chemical results which would at once relieve the parts of their unnatural distension. Sulphuric ether, if

---

* Messrs. Youatt and Spooner mention such an instrument (having a stylet within it which is withdrawn after its insertion into the stomach,) invented by Dr. Munro. I have never seen it. Both writers state that its use is difficult and dangerous in unpracticed hands; and Mr. Youatt expresses a preference for the trocar.

more accessible, would also, in doses of two drachms, condense the inflating gases. A remedy in use among farmers, but which I have never seen tried, is composed of four ounces of lard and a pint of well-boiled milk, poured down at blood heat — half at once, and the remainder soon after. Others administer a gill of urine with as much salt as it will dissolve.* Some give milk with a small quantity of soft soap.

Poisons.—The effect of St. John's-Wort has been adverted to at page 269. The most ordinary poison which the sheep partakes of in the regions with which I am familiar, is the narrow-leaved or low laurel, (*Kalmia angustifolia*.) Sheep unused to this plant, and driven hungry along roadsides where it abounds, consume it and it acts as a virulent poison on them. I never saw a sheep laboring under its effects. Mr. Morrell says:—"The animal appears to be dull and stupid, swells a little, and is constantly gulping a greenish fluid which it swallows down; a part of it will trickle out of its mouth and discolor its lips." He adds:—"In the early stages, if the greenish fluid be suffered to escape from the stomach, the animal most generally recovers. To effect this, gag the sheep, which may be done in this manner: Take a stick of the size of your wrist and six inches long — place it in the animal's mouth — tie a string to one end of it, pass it over the head and down to the other end, and there make it fast. The fluid will then run from the mouth as fast as thrown up from the stomach. In addition to this, give roasted onions and sweetened milk freely."

Mr. Grennell, in the Massachusetts Agricultural Report, 1860, states that the broad-leaved laurel, "calico bush" or "spoonwood," (*Kalmia latifolia*,) is equally fatal, and that "a farmer in Franklin county, Mass., lost sixty from a flock of two hundred, in the fall of 1860, which strayed from a good feed of aftermath grass, into an adjoining pasture, to eat laurel and die." Several plants growing in our Western States are thought occasionally to poison sheep, but I do not know their names or the facts. Mr. Youatt enumerates among the vegetable sheep poisons, the yew (*Taxus baccata*,) and the corn-crowfoot (*Ranunculus arvensis*.) Mr. Spooner mentions that soot, when applied as a top-dressing on wheat on which sheep were soon afterwards turned, acted as a poison, producing palsy of the limbs and death.

---
* American Agriculturist, Vol. 3, p. 66.

M. Brugnone very successfully administered diluted white wine vinegar to sheep poisoned with corn-crowfoot. They were all comparatively well on the succeeding day. Mr. Spooner quotes Mr. Coates, of Gainsborough, England, as saying that three or four score of sheep poisoned by soot—all those which had not become paralyzed—recovered on the administration of cathartics until their bowels were well acted upon. "They were then fed on linseed cake, and ultimately did well." It is a popular impression in this country, how well founded I do not know, that a strong decoction of the white ash, made by boiling the bruised twigs for an hour, and administered from half a gill to a gill, repeating the dose if necessary, is a sure antidote to the effects of laurel, if taken within a day of the poisoning.* Drenches of milk and castor oil are also resorted to for the same object.

Active aperient medicine, so administered as to have its full effect, (see page 299,) is usually given to poisoned sheep. Mr. Youatt recommends, with obvious propriety, that "warm water be injected into the paunch by means of Read's apparatus, pumped out again, and this repeated until either vomiting is excited or the poison has been rendered harmless by dilution." There is a simple and inexpensive stomach pump composed of a hollow ball and perforated tube of India rubber worked by alternately squeezing out the air and fluid from the ball, for sale in all our American drug stores, which would answer for the above purpose admirably. Every analogy goes to show that cathartics are not rapid enough in their effects; and they do not, at best, avert the destructive results of virulent poisons which have been received into the stomach in quantities sufficient to produce death.

INFLAMMATION OF THE RUMEN OR PAUNCH.—This is unknown in the United States, and as it is of very rare occurrence even in England, where ovine maladies generally flourish so vigorously, we have not much reason to fear its future advent—accordingly space will not be consumed here in describing it.

OBSTRUCTION OF THE MANIPLUS.—It would appear from Mr. Youatt's statements (page 435,) that this is more common

---

* Since the above was written I find this remedy given in Allen's "Domestic Animals," and also the following:—"Pour a gill of melted lard down the throat."

than the preceding in England, and is a very serious malady; but I do not consider it necessary to describe it here.

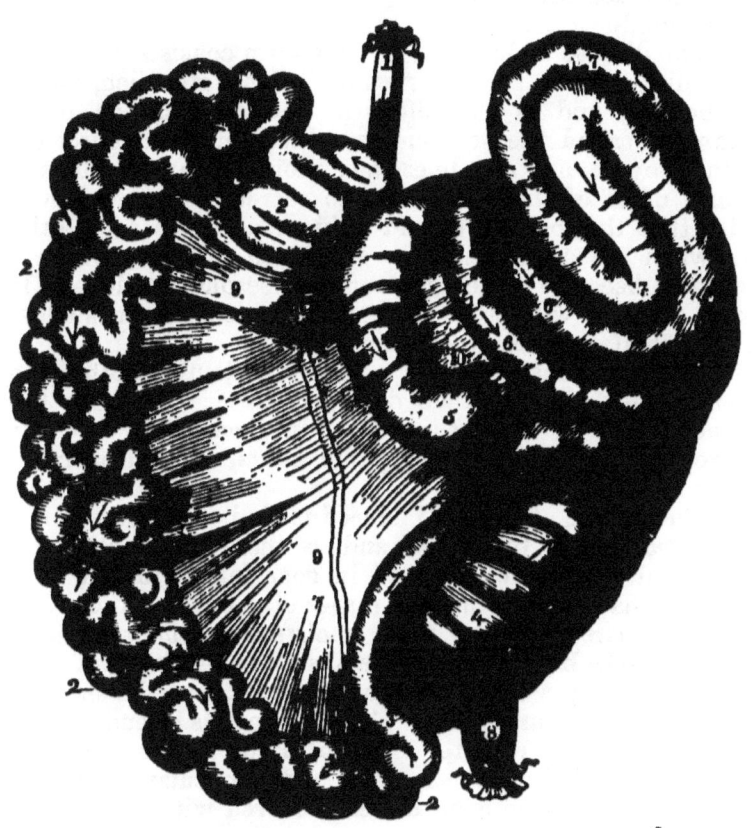

THE INTESTINES AND MESENTARY.

1. The duodenum.
2. The jejunum.
3. The ileum.
4. The cœcum, being the anterior prolongation of the colon, or first large intestine. The ileum opens into this (on the back side as presented in the cut,) about twelve inches from its extremity—the opening being defended by a valve.
5. The large anterior portion of the colon, retaining its size (about three times that of the smaller intestines) for about two feet.
6, 6. The colon tending toward the center.
7, 7. The returning convolutions of the colon.
8. The rectum or straight gut, communicating with the anus.
9, 9. The mesentary, or that portion of the peritoneum which retains the intestines in their places.
10. The portion of the mesentary supporting the colon, &c.
The united length of these intestines is upward of sixty feet.

ACUTE DROPSY, OR RED-WATER. — I have never seen this disease in our country, but as others think they have, I will introduce Mr. Spooner's description of it and of the proper mode of treating it. He says:

"The disease understood by this term consists of effusion of reddish-colored serum or water in the abdomen, outside the bowels, and is the effect of increased action of the membrane called the peritoneum, which forms the outer coat of the bowels, and also lines the abdominal cavity. It is the natural office of this membrane to secrete a watery fluid, in order that the bowels should glide readily on each other, but when diseased action is set up in this membrane its secretion becomes excessive, and the serous portion of the blood, mingled with some of the red portion, becomes effused in this cavity, where it can not escape.

"The disease is extremely common to lambs, both during the time they are with their dams, and after they have been weaned; and in them, as well as in sheep, it is very fatal, destroying the latter in twenty-four hours, and the former in less time.

"The nature of the fluid effused is similar to the serum or watery portion of the blood, and as there is no active pain manifested, we are not justified in considering that it is the effect of inflammation, but one rather of debility of the vessels, and the existence of too much moisture in the system. It usually attacks both sheep and lambs when feeding on turnips, and particularly when there is a hoar-frost, and the sheep are folded on them during the night. From this circumstance it has been attributed to the effect of lying on the cold, damp ground, thus chilling the system, and particularly the abdomen. But the sheep is an animal covered with wool, which can readily bear this exposure, and it is more likely to be produced by an excess of this cold, watery frost taken into the system, though perhaps assisted by cold lairs.

"This view of the matter, too, is borne out by the fact, that when ewes in lamb are kept too much or too long on turnips, they often cast their lambs, which are found dead and water-bellied, as it is termed, that is, the abdomen is found distended with the same description of watery fluid as we find in red-water. Now, in this case, the ewe generally escapes disease, therefore it cannot be from external cold, but from the nature of the food; so likewise it is most probable that such is the case with red-water.

"The symptoms usually observed in sheep, are refusal to

feed or ruminate, a dull, heavy appearance, often attended with giddiness, a staring eye, obstinate costiveness, and sometimes the head is carried on one side. In lambs these symptoms are less decidedly marked, but the little animal lags behind its fellows, is unwilling to move, and is very dull, and dies in a shorter time than the sheep. Acute pain is rarely manifested in either sheep or lamb, but they are generally carried off in a short time. It is not at all uncommon for the shepherd to leave them apparently well over night, and to find one or more dead in the morning.

"The treatment of the cases where the symptoms have fully manifested themselves will generally be unsuccessful; but in the earliest stages, and before the disease has actually been manifested, much can be done. The sheep should be removed to a drier situation, and pasture or seeds or stubble should be substituted for the turnips, and the following medicine administered:— Sulphate of magnesia, one pound; gentian, powdered, one ounce; ginger, dissolved in warm water, one ounce. This is sufficient for eight or ten sheep, or double or treble the number of lambs.

"Above all, it is desirable, by way of prevention, to remove the healthy sheep to some dry pasture, giving them good, sound hay, a little corn and turnips, only in moderation. Such, however, is the fatality of the disease, that it is a question whether it will not be more prudent to kill the sheep or lambs affected; that is, if they are in any condition for the table, or unless from any particular reason it is very desirable to preserve them. Bleeding in these cases will not be prudent unless we are sure that inflammation is present, which we may expect if active pain is manifested.

"Mr. W. Greaves advises the employment of tar as a preventive, and adduces the following instance of its successful employment. He says:—'This disease is very prevalent in this part of Derbyshire, and a friend of mine, Mr. Cooper, of Ashford, for many years lost one-fifth of his hoggets from red-water. Three or four years ago he was advised to bring them into a yard, and give each hogget a tablespoonful of common tar every fortnight, and the consequence has been, that although they are kept in every respect in the same way as before, and on the same ground, he has not lost one sheep since the adoption of this treatment.'

"We give the above on the responsibility of the advisers, in case any farmers may be desirous of trying it, but we can give no opinion in favor of its efficacy."

ENTERITIS, OR INFLAMMATION OF THE COATS OF THE INTESTINES.—For the same reasons which are given in regard to red-water, I present Mr. Youatt's account of this malady. He says:

"By this term is understood inflammation of most, if not all, of the coats of the intestines. * * Its early symptoms are not to be distinguished from those of colic: possibly it is simple colic which then exists; but the disease does not yield to common remedies. The symptoms continue—they become more aggravated—the animal stamps the ground with his feet—he scratches it—he attempts to strike his belly with his hind-legs—he bends his knees as if he would lie down, but he dreads the pain resulting from the consentaneous action of the muscles of the belly, and their pressure on the contents of the belly; he looks round at his sides: at length he comes suddenly down—he rolls on his back:—he maintains this position for some seconds, and then he suddenly starts and scrambles up again. The muzzle, the horns, and the feet are cold. The pulse is quick but small—the bowels are usually confined—obstinately so—the strength of the animal rapidly wastes away. Sometimes there is a determination of blood to the head; the animal is heedless of all around it, the pupil is widely dilated—and to this delirium occasionally supervenes. * * *

"The causes belong almost exclusively to the food or the locality. Enteritis is produced by stimulating and acrimonious nutriment—by an excess of that which is healthful—by the injudicious administration of purgatives, by exposure to cold, and, more particularly, by the mingled influence of cold and wet.

"The treatment is sufficiently plain—bleeding according to the age and condition of the animal, and the urgency of the symptoms—purgatives perseveringly administered until the bowels are opened, and the purging being afterwards kept up; the Epsom salts being employed to produce the first effect, and sulphur the second. The food to consist of mashes or gruel. No tonic to be allowed until the febrile stage is passed, or until violent diarrhea, difficult to check, has succeeded to the constipation."

DIARRHEA.—This disease is often more properly a nervous than a febrile one—in the former case, a morbid increase of the peristaltic motion of the bowels—in the latter, an inflammation of the mucous coat of the smaller intestines. But for the purpose of viewing it in connection with dysentery, to which it is sometimes closely allied, and into which it often runs — and which is clearly a febrile disease — it will be described here.

Common diarrhea, purging, or scours, manifests itself simply by the copiousness and fluidity of the evacuations of dung. It is brought on by a sudden change from dry feed to green, or by the introduction of improper substances into the stomach. It is important clearly to distinguish this disease from dysentery. In diarrhea there is no apparent general fever; the appetite remains good; the stools are thin and watery, but unaccompanied with mucus (slime) and blood; the odor of the dung is far less offensive than in dysentery; the general condition of the animal is but little changed.

Confinement to dry food for a day or two, and a gradual return to it, oftentimes suffice for its cure. I have rarely administered anything to grown sheep, and never have lost one from this disease. To lambs, especially if attacked in the fall, the disease is more serious. If the purging is severe, and especially if any mucus (slime) is observed with the dung, the feculent matter should be removed from the bowels by a gentle cathartic — as half a drachm of rhubarb, or an ounce of linseed-oil, or half an ounce of Epsom salts to a lamb. This should always be followed by an astringent, and in nine cases out of ten, the latter will serve in the first instance. I generally administer, say, $\frac{1}{4}$ oz. of prepared chalk in half a pint of tepid milk, once a day for two or three days, at the end of which, and frequently after the first dose, the purging will ordinarily have abated or entirely ceased.*

The following is the formula of the English "sheep's cordial," usually prescribed in cases of diarrhea by the English veterinarians, and there can be no doubt it is a safe and excellent remedy — better probably than simple chalk and milk in severe cases: Take of prepared chalk one ounce, powdered catechu half an ounce, powdered ginger two drachms, and powdered opium half a drachm; mix them with half a pint of peppermint water — give two or three table-spoonfuls morning and night to a grown sheep, and half that quantity to a lamb.

Mr. Spooner says: — "If the cases are not severe, and entirely confined to diarrhea, astringents alone may be given; but if any mucus is perceived, it will be proper to administer

---

* This, and the two preceding paragraphs, are quoted with the alteration of a few words from my Sheep Husbandry in the South.

a laxative in the first instance. * * * In cases of simple diarrhea the following astringent medicine will be found very useful:—Powdered chalk, one ounce; catechu, four drachms; ginger, two drachms; opium, half drachm; to be mixed carefully with half a pint of peppermint water, and two or three tablespoonfuls given morning and night to a sheep, and half this quantity to a lamb."

The following remedy for diarrhea appears in Mr. Robert Smith's prize essay On the Management of Sheep, already several times cited. He says:— "When the disease is observed to be coming on, the animals should be instantly changed to older or dry keeping. If the disease has advanced unnoticed, they should be taken up, kept warm, supplied with dry food, and given one ounce of castor oil in half a pint of gruel; if the animal has much pain or straining, add twenty drops of laudanum, with rather more gruel; if the discharge still continues, and the bowels have been cleared by this dose, it will be proper to check it by astringents. The following is found to be an excellent medicine and rarely fails:— Four ounces logwood, one drachm of the extract of catechu, and two drachms of cinnamon, mixed with three pints of water, boiled for a quarter of an hour; strain it off, then add sixty drops of laudanum. Give a pint night and morning as long as the flux continues."

DYSENTERY.— Dysentery is caused by an inflammation of the mucous or inner coat of the larger intestines, causing a preternatural increase in their secretions, and a morbid alteration in the character of those secretions. It is frequently consequent on that form of diarrhea which is caused by an inflammation of the mucous coat of the smaller intestines. The inflammation extends throughout the whole alimentary canal, increases in virulence, and it becomes dysentery — a disease frequently dangerous and obstinate in its character, but fortunately not common among sheep in the United States. It differs from diarrhea in several readily observed particulars. There is evident fever; the appetite is capricious, ordinarily very feeble; the stools are as thin or even thinner than in diarrhea, but much more slimy and sticky. As the erosion of the intestines advances, the dung is tinged with blood; its odor is intolerably offensive; and the animal rapidly wastes away. The course of the disease extends from a few days to several weeks.

I have seen but a few well-defined cases of dysentery, and in the half-dozen instances which have occurred in my own flock, I have usually administered a couple of purges of linseed oil, followed by chalk and milk as in diarrhea (only doubling the dose of chalk,) and a few drops of laudanum, say twenty or thirty — with ginger and gentian. According to my recollection, about one-third of the cases have proved fatal, but they have usually been old and feeble sheep.*

Mr. Youatt prescribes bleeding as indispensable, cathartics, mashes, gruel, etc. He adds:

"Two doses of physic having been administered, the practitioner will probably have recourse to astringents. The sheep's cordial will probably supply him with the best; and to this, tonics may soon begin to be added — an additional quantity of ginger may enter into the composition of the cordial, and gentian powder will be a useful auxiliary. With this — as an excellent stimulus to cause the sphincter of the anus to contract, and also the mouths of the innumerable secretory and exhalent vessels which open on the inner surface of the intestine — a half grain of strychnine may be combined. Smaller doses should be given for three or four days."

The following remarks on dysentery and its treatment, occur in Mr. Robert Smith's prize essay:—"If the disease has only just commenced, bleeding is highly necessary; but if advanced, great caution should be observed, and the pulse attended to, to avoid lowering the system too much. To effect a cure, a reaction or perfect change in the system is necessary, and may be best produced by exciting the action of the skin. To effect this the animal should be immersed in a tub of hot water for fifteen minutes, then given one ounce of castor oil, with thirty drops of laudanum, in a little gruel, taking care that the animal be kept warm by wrapping, and placed in a warm shed. As the animal recovers, give gruel freely, with a more moderate dose of the above; when the appetite returns, give mixed feed, such as hay and vegetables. During this disease care should be taken not to pull the wool, as it frequently falls off; a change of pasture, and not

---

* This is also from Sheep Husbandry in the South, with a change of a few lines. Since it was written, I have had sheep die where one symptom of the fatal malady was dysentery. I have ceased to administer more than one purge — and the sheep which I have had thus affected have been in such a situation that I dared not resort to bleeding — notwithstanding the universal tide of modern authority in that direction, when it can be resorted to in an early stage of the disease. What I have mentioned as "hunger-rot" on page 204, frequently closes with dysentery; but the poverty and debility have reached an advanced stage before that sets in,— so that dysentery can not be considered the primary disease.

run too thick, is the best preventive. I have also found either of the following recipes to stay its ravages when given in time; they may be adopted, where parties reject the hot water plan, with equal success:

"No. 1. Four tablespoonfuls of common salt, one teaspoonful of turpentine, mixed with a little water, and repeated in a milder dose when necessary.

"No. 2. One teaspoonful of laudanum, one tablespoonful either gin or rum, well mixed and given; repeat the dose if necessary, or in a milder form.

"No. 3. One ounce of alum in half a pint of warm water. The above three recipes will also stay the progress of diarrhea in lambs."

CONSTIPATION.—There is a tendency toward this in pregnant ewes confined too long to dry feed, as has been already mentioned: and the appropriate remedy is to give a portion of green feed, (see pages 221–228.) Long confinement to dry feed produces a degree of costiveness in all sheep, which occasionally results in colic. The preventive is the same. The constipation of young lambs and its proper treatment have been sufficiently described at page 149.

COLIC, OR STRETCHES.—The cause of this disease is given under preceding head. The paroxysms recur at intervals. During the continuance of them the sheep stretches itself incessantly and often twists about its head as if in severe pain. It lies down and rises frequently. The termination is occasionally fatal, unless the bowels are promptly opened by medicine. An ounce of Epsom salts dissolved in warm water, with a drachm of ginger and a teaspoonful of the essence of peppermint should be administered to a sheep and half as much to a lamb.* Three very excellent practical shepherds† write me—the first, that "he gives Epsom salts successfully for stretches:" the second, that he "uses a decoction of thoroughwort or boneset—that warm tea is also good:" the third, that he "employs castor oil, and if the case is obstinate, a moderate dose of aloes." Attacks of this disease become habitual to some sheep. It can always be prevented by giving green feed daily, or even once or twice a week.

---

* Some farmers lift up the sheep by its hind-legs and shake it a little in that position, under the belief that it cures stretches. I have never tried it. Others "drag it about by the hind-legs!"

† Nelson A. Saxton, of Vergennes, Vermont; William R. Sanford, of Orwell, Vermont; and Prosper Ellithorp, of Bridport, Vermont.

BRAXY, OR INFLAMMATION OF THE BOWELS. — Braxy is one of the formidable diseases of Europe, which I have never met with in this country, though Mr. Morrell says "it is not unusual to sheep kept in the latitude of ours." It is stated in the Mountain Shepherd's Manual that it chiefly attacks lambs about the end of autumn and beginning of winter, and that "inflammation of the bowels seems to be the most common form" of it. "When a sheep is observed to be restless, lying down and rising up frequently, and at intervals standing with its head down and its back raised, and when it appears to move with pain, inflammation may be suspected. The progress of the inflammation excites great pain; but when mortification comes on the pain ceases; and thus we may sometimes account for an animal dying suddenly when apparently well." "The causes of the inflammation," continues the same authority, "may be various. Costiveness from eating hard, dry food, drinking cold water while the body is over-heated, or being plunged into cold water while in that state, or suddenly chilled by a shower of rain or snow, may bring on this destructive malady. Feeding on strong, rank grass is also strongly suspected of inducing braxy. * * Along with long, rank leaves, others that are decayed and rotten or flaccid, may be eaten, and together with the too large quantity of such rank food, which young sheep are apt to swallow, contribute to excite fermentation; and this, from the extrication of air, swells out the intestines, preventing due rumination; and thus, while the food itself is vitiated and does mischief, the over-stretching of the bowels causes inflammation." Mr. Spooner thus gives the post mortem appearances where death has been produced by inflammation of the bowels:—"After death, the paunch is found distended with gas and with food—the latter in a state of putrid fermentation, and necessarily producing the former. The small intestines are in a gangrenous state, the liver is partly decomposed, and filled with vitiated bile; but, most of all, the spleen is gorged with blood, softened, enlarged, not unfrequently ruptured, and filled with tubercles and ulcers, with, in short, various appearances of disease, but all of them the consequence of inflammation principally belonging to this gland, and of the most serious character."

Mr. Spooner recommends the following treatment:—"It should be met with very active treatment. Bleeding from the neck in the early stage, mild aperients, setons, and blisters appear to be called for; but depletion should not be persisted in long, and should be followed by plenty of gruel, vegetable tonics and good nursing."

WORMS.—Sheep, says Mr. Spooner, are subject in rare instances, in England, to a disease arising from the presence of worms in the intestines. Mr. Copeman, of Suffolk, found fifty lambs laboring under violent diarrhea. On examining some which died, he found large patches of inflammation on the villous membrane of the fourth stomach. "The small intestines contained thousands of the folded tape-worm (*Toenia plicata*,) and about twenty-five of the large round worms, (*Ascaris lumbricoides*) with a large quantity—several ounces —of sand. The villous membrane was in a stage approaching to mortification." He ordered a total change in the diet, and the following medicine: Castor oil, 1 oz.; powdered opium, 3 grs.; starch, 1 oz.; boiling water sufficient to make a draught. Thin starch was given night and morning. The lambs improved. After administering this medicine, for four or five days, a stimulant was administered to destroy the parasites: linseed oil, 2 oz.; oil of turpentine, 4 drachms. "One dose only was given to some of them, others required two, and a few had three or four in the course of the following month, and then all were well." I never heard but of a single alledged case, in the United States, of worms proving injurious in the intestines of sheep.

PINING.—Under this name Mr. Spooner describes a very destructive malady in certain districts of Scotland, and particularly on the Cheviot Mountains. Mr. James Hogg, the "Ettrick Shepherd," lost upward of nine hundred sheep by its ravages, within the space of nine years. I do not think this peculiar disease, or anything analagous to it, has yet appeared in the United States, but as the limits of sheep breeding are rapidly extending to fresh regions, embracing new varieties and combinations of climate, soil and verdure, it may be erring on the safe side to include it in this catalogue of maladies. Mr. Hogg says:—"The distemper is a strange one; it may effect a whole flock at once. The first symptoms to a practiced eye are lassitude of motion, and a heaviness about the pupil of the eye, indicating febrile action. On attempting to bleed the animal, the blood is thick and dark colored, and cannot be made to spring; and when dead there is found but little blood in the carcass, and even the ventricles of the heart become as dry and pale as its skin. On the genuine *pining* farms, the disease is more fatal in dry than in wet seasons;

and most so at that season when, by the influence of the sun, the plants are less juicy, or early in autumn, when the grasses which have pushed to seed become less succulent. Consequently, June and September are the most deadly months. If ever a farmer perceives a flock on such a farm, having a flushed appearance of more than ordinary rapid thriving, he is gone.* By that day eight days, when he goes out to look at them again, he will find them all lying, hanging their ears, running at the eyes, and looking at him like so many condemned criminals. As the disease proceeds, the hair of the animal's face becomes dry, the wool assumes a bluish cast, and if the shepherd has not the means of changing the pasture, all those affected will fall in the course of a month."

Pining is thought to proceed "from an enervated and costive habit, producible by want of proper exercise and eating astringent food." "The farms most liable to this disease are those dry, grassy farms, abounding in flats and ridges of white and flying bent. * * The lands which are now most subject to this disease were once in the same manner liable to the rot. As the draining of the sheep pastures proceeded, the rot gradually became extinct, and was ultimately superseded by the pining." Mr. Hogg and Mr. Laidlaw are of the opinion that the primary cause of the pining of sheep was the extirpation of the ground moles from their ranges. These, by throwing up the fresh earth on the surface, preserved the soft, succulent herbage: on their disappearance, it became coarse, harsh and unpalatable.†
"In dripping seasons, shepherds, by strict attention in changing the sheep's pasture every day, may, in great measure, prevent its ravages; but in a dry one, without infield land sown with succulent grasses or limed, it is impossible to prevent it."

Mr. Spooner, after recommending the preparation of more succulent pasturage, and suggesting the culture of some plants in them having laxative qualities, such as the purging flax, adds:—"With regard to medicine, the Epsom and Glauber salts offer themselves as the most suitable, and the employment of common salt will also be found of much service."

I feel constrained to say that the explanation above given of the nature and causes of the malady termed pining, are wholly unsatisfactory to my mind.

---

* In this and the succeeding sentences I think we may suspect a little *poetic* exaggeration — rather a *habitual* tendency in the mind of the author of The Queens' Wake.

† This cause for so general a result appears to me inadequate, not to say fanciful.

CHAPTER XXV.

DISEASES OF THE CIRCULATORY AND THE RESPIRATORY SYSTEM.

THE PULSE — PLACE AND MODE OF BLEEDING — FEVER — INFLAMMATORY FEVER — MALIGNANT INFLAMMATORY FEVER — TYPHUS FEVER — CATARRH — MALIGNANT EPIZOOTIC CATARRH — PNEUMONIA, OR INFLAMMATION OF THE LUNGS — PLEURITIS OR PLEURISY — CONSUMPTION.

The Circulatory System consists of the heart, arteries and veins. It does not enter within the scope of this work to describe their functions and action.

The Pulse.— The pulse in a healthy, full-grown sheep beats according to Gasparin sixty-five, according to Youatt about seventy, and according to Hurtel d' Arboval seventy-five times per minute. To ascertain the number of pulsations, the hand is placed on the left side where the beatings of the heart can be felt. When it is necessary to judge of the character of the pulse, it is felt at about the middle of the inside of the thigh, where the femoral artery passes obliquely across it.

Place and Mode of Bleeding.— Bleeding from the ears or tail, as is commonly practiced, rarely extracts a quantity of blood sufficient to do any good where bleeding is indicated. To bleed from the eye-vein, the point of a knife is usually inserted near the lower extremity of the pouch below the eye, pressed down, and then a cut made inward toward the middle of the face. Daubenton recommends bleeding from the angular or cheek vein — "in the lower part of the cheek, at the spot where the root of the fourth tooth is placed, which is the thickest part of the cheek, and is marked on the external surface of the bone of the upper jaw by a tubercle, sufficiently prominent to be very sensible to the finger when the skin of the cheek is touched. This tubercle is a certain index to the angular vein which is placed below. * * The shepherd takes the sheep between his legs; his left hand more advanced than his right, which he places under the head, and grasps the under jaw near to the hinder extremity, in order to press the angular vein, which passes in that place, to make it swell; he touches the right cheek at the spot nearly equi-distant from the eye and mouth, and there finds the tubercle which is to guide him, and also feels the angular vein swelled below this tubercle; he then makes the incision from below upward, half a finger's breadth below the middle of the tubercle."

When the vein is no longer pressed upon, the bleeding will ordinarily cease. If not, a pin may be passed through the lips of the orifice, and a lock of wool tied round them.

Mr. Youatt says:—" In cases of rheumatism, or garget, or local inflammation referable to the hind-quarters, it may sometimes be advisable to bleed from the saphena or thigh vein. The assistance of another person is required here. The sheep must be laid on his side, on a table, or on some straw, the thigh from which it is intended to extract the blood being undermost. The other three legs must then be tied together, and the assistant must draw out and firmly hold the fourth, while the operator cuts away the hair from that portion of the thigh at which he intends to operate. A person acquainted with the anatomy of the part will at once put his finger on the course of the vein on the upper part of the thigh, and compress it, and thus cause it to become larger below the pressure; but he who is not so much used to the operation will do right to pass a ligature (a piece of coarse tape will constitute the best,) round the hinder part of the thigh, which will render the vein sufficiently evident. It must be opened and afterward secured in the same manner as the cheek vein."

But for thorough bleeding, the jugular vein is generally to be preferred. The sheep should be firmly held by the head by an assistant, and the body confined between his knees, with its rump against a wall. Some of the wool is then cut away from the middle of the neck over the jugular vein, and a ligature, brought in contact with the neck by opening the wool, is tied around it below the shorn spot near the shoulder. The vein will soon rise. The orifice may be secured, after bleeding, as described in the first of the preceding methods.

The good effects of bleeding depend almost as much on the *rapidity* with which the blood is abstracted, as on the *amount* taken. This is especially true in acute disorders. Blacklock tersely remarks:—"Either bleed rapidly or bleed not at all." The orifice in the vein, therefore, should be of some length, and I need not inform the least experienced practitioner that it should be made lengthwise with the vein. A lancet is by far the best implement, and even a sharp-pointed pen-knife is preferable to the bungling fleam. Another important rule in bleeding is that, when indicated at all, it should always be resorted to as nearly as possible to the *commencement* of the malady.

The amount of blood drawn should never be determined by admeasurement, but by constitutional effect—the lowering of the pulse, and indications of weakness. In urgent cases, as, for example, apoplexy or cerebral inflammation, it would be proper to bleed until the sheep staggers or falls. The amount of blood in the sheep is less, in comparison, than that in the horse or ox. The blood of the horse constitutes about one-eighteenth part of his weight, that of the ox at least one-twentieth, while the sheep, in ordinary condition, is one-twenty-second. For this reason, we should be more cautious in bleeding the latter, especially in frequently resorting to it. Otherwise the vital powers will be rapidly and fatally prostrated. Many a sheep is destroyed by bleeding freely in disorders not requiring it, and in disorders which did require it at the commencement, but of which the inflammatory stage has passed.

FEVER.—Fever, without any particular local disease, is very rare in the United States. I never saw a case which I believed came strictly within this class. The sheep suffering from it is without appetite, retreats to a shady place and lies on the ground, pants if it is driven, has a high pulse, a clammy mouth, a dry, hot nose, hot feet, red eyes, and a dull, anxious countenance. On examination, the disease has not yet fastened upon any organ; it is simple fever. At this stage it yields readily to moderate depletion—the abstraction of a small amount of blood and a dose of cathartic medicine.

INFLAMMATORY FEVER.—Mr. Price, an English writer on Sheep Grazing and Management, gives the following account of this disease:—"The number of animals that die of this disorder in Romney Marsh is truly astonishing: I should suppose nearly four in a hundred yearly in some soils and situations, and at peculiar seasons, although every precaution in stocking is taken to prevent it; which if the graziers did not, they would lose half their flock annually. My opinion is that the soil of Romney Marsh, being very rich, consequently the clover and grasses equally so, that sheep feeding on these rich pastures must be more subject to inflammation than those fed on poorer soils, particularly in the spring, when the young shoots of the grasses and natural clover are full of juices: besides, when in this state they are greedily eaten by the animals, which often proves fatal, particularly after a warm day or two.

"On opening them the contents of the abdomen are more or less inflamed, and some parts are very dark colored, and emit a very offensive smell. Sometimes the heart or lungs appear to be primarily affected; and sometimes the liver, bowels, and stomach, which is very easily perceived by the dark and livid appearance of the part. It is said that bad-mouthed sheep never die of this disease, because they can not feed on short, nutritious grass, but on coarse long herbage which does not enrich the blood. I am of opinion that it is an inflammatory disease, and that the only remedy is large bleedings, so as rapidly to lower the system."

MALIGNANT INFLAMMATORY FEVER. — This malady appears occasionally in England, but is common as a very destructive epizootic in France, where it is termed *La Maladie de Sologne*. It prevails in low, marshy districts where the sheep are wintered very poorly, folded in close, damp stables, and turned out in the spring to gorge themselves on the watery, rapidly-growing vegetation. It appears toward the close of spring, and rages until August. Its early symptoms are, says Mr. Youatt, "suspension of rumination, loss of appetite, dullness, weeping from the eye, coldness of the ears, alternate shiverings and flashings of heat. Soon afterward the mouth and the breath become hot — the eyes are red — the pulse is accelerated, and weak and irregular — and there is a mucous discharge from the nostrils, to which succeeds bloody mucus, and then a mixture of purulent matter and blood. By degrees, the urine becomes bloody and the excrements are covered with grumous blood — the head and legs are swelled — the debility is extreme, and the animal dies in the course of eight or ten days. The greater part of the animals attacked by this disease perish. The sheep in the finest condition die soonest, and with greatest certainty."

Dry food, salt, camphorated drinks and vegetable tonics, are usually administered. Bleedings are sometimes resorted to in the very earliest stage of the malady. Tessier, one of the ablest agricultural writers of France, suggests the following modes of prevention:—"To keep the flock more in the sheep house during the rainy season; to feed better the ewes that are pregnant, or that are giving suck; never to milk the ewes;* not to turn the young lambs on those marshy situations on which the danger of being infected by the rot makes them afraid to place the mothers; to keep salt within the reach both of the lambs and the ewes; not to send the sheep to the field when the weather is cold, and to drive them back when storms threaten; not to shear the sheep so early as they are accustomed to perform that operation; and to endeavor by every possible means to drain the ponds and marshes with which that [La Sologne] and so many other districts of France abounds."†

This formidable malady has never yet appeared in the United States.

TYPHUS FEVER.— Mr. Youatt expresses the opinion that this disease often destroys thousands of sheep in Great Britain, and that many of the diseases recognized as braxy are really of this class. I do not know that it ever occurs, as a distinct or idiopathic malady, in the United States; but I scarcely ever saw any febrile symptoms attend any form of ovine disease in our country which were not, or did not very soon become typhoid in their character. (See page 262.) English practitioners recommend "the lancet and Epsom salts," at the "very commencement" of the disease. In any stage of any malady attended by the characteristic low form of fever, where I have seen bleeding and purgative medicines both resorted to, to any serious extent, their apparent effect has been uniformly to accelerate the fatal result.

CATARRH.— Catarrh is an inflammation of the mucous membrane which lines the nasal passages — and it sometimes extends to the larynx and pharynx. In the first instance — where the lining of the nasal passages is alone and not very violently affected — it is merely accompanied by an increased discharge of mucus, and is rarely attended with much danger.

---

* The French, in many districts, milk their ewes and manufacture the milk into cheese.

† Quoted by Youatt, at p. 481.

In this form it is usually termed snuffles, and high-bred English mutton sheep, in this country, are apt to manifest more or less of it, after every sudden change of weather. When the inflammation extends to the mucous lining of the larynx and pharynx, some degree of fever usually supervenes, accompanied by cough, and some loss of appetite. At this point the English veterinarians usually recommend bleeding and purging. Catarrh rarely attacks the American fine-wooled sheep with sufficient violence, in summer, to require the exhibition of remedies. I early found that depletion, in catarrh, in our severe winter months, rapidly produced that fatal prostration from which it is next to impossible to recover the sheep — entirely impossible, without bestowing an amount of time and care on it, costing far more than the price of any ordinary sheep.

The best course is to prevent the disease by judicious precautions. With that amount of attention which every prudent flock-master should bestow on his sheep, the hardy American Merino is little subject to it. Good, comfortable, but well-ventilated shelters, constantly accessible to the sheep in winter, with a proper supply of food regularly administered, is usually a sufficient safeguard; and after some years of experience, during which I have tried a variety of experiments on this disease, I resort to no other remedies — in other words, I do nothing for those occasional cases of ordinary catarrh which arise in my flock; and they never prove fatal.

MALIGNANT EPIZOOTIC CATARRH.—In "Sheep Husbandry in the South," from which the preceding paragraph is transferred, I give an extended account of a disease which prevailed with destructive violence in the State of New York in the winter of 1846-47. Some flock-masters lost half, others three-quarters, and a few seven-eighths of their flocks. One individual within a few miles of me lost five hundred out of eight hundred — another nine hundred out of one thousand. But those severe losses fell mainly on the holders of the delicate Saxon sheep, and perhaps, generally, on those possessing neither the best accommodations, nor the greatest degree of energy and skill.

I lost about fifty sheep by the disease. Up to February, my sheep remained apparently perfectly sound, and they were in good flesh. Each flock had excellent shelters, were fed regularly, etc., and although sheep were beginning to perish about the country, my uniform previous impunity in those "bad winters" led me to entertain no apprehensions of the prevailing epizootic. About the first of February my sheep went into the charge of a new man, hired upon the highest recommendations. A few days after I was called away from home for a week. The weather during my absence was, a part of the time, very severe. The sheep house occupied by one flock containing one hundred sheep, was, with the exception of two doors, as close a room as can be made by nailing on the wall-boards vertically and without lapping, as is common on our Northern barns. One of the doors was always left open, to permit the free ingress and egress of the sheep, and for necessary ventilation. A half dozen ewes, which had been untimely impregnated by a neighbor's ram, were on the point of lambing, and it being safer to confine the ewes in a warm room over night, the shepherd, instead of removing them to such a room, confined the whole flock in the sheep house every night, and rendered it warm by closing both doors. After two or three hours, the air must have become excessively impure. On entering the sheep house, on my return, I was at once struck with its highly offensive smell. A change, too, slight but ominous, had taken place in the appearance of a part of the flock. They showed no signs of violent colds, I heard no coughing, sneezing, or labored respiration — and the only indication of catarrh which I noticed, was a nasal discharge, by a few sheep. But those having this nasal discharge, and some others, looked dull and drooping; their eyes ran a little — were partially closed, the lachrymal caruncle and lids looked pale — their movements were languid — and the shepherd complained that they did not eat quite so well as the others. The pulse was nearly natural — though I thought a trifle too languid.

Not knowing what the disease was — and fully believing that depletion by bleeding or physic was not called for, I contented myself with thoroughly purifying the sheep house — seeing that the feeding, etc.,* was managed with the greatest regularity — and closely watching the further symptoms of disease in the flock. In about a week, the above described symptoms were evidently aggravated, and there had been a rapid emaciation, accompanied with debility, in the sheep first attacked. The countenance was exceedingly dull and

---

* They had been fed with bright hay three times a day and turnips. As those affected did not eat their turnips well, I commenced feeding some oats in addition to the turnips. I believed that a generous diet was called for and I gave it.

drooping — the eye kept more than half closed — the caruncle, lids, etc., almost bloodless — a gummy, yellow secretion below the eye — thick glutinous mucus adhering in and about the nostrils — appetite feeble — pulse languid — and the muscular energy greatly prostrated. Nothing unusual was yet noticed about their stools or urine.

I now had all the diseased sheep removed from the flock, and placed in rooms the temperature of which could easily be regulated. I commenced giving slight tonics and stimulants, such as gentian, ginger, etc., but apparently with no material effect. They rapidly grew weaker, stumbled and fell as they walked, and soon became unable to rise. The appetite grew feebler — the mucus at the nose, in some instances tinged with dark grumous blood — the respiration oppressed, and they died within a day or two after they became unable to rise.

I proceeded to make post mortem examinations with great care and deliberation — aided by Dr. Frederick Hyde, now Professor of Anatomy in the Geneva Medical College.* My minutes of those examinations have already been partially published in "Sheep Husbandry in the South;" and they are quite too long for insertion here.

Laboring very strongly under the impression that the seat of the disease would be found in the lungs, or some of the abdominal viscera, no examination was made, in the first six cases, of the interior organs of the head and neck. But failing to discover any sufficient indications of primary disease among the latter to account for the results, I, in the next case, examined the bronchial tubes, the lower portions of the windpipe, esophagus, &c., and found them all in an apparently healthy condition. Before tracing these passages to the throat, I removed the upper portion of the skull and carefully examined the brain and its investing membranes. All seemed in a perfectly normal state. I then made a longitudinal section down through the middle part of the whole head, and the seat and character of the fatal malady stood at once revealed. The mucous membrane lining the whole nasal cavity, highly congested and thickened throughout its whole extent, betrayed the most intense inflammation. At the junction of the cellular ethmoid bones with the cribri-

---

* To guard against any misapprehensions on this point, I may be permitted to say that we had a well warmed room — all the proper instruments for making such an examination — and several hours were usually devoted to each case.

form plate—in the ethmoidal cells—slight ulcers were forming on the membraneous lining. The inflammation also extended to the mucous membrane of the pharynx, and say three inches of the upper portion of the esophagus. Here it rather abruptly terminated. Fifteen or twenty more cases were examined, and so far as the seat and character of the disease of the mucous membrane was concerned, the appearances were uniform in every instance.

This was obviously a species of catarrh—though the feverish symptoms which ordinarily accompany a severe attack of that disease were wanting. From the very outset, and in every case, the type of the disease was typhoid—sinking—and rapidly tending to fatal prostration.

I was anxious, of course, to reduce the local inflammation of the membranes lining the head, but felt perfectly satisfied there was too much debility to admit of depletory treatment. Nevertheless, to make myself sure, and to gratify the curiosity of others, I bled in three or four instances, as near as possible to the commencement of the attack. As anticipated, it evidently hastened the fatal termination. Blisters not being regarded as available under all the circumstances, I blew Scotch snuff (through paper tubes) up the nostrils of some of the sheep, to cause the removal, by sneezing, of the mucus which seriously obstructed respiration, and in the faint hope that it might produce a new action, by which an increased mucous secretion would be excited and the congested membrane relieved. This was the only local treatment resorted to.

The next step was to fix on the constitutional treatment. The liver had been shown to be in a torpid state. There was a functional derangement in the mesenteric and probably other glands, and a want of activity in the general secretory system. What medicine would stimulate the liver, cause it to secrete the proper quantity as well as quality of bile, change the morbid action of the glands and secretory system, and restore activity and health to the vital functions generally? In my judgment, nothing promised so well as mercury; and by its well known effect on the entire secretory system, it would powerfully tend to relieve the congested membranes of the head. The proto-chloride of mercury (calomel) was supposed to possess too much specific gravity to reach the fourth stomach, with any certainty, administered in a liquid; and if administered as a ball or pill, it would be almost sure not to reach that stomach. The dissolved bi-chloride of mercury (corrosive sublimate) was therefore hit upon. One grain was dissolved in two ounces of water, and one-half ounce of the water (or one-eighth of a grain of corrosive sublimate) was exhibited in a day, in two doses.

As constipation existed in most of the cases, it was thought that the bowels required to be stimulated into action, and slightly evacuated by a mild laxative. Having noticed in similar cases of debility and torpor of the intestinal canal, that purgation is often followed by a serious diarrhea, difficult to correct, and leading to rapid prostration, and there being no intestinal irritation to suffer additional excitement, I thought that rhubarb — from its well known tendency to give tone to the bowels, and its secondary effect as a mild astringent—was particularly indicated. It was given in a decoction—the equivalent of ten or fifteen grains at a dose—accompanied with ginger and gentian, in infusion. To a portion of the sheep I administered the rhubarb and its adjuvants alone; to others I gave the bi-chloride of mercury in addition.

Not a single sheep recovered after the emaciation and debility had proceeded to any great extent. One such only lingered along until shearing. Its wool gradually dropped off: it seemed to rally a little once or twice and then relapse; and it perished one night in a rain-storm. In the generality of instances the time from the first observed symptoms until death, varied from ten to fifteen days. A few died in a shorter time. I thought that the treatment produced favorable effects in some instances — particularly when resorted to at the commencement of the disease. At all events, some of the sheep recovered under the treatment — particularly under that including the exhibition of the bi-chloride of mercury—and very few, if any, recovered without any treatment. Candor compels me to say, however, that the results of the treatment were far from being satisfactory — that the cases of recovery were much fewer than the deaths. I have merely stated what I believe to be the facts in the premises; I do not feel prepared to make any recommendations. As I now look back on, and quote from my records written seventeen years ago, I feel greatly disposed to doubt whether more recovered under my treatment than would have recovered without it. At all events, I prefer that view of the case should be taken, so that if a similar epizootic should recur, those called upon to combat it will start without any misconceptions derived from me. I have given my treatment because it constitutes part of the true history of the case; and because records of *failures* are not without their value.

The epizootic gradually abated towards spring, and my flock regained its perfect health. Near spring, many farmers found what seemed to them an unusual number of "grubs" in the heads of their sheep which died of the prevailing epizootic, and therefore they attributed the disease to this cause — and this seemed to be the prevailing popular opinion. In some of the latest cases in my flock, I discovered more or less grubs; and, in two or three instances, an unusual number. In other cases, where the external symptoms and the post mortem appearances were almost identical, no grubs were to be seen — convincing proof that they had nothing to do in originating this destructive disease.

The whole value of the preceding records, in connection with the omitted post mortem examinations,—if they have any—is in enabling us to determine what the sheep epizootic of 1846–47 was, and what it was not. I am not prepared to aver that it was identically the same with the "distemper" which used to sweep off from twenty to forty or fifty per cent. of carelessly managed flocks as often as once in five or six winters — and which, though greatly mitigated in the frequency and severity of its visitations, continues to destroy more American sheep than all other maladies combined. It is, indeed, the only malady which proves *mortal* on a large scale. But, except that the "distemper" of the "bad winters" sometimes closes with dysentery, I never saw any difference between its general external symptoms and that of the epizootic of 1846–47. If their identity should be established, it would be a most important point gained; for then we should know against *what* enemy to concentrate our efforts, instead of "doctoring" for rot, inflammation of the lungs, braxy, consumption, grub in the head, etc., etc.— each of which maladies the winter "distemper" of this climate has often been pronounced to be. Unfortunately, I have had no opportunities to make post mortem examinations of sheep dying of that disease since 1846–47. Without this, all other observations are uncertain and comparatively valueless. The farmer who finds a prevailing and mortal disease among his sheep, and who is not sufficiently familiar with the internal structures and appearances of the animal to make an intelligent examination of them after death, should always avail himself of the services of a well educated physician.* How-

---

* I never knew *such* a physician who disdained to bestow his skill, on proper occasion, on a brute. "Where Allah hath deigned to bestow life, and a sense of pain and pleasure," said Adonbec el Hakim, "it were sinful pride should the sage, whom he has enlightened, refuse to prolong existence or assuage agony."

ever little the latter may be acquainted with veterinary practice, he will be entirely competent to decide, in a great majority of cases, what organ is the seat of a mortal malady:* and it will be far safer to rely on his general directions, founded on established principles and on a knowledge of the properties of remedial agents, than to make experiments at random, or what is equally dangerous, call in the aid of an ignorant quack.

PNEUMONIA, OR INFLAMMATION OF THE LUNGS. — Pneumonia, or inflammation of the lungs, is not a common disease in the Northern States, but undoubted cases of it sometimes occur, after sheep have been exposed to sudden cold — particularly when recently shorn. The adhesions occasionally witnessed between the lungs and pleura of slaughtered sheep, betray the former existence of this disease — though in many instances it was so slight as to be mistaken, in the time of it, for a hard cold. The sheep laboring under pneumonia is dull, ceases to ruminate, neglects its food, drinks frequently and largely, and its breathing is rapid and laborious. The eye is clouded — the nose discharges a tenacious, fetid matter — the teeth are ground frequently, so that the sound is audible to some distance. The pulse is at first hard and rapid — sometimes intermittent; but before death it becomes weak. During the height of the fever, the flanks heave violently. There is a hard, painful cough during the first stages of the disease. This becomes weaker, and seems to be accompanied with more pain as death approaches.

After death, the lungs are found more or less *hepatized*, i. e., permanently condensed, and engorged with blood, so that their structure resembles that of the *hepar*, or liver — and they have so far lost their integrity that they are torn asunder by the slightest force.

It may be well in this place to remark that when sheep die from any cause with their blood in them, the lungs have a dark hepatized appearance. But it can be readily decided whether they are actually hepatized or not, by compressing

---

* A healthy situation of the lungs, bronchial tubes, &c., would at once show the absence of pneumonia, consumption, bronchitis, etc. The healthy condition of the liver would show the absence of rot — the healthy condition of the intestines, the absence of braxy, etc. Were any of those organs found diseased, it might not be so easy in all instances to decide on the precise character of the malady, — but enough at least would be learned to furnish a guide to the general treatment in subsequent cases; and at all events, to avoid exasperating the disease by entirely improper remedies. It is much to be hoped that a professional body of educated and learned veterinarians will soon be spread throughout our country.

the windpipe, so that air can not escape through it, and then between such compression and the body of the lungs, in a closely fitting orifice, insert a goose quill or other tube, and continue to blow until the lungs are inflated so far as they can be. As they inflate they will become lighter colored, and plainly manifest their cellular structure. If any portions of them can not be inflated, and retain their dark, liver-like consistency and color, they exhibit hepatization — the result of high inflammatory action — and a state utterly incompatible, in the living animal, with the discharge of the natural functions of the lungs.

With the treatment of pneumonia, I have but little personal experience. In the first or inflammatory stages of the disease, bleeding and aperients are clearly called for. Mr. Spooner recommends "early and copious bleeding, repeated, if necessary, in a few hours — this followed by aperient medicines, such as two ounces of Epsom salts, which may be repeated in smaller doses if the bowels are not sufficiently relaxed. The following sedative may also be given with gruel twice a day:—nitrate of potash, one drachm; digitalis powdered, one scruple; tartarized antimony, one scruple."

The few cases I have seen have been of a sub-acute character, and would not bear treatment so decided. Mr. Youatt remarks:—"Depletion may be of inestimable value during the continuance — the short continuance — of the febrile state; but excitation like this will soon be followed by corresponding exhaustion, and then the bleeding and the purging would be murderous expedients, and gentian, ginger, and the spirit of nitrous ether will afford the only hope of cure."

BRONCHITIS.— It would be difficult to suppose that where sheep are subject to pneumonia they would not also be subject to bronchitis — which is an inflammation of the mucous membrane which lines the bronchial tubes—the air-passages of the lungs. I have seen no cases, however, which I have been able to identify as bronchitis, and have examined no subjects, after death, which exhibited its characteristic lesions. Its symptoms are those of an ordinary cold, but attended with more fever and a tenderness of the throat and belly when pressed upon. Treatment: Administer salt in doses from 1½ to 2 oz., with 6 or 8 oz. of lime-water, given in some other part of the day. This is Mr. Youatt's prescription.

PLEURITIS OR PLEURISY.— I have seen no instance of this disease. Mr. Spooner says of it:—"This disease consists of inflammation of the pleura or membrane lining the chest. It is produced by the same causes as inflammation of the lungs, with which it may be accompanied, and particularly by any sudden changes that may chill the whole system. It often occurs from this cause after sheep washing, when it is very common to find a few sheep failing and in proportion to the want of care exercised. It is not unusual, in examining the bodies of sheep, to find the lungs in part adhering to the sides of the chest, and the animal thus affected generally loses flesh. This adhesion is the effect of pleurisy, and another and still more dangerous result is water in the chest.

"The symptoms of this disease are in many respects like those of inflammation of the lungs, but it is attended occasionally by severe pain and by a variation of the symptoms generally, such as a harder and more defined pulse and more warmth of the body. The treatment must consist of active bleeding in the first instance; and in this disease the sheep can bear blood-letting to a greater extent than in most diseases. The bleeding may be repeated if necessary, setons may be inserted in the brisket, the bowels moderately relaxed, and in other respects the same treatment observed as advised for inflamed lungs."

CONSUMPTION.— This has never, so far as my knowledge extends, appeared in American flocks. Mr. Youatt thus describes it:— "There is another and still more frequent and equally fatal disease of the lungs, [with acute inflammation,] but it assumes an insidious character, and is not recognized until irreparable mischief is effected, viz., sub-acute, or chronic inflammation of the lungs, and leading on to disorganization of a peculiar character — tubercles in the lungs, and terminating in phthisis [consumption.] The sheep is observed to cough — he feeds well and he is in tolerable condition — if he does not improve quite so fast as his companions, still he is not losing ground, and the farmer takes little or no notice of his ailment. * * * He is driven to the market and he is slaughtered, and the meat looks and sells well; but in what state are the lungs? Let him who is in the habit of observing the plucks of the sheep, as they hang by the butcher's door, answer the question. He sees plenty of sound lungs from oxen — he sees the lungs of the calf in a beautifully healthy state; but he does not see one lung in three belonging to the sheep that is unscathed by disease.— whose mottled surface does not betray inflammation of the investing membranes, and in the substance of which there are not numerous minute concretions — tubercles.

"Perhaps these lesions quickly follow sub-acute inflammation of the lungs, but they do not rapidly increase afterwards. Their existence produces a slight cough which scarcely interferes with health. * * * But what is the case, and that not unfrequently, with the ram and the ewe when they get three or four years old? The cough continues — it increases — a pallidness of the lips, or of the conjunctiva, is observed — a gradual loss of flesh — an occasional or constant diarrhea, which yields for a while to proper medicine, but returns again and again until it wears the animal away. Of how many diseases is this cough and gradual wasting the termination? It is the frequent winding up of turnsick; it is the companion and child of rot.

"This disease is especially prevalent in low and moist pastures, and it is of most frequent occurrence in spring and in autumn, and when the weather at those seasons is unusually cold and changeable. It is almost useless to enter into the consideration of treatment. It would consist in a change to dry and wholesome and somewhat abundant pasture — the placing of salt within the reach of the animal, and, if he was valued, the administration of the hydriodate of potash, in doses of three grains, morning and night, and gradually increasing the dose to twelve grains. With regard, however, to the common run of sheep—when wasting has commenced, and is accompanied by cough or dysentery, the most honest and profitable advice which the surgeon could give to the farmer would be, to send the animal to the butcher while the carcass will readily sell."

Some American writers appear to think they have recognized this disease among the sheep of our country. Consumption is considered distinctly hereditary in almost all domestic animals.

CHAPTER XXVI.

DISEASES OF THE GENERATIVE AND URINARY ORGANS.

ABORTION — INVERSION OF THE WOMB — GARGET — PARTURIENT OR PUERPERAL FEVER—CYSTITIS, OR INFLAMMATION OF THE BLADDER.

A PORTION of the more ordinary diseases of the generative system have been described in the Chapters devoted to the treatment of sheep during the different seasons of the year.

ABORTION.—Abortion is unusual among sheep in our country; and when it occurs, is usually produced by some violence, such as the hooking of a cow, the kick of a wanton colt, the heavy sidewise blow inflicted by the horns of a cross ram as he forces his way impatiently up to the rack or feeding-trough, or the like. Severe running, leaping, or the rough, careless handling of the operator for hoof-rot, sometimes produces it. There seems to be an occasional ewe which is habitually subject to it from some unknown cause. Mr. Youatt and Mr. Spooner both mention that it is thought sometimes to occur in England in consequence of eating salt. The constant habit of feeding salt freely at all periods of the year, during my whole life, without, so far as could be reasonably judged, producing such an effect in a single instance, leads one wholly to discredit this hypothesis. Mr. Spooner says:—" But what causes it more than anything else is the unlimited use of turnips and succulent food." I have no experience in the "unlimited" winter feeding of any green food; but I have fed breeding ewes about a pound of turnips per head, sometimes a trifle more, daily, during their entire pregnancy for many years; and by comparing them with flocks about me restricted to dry feed, I have always been satisfied that a *moderate* supply of green feed tended decidedly to prevent abortion.*

---

* Mr. Youatt gives another singular cause of abortion—"continued *intercourse with the ram* after the period of gestation has considerably advanced;" and he says:

So far as this has fallen under my observation, it has occurred oftenest about the close of the third or the beginning of the fourth month of pregnancy. I have never known it to assume that semi-infectious or enzootical character which it occasionally takes in our great dairies of cows—though, as a matter of precaution, as well as to give her a better chance, I always prefer to have the ewe that has miscarried, drawn from the breeding flock and put in "the hospital." The aborted lamb and everything that comes with it from the vagina, is also removed from the sheep yard. The lamb is almost invariably dead at birth. I have not been in the habit of administering any medicine to the ewe.* She usually becomes poor and weak unless nursed with great care—her wool ceases to grow, and is very apt to be shed off. Sometimes she scarcely recovers her condition during the ensuing summer. It is a very great injury to a ewe to abort, and if she does so the second time, she should invariably be excluded from the breeding flock.

INVERSION OF THE WOMB.—This has been sufficiently noticed at page 145 of this work.

GARGET.—This has also been noticed at page 157 under the head of Inflamed Udder. In high-fed English ewes it assumes a more acute and dangerous form than is there described. Hard kernels or tumors form in the udder. The udder itself becomes much swollen, with great heat and tenderness. An ounce or two of Epsom salts with a drachm of ginger, should be administered. If matter forms in any part of the udder, a deep incision should at once be made, the pus squeezed out, the parts well fomented, and if any offensive smell proceeds from the wound it should be bathed or syringed two or three times a day with a weak solution of chloride of lime, until it assumes a healthy action.

In the place of the iodine ointment recommended by me (at page 158) as an application to the udder from the earliest

---

"This is frequently the case among the mountain and the moor sheep." American sheep are more modest! I will not undertake to say such a thing never occurs, but I never yet saw or heard of one of our sheep taking the ram after the beginning of pregnancy, though nothing is more common than to allow rams to run with "in-lambed" ewes the entire winter.

* Mr. Spooner recommends giving Epsom salts ½ oz., tincture of opium 1 drachm, powdered camphor ½ drachm, with nourishing gruel: the two latter medicines to be repeated the next day, but not the salts unless the bowels are constipated. Mr. Youatt says, "If the fœtus has been long dead—shown by the fetid smell and the vaginal discharge—the parts should be washed with a weak solution of the chloride of lime; and some of it also injected into the womb."

stage of the disease, Mr. Spooner recommends camphor ointment, (see List of Medicines,) and Mr. Youatt one drachm of camphor ointment, one drachm of mercurial ointment, and one ounce of elder ointment, well incorporated together. Both also rely greatly on constant fomentation with *hot* water, without the ingredients which I mentioned as proper to mix with it. (See page 158.) But those ingredients must add to its salutary effects.

PARTURIENT OR PUERPERAL FEVER.—This disease, as already remarked, is very unusual in this country, and is, so far as I have learned, confined exclusively to English sheep. I have never seen a case of it. I shall therefore present the following account of its symptoms and treatment from a Prize Essay on the subject, prepared for the Royal Agricultural Society of England by Mr. Isaac Seaman. He says:

"Parturient fever may be defined a disease of low inflammatory character, involving more or less extensively the organs of reproduction, digestion and respiration; the brain and spinal marrow are also involved. There is generally a greater determination of blood to some organs than to others; mostly the uterus is first and principally affected, in some the bowels and lining membrane of the abdomen (peritoneum,) in others the lungs; the brain and spinal marrow are often very much affected. It shows itself generally during the last twenty days' gestation, and within the first six days after parturition: the average duration of the disease is from seven to fourteen days; some die in two days while others linger a month.

"*Causes.*—Any circumstance or agency which depresses the power of the system, insufficient or improper food, close folding, exposure to fatigue, to cold, and moisture, may be considered causes of the affection. I have repeatedly noticed, where ewes about a month before lambing have been removed from a sufficiency of wholesome food to other possessing less nutritive qualities, they have suffered greatly from parturient fever. The practice of fattening sheep and ewes being fed upon the same piece of turnips, (the best parts of which are consumed by the former, whilst the roots and other inferior parts are consumed by the latter,) ought to be abandoned; *a small fold, too — a circumstance so essential to the development of fat in the one, whilst highly injurious to the pregnant ewe, to whom exercise is of the greatest importance for the maintenance of health.** Moist and warm seasons,

---

* I have italicised these words, so strongly confirmatory of the views expressed in the closing portion of Chapter XIX—extending from page 221 to 228 of this volume.

vegetables growing luxuriantly, and the non-supply of dry, farinaceous food, are alike productive of the affection. Fat condition is thought to be a grand cause of the disease. I certainly have noticed that the Sussex Downs (a breed most disposed to collect fat,) suffer most; and, as I before stated, a delicate sheep; but losses have been sustained from the fact that the breeder, thinking them too fat, a short time before the full period of gestation lessens the supply of food, which is plentiful and nutritious, and substitutes that of a poorer nature. * * * * * * * * *

"*Symptoms.*—The most early symptom that marks the commencement of this disease — first the ewe suddenly leaves her food, twitches both hind-legs and ears, and returns again to her food; during the next two or three days she eats but little, appears dull and stupid; after this time there is a degree of general weakness, loss of appetite and giddiness, and a discharge of dark color from the vagina; whilst the flock is driven from fold to fold the affected sheep loiters behind and staggers in her gait, the head is carried downward, and the eyelids partly closed. If parturition takes place during this stage of the disease, and the animal is kept warm and carefully nursed, recovery will frequently take place in two or three days; if, on the contrary, no relief is afforded, symptoms of a typhoid character present themselves; the animal is found in one corner of the fold, the head down, and extremely uneasy, the body is frequently struck with the hind feet, a dark colored fetid discharge continues to flow from the vagina, and there is great prostration of strength. A pair of lambs are now often expelled in a high state of putrefaction; and the ewe down and unable to rise, the head is crouching upon the ground, and there is extreme insensibility; the skin may be punctured and the finger placed under the eyelids without giving any evidence of pain; the animal now rapidly sinks and dies, often in three or four days from the commencement of the attack. Ewes that recover suffer afterward for some time great weakness, and many parts of the body become denuded of wool.

"*Treatment.*—The ewe immediately noticed ill should be removed from the flock to a warm fold apart from all other sheep, and be fed with oatmeal gruel, bruised oats and cut hay, with a little linseed cake. If in two or three days the patient continues ill, is dull and weak, a dark colored fetid discharge from the vagina, and apparently uneasy, an attempt to remove the lambs should be made. The lambs in a great majority of cases at this period are dead, and their decomposition (that is, giving off putrid matter,) is a frequent cause of giddiness and stupor in the ewe. If the *os uteri* (the entrance into the womb) is not sufficiently dilated to admit of the hand of the operator, the vaginal cavity and *os uteri* should be smeared every three hours with the extract of belladonna, and medicine as follows, given: — Calomel eight grains, extract hyoscyamus one drachm, oatmeal gruel eight ounces — mix and give two tablespoonfuls twice a day. Epsom salts two ounces, nitre half ounce, carbonate of soda two ounces, water one pint — mix and give two wine-glassfuls at the same time the former mixture is given. Let both mixtures be kept in separate bottles, and well shaken before given. The bowels being operated upon, omit both former prescriptions, and give the following: — Nitre half ounce, carbonate of soda one ounce, camphor one drachm, water eight ounces — a wine-glassful to be given twice a day. Feed the ewe principally upon gruel and milk, or linseed porridge. Parturition having taken place, the uterus should be injected with a solution of chloride of lime, in the proportion of a drachm to a pint of water, and repeated twice a day whilst any fetid discharge from the vagina remains. * * *

"*Post Mortem Appearances.*—On opening the body of an ewe in which parturient fever has existed, and has been the cause of death, a great variety of appearances are presented. In some cases a degree of redness, varying from clear vermillion to a reddish brown, is variously disposed over the coats of the intestines and lining membrane of the abdomen (peritoneum) and the cavity of the abdomen, invariably containing a great quantity of reddish serum (red-water.) The liver mottled, its structure soft, and the bile appearing dark and viscid. The cavity of the womb containing much dark colored putrid matter, emitting a most horrible stench, its structure soft and almost black. The blood in the heart and large blood vessels frequently found black, would not coagulate, and destitute of tenacity. The lungs frequently found gorged with a reddish serosity [fluid] and of a deeply red or brown color, and as soft as pulp, the cavity of the chest containing much red serum. Dark colored spots variously disposed over the surface of the brain, and within the sheath of the spinal marrow.

"*Prevention.*—The most important feature connected with our subject is the prevention of the disease, for it most interests the breeder in a pecuniary point of view. I would recommend as most important during the last five or six weeks' gestation, regular and nutritious feeding, *regular exercise*, dry and extensive folding. If turnips be the article of food, let there be given in addition a few oats, linseed cake, with hay and straw chaff; let a well sheltered and dry fold be arranged at a short distance from where the ewes are fed during the day, wherein to lodge for the night; *the driving to and from these folds will give exercise — a circumstance tending much to promote health in the pregnant ewe;*[*] if the system of heath or pasture feeding is practiced, night folding is then equally necessary. The night fold in common use — that formed by building straw and stubble walls, with sheds attached, the front of which has a southern aspect — answers admirably. Further explaining the comforts of the pregnant ewe, I will add in the words of the poet,

> "First with assiduous care from winter keep,
> Well foddered in the stalls, thy tender sheep:
> Then spread with straw the bedding of thy fold,
> With fern beneath, to 'fend the bitter cold."

These statements scarcely need addition; but as there is a strong probability that this formidable malady will become more common in the United States as the high bred English sheep, and English systems of keeping are introduced, I will append to it the following letter addressed to me by Mr. Thorne:

"THORNDALE, WASHINGTON HOLLOW, N. Y., April 13, 1863.

DEAR SIR:— * * The puerperal fever has been known in this neighborhood since I first came here, though only to a limited extent during the last two seasons.[†] * * * The disease more generally affects middle aged ewes, and ewes producing or carrying twins. It does not select those lowest in flesh; hence the farmers, as a class, are unwilling to believe that feed can remedy it. It generally shows itself from four or five to ten days before lambing. The symptoms you will find fully described in Seaman's Essay, in Vol. XV, of the Journal of the Royal Agricultural Society. The treatment which my shepherd has followed, and with good success — saving sixteen out of twenty, sick in 1859 — has been to separate the sick ewe at once from the flock and give a dose of 2 ozs. Epsom salts, 2 to 3 ozs. molasses, 1 drachm of nitre, mixed with a pint of warm linseed gruel. The

---

[*] I placed these words in italics,—and also the words "regular exercise" above.

[†] Mr. Thorne's statements of his losses, which here follow, have already been mentioned at page 59.

object is to open the bowels, and should the above not operate in eight or ten hours, it should be repeated. After that, the nitre and molasses are given night and morning in an ordinary quart bottle of gruel until there is an abatement of the fever, when the nitre is discontinued. Frequently, in fact generally, after they have been down three or four days — if they live so long — the brown discharge which has been noticed passing from the vagina becomes putrid, showing that the fœtus is dead. In such cases a small quantity of belladona — applied dry on the end of the finger — is applied to the mouth of the womb every hour until it is sufficiently relaxed to allow of the removal of the decaying mass. After that has been done, the womb is thoroughly syringed with warm water, to which milk is sometimes added. The ewes' position is made as comfortable as possible, and always changed once or twice a day. Where the ewe brings forth her young alive she recovers more rapidly. The remedies and treatment, as you will see, are perfectly simple and easily tried by any flock owner. The great secret of success with it, as with a large majority of diseases, I believe is good nursing. * * * Since my flock have received a small quantity of grain, say half a pint per head daily, before lambing,* they have been quite free from any signs of that trouble. As an illustration that a small quantity of feed is a preventive, a flock belonging to one of my friends was divided, upon going into winter quarters, into two lots,—one of sixty old ewes, the other of thirty two-year-olds. The former received a very small quantity of corn daily — the latter only hay. His losses from the former lot was two — from the latter fourteen head; though the younger ones generally escaped. * * *

Yours faithfully, SAM'L THORNE."

While an over-fleshy, plethoric condition is obviously improper for breeding ewes, there is not a particle of doubt that both Mr. Seaman and Mr. Thorne are correct in the position — not only as respects the attack of parturient fever, but all other maladies and difficulties connected with parturition — that ewes should not be suffered to fall off seriously in flesh during the period of gestation. Even if the ewe enters that period in too high condition, it is safer to keep her there than it is to reduce her. It would be better, if we could have

---

* In a subsequent letter Mr. Thorne says :—"I commence with a small quantity of grain eight weeks before lambing, which is soon increased to half a pint each."

things exactly according to our wishes, to have the ewe enter the term of gestation in moderate order, and then *gain a little* — almost imperceptibly — to the time of lambing. But let me not be mistaken. This is no time to fatten or to stimulate — no time to over-feed, as many do, on the wholly unfounded hypothesis that it is necessary for the support of the fœtus. On this last point let me corroborate my opinions by much more authoritative ones. The well known Dr. Dewees, speaking of pregnant human females, says:

"Errors in diet are almost constantly committed during pregnancy, than which few things are more mischievous. We have already adverted to the tending of the system to plethora, during this condition of the female: on this account it can not fail to be injurious to overcharge, or to overstimulate the stomach. No one circumstance has contributed so certainly to fix this error, as the vulgar speculation on this subject; namely, the necessity the female is under to prepare nourishment for two beings, at one and the same time; that is, for herself and the child within her. It is, therefore, constantly recommended, to eat and drink heartily; and this she often does, until the system is goaded to fever; and sometimes to more sudden and greater evils, as convulsions or apoplexy."[*]

Mr. Youatt says:—"It has been supposed by some breeders that, because the ewe is with lamb, an additional quantity of food, of more nutritive food, should be allowed; nothing can be more erroneous or dangerous both to the mother and the offspring. There will be too many causes of inflammation ready to act, and to act powerfully, during the time of going with lamb, to prevent the least approach to excess of food."[†]

According to eminent British medical writers, like Dr. Hey, Dr. Gordon, Dr. Joseph Clarke, Dr. John Clarke, etc., puerperal fever in the human subject often assumes an epidemic and highly destructive character in Europe, and particularly in England. According to Dr. Dewees, it is very rare in the United States.[‡]

The history of the disease thus far seems to run parallel between human and ovine subjects, in this country and Europe. It would seem that it assumed an epizootic, or rather enzootic character in Mr. Thorne's neighborhood.

---

[*] Treatise on the Diseases of Females, by William P. Dewees, M. D., late Professor of Midwifery in the University of Pennsylvania, &c., &c., 1840.

[†] Youatt on Sheep, p. 497. He repeats these views again and again.

[‡] Dewees on the Diseases of Females, p. 380.

Those desirous of reading a more elaborate paper on this subject than that of Mr. Seaman, the important parts of which I have quoted, will find it in a "Prize Report" by Mr. W. C. Sibbald, in XII Volume of the Journal of the Royal Agricultural Society of England, 1851, (page 554.)

CYSTITIS, OR INFLAMMATION OF THE BLADDER.— Mr. Spooner says:—"Inflammation of the bladder is a rather rare disease with sheep, and is chiefly confined to such as are kept on artificial food, such as oil-cake, beans, &c., though clover that has been mown, it is said, will produce it. There are more losses from this cause than farmers are aware, it being generally this disease when a sheep is said to drop with water. It is mostly confined to the male sex, and principally to rams, and such as are highly fed. The state of the bladder appears to be that of fullness, which shows its neck is involved in inflammation, and thus becomes contracted and loses the cavity. In horses, cystitis is generally attended with constant staling, the bladder being so irritable as scarcely to retain a drop of urine. In sheep there is the same predisposition to stale, but an incapability of performing the act." Mr. Dickens abstracted three pints of blood from the neck of a "highly fed tup," laboring under this disease, which produced fainting. "He soon rallied, and an oleagenous draught, accompanied by an opiate, was given twice during the day. Toward night he appeared much better, ate a little, and was seen to void some very highly colored urine. His medicine acted well during the night, but on the next day his straining came on at times. He again bled him from the other side of his neck to the amount of two pints. From this time he continued mending."

CHAPTER XXVII.

DISEASES OF THE SKIN.

THE SCAB — ERYSIPELATOUS SCAB — WILD FIRE AND IGNIS SACER — OTHER CUTANEOUS ERUPTIONS — SMALL POX, OR VARIOLA OVINA.

THE SCAB.— The scab is a cutaneous disease, analogous to the mange in horses and the itch in men. It is caused and propagated by a minute insect, the *acarus*. M. Walz, a German veterinarian, who has thrown great light on the habits of these parasites, says:

"If one or more female acari are placed on the wool of a sound sheep, they quickly travel to the root of it, and bury themselves in the skin, the place at which they penetrated being scarcely visible, or only distinguished by a minute red point. On the tenth or twelfth day a little swelling may be detected with the finger, and the skin changes its color, and has a greenish blue tint. The pustule is now rapidly formed, and about the sixteenth day breaks, and the mothers again appear, with their little ones attached to their feet, and covered by a portion of the shell of the egg from which they have just escaped. These little ones immediately set to work and penetrate the neighboring skin, and bury themselves beneath it, and find their proper nourishment, and grow and propagate, until the poor animal has myriads of them to prey on him, and it is not wonderful that he should speedily sink. Some of the male acari were placed on the sound skin of a sheep, and they, too, burrowed their way and disappeared for awhile, and the pustule in due time arose, but the itching and the scab soon disappeared without the employment of any remedy."

The figures on the next page are copied from M. Walz's work.

The female acarus brings forth from eight to fifteen young at a litter.

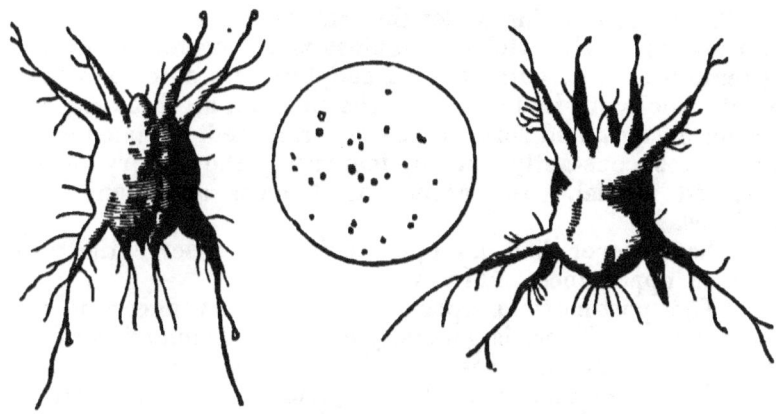

THE ACARUS WHICH CAUSES SCAB.

The central figure represents the acari of their natural size on a white ground. The left hand figure represents the male on its back, magnified to 306 times the natural size. The right figure represents the female seen by the same magnifying power. The heads or suckers of both point upward, between the inner pair of legs. The legs have trumpet-like appendices. Hairs are seen on them to which the young ones adhere when they first escape from the pustule.

The scab is thought often to be produced spontaneously, in England, by mismanagement of various kinds, such as "bad keep, starvation, hasty driving, dogging, and exposure afterward to cold and wet;" and it spreads rapidly by contagion. It is very prevalent there, and annually causes an immense loss in the wool and flesh of the British flocks. In the United States it is comparatively little known, and so far as I am able to learn, never originates spontaneously. It is a singular fact that short-wooled sheep, like the Merino, are much less subject to its attacks, and this is probably *one* reason for its little comparative prevalence in the United States. Mr. Youatt observes:

"The old and unhealthy sheep are first attacked, and long-wooled sheep in preference to the short; a healthy, short-wooled sheep will long bid defiance to the contagion, or probably escape it altogether."

It spreads from individual to individual, and from flock to flock, not only by means of direct contact, but by the acari left on posts, stones, and other substances against which diseased sheep have rubbed themselves. Healthy sheep are therefore liable to contract the malady if turned on pastures previously occupied by scabby sheep, though some considerable time may have elapsed since the departure of the latter.

The sheep laboring under the scab is exceedingly restless. It rubs itself with violence against trees, stones, fences, &c. It scratches itself with its feet, and bites its sores and tears off its wool with its teeth. As the pustules are broken, their matter escapes, and forms scabs covering red, inflamed sores. The sores constantly extend, increasing the misery of the tortured animal. If unrelieved, it pines away and soon perishes.

I have never had an opportunity to observe the post-mortem appearances. Mr. Youatt says:

"The post-mortem appearances are very uncertain and inconclusive. There is generally chronic inflammation of the intestines, with the presence of a great number of worms. The liver is occasionally schirrous, and the spleen enlarged; and there are frequently serous effusions in the belly, and sometimes in the chest. There has been evident sympathy between the digestive and the cutaneous systems."

Twenty-seven years since, I purchased one hundred and fifty fine-wooled sheep just driven into the county from a considerable distance. I placed them on a farm then owned by me, in another town, and did not see them for about three weeks. One of my men then reported to me that the sheep were amiss — that they were shedding off their wool — that sore spots were beginning to show on them — and that they rubbed themselves against the fence-corners, &c. Though I had never seen the scab, I took it for granted that this was the disease. No time was to be lost, as I had seven hundred other sheep on the farm — though fortunately, thus far, the new comers had been kept entirely separate from them. Barely looking into Mr. Livingston's work for a remedy, I provided myself with an ample supply of tobacco and set out. The sheep had been shorn, and their backs were covered with scabs and sores. They evidently had the scab. I had a large potash kettle sunk partly in the ground as an extempore vat, and an unweighed quantity of tobacco put to boiling in several other kettles. The only care was to have enough of the decoction, as it was rapidly wasted, and to have it strong enough. A little spirits of turpentine was occasionally thrown on the decoction, say to every third or fourth sheep dipped. It was necessary to use it sparingly, as, not mixing with the fluid and floating on the surface, too much of it otherwise came in contact with the sheep. Not attending to this at first, two or three of the sheep were thrown into great agony, and appeared to be on the point of dying. I had each sheep caught and its scabs scoured off, by two men who rubbed them with stiff shoe-brushes, dipped in a suds of tobacco-water and soft soap. The two men then dipped the sheep all over in the large kettle of tobacco-water, rubbing and kneading the sore spots with their hands while immersed in the fluid. The decoction was so strong that many of the sheep appeared to be sickened either by immersion or by its fumes; and one of the men who dipped, though a tobacco-chewer, vomited, and became so sick that his place had to be supplied by another.

The effect on the sheep was almost magical. The sores rapidly healed, the sheep gained in condition, the new wool immediately started, and I never had a more perfectly healthy flock on my farm. Though administered with little reference to economy, the remedy was a decisive one. With a vat like figure on page 187, this would not necessarily be a very expensive method, with sheep recently sheared. But the assaults of the scab usually come on in the spring before shearing time, and it would require an immense quantity of tobacco decoction to dip sheep with their fleeces on, however carefully it might be pressed out.

The following is the remedy recommended by Chancellor Livingston:—"First, I separate the sheep (for it is very infectious;) I then cut off the wool as far as the skin feels hard to the finger; the scab is then washed with soap-suds, and rubbed hard with a shoe-brush, so as to cleanse and break the scab. I always keep for this use a decoction of tobacco, to which I add one-third by measure of the lye of wood ashes, as much hog's lard as will be dissolved by the lye, a small quantity of tar from the tar-bucket, which contains grease, and about one-eighth of the whole by measure of spirits of turpentine. This liquor is rubbed upon the part infected, and spread to a little distance round it, in three washings, with an interval of three days each. I have never failed in this way to effect a cure when the disorder was only partial. * * * I can not say whether it would cure a sheep infected so as to lose half its fleece."*

The following remedies are much used in Great Britain:

No. 1.—Dip the sheep in an infusion of arsenic, in the proportion of half a pound of arsenic to twelve gallons of water. The sheep should previously be washed in soap and water. The infusion must not be permitted to enter the mouth or nostrils.

* Livingston's Essay. Appendix, p. 177.

No. 2.—Take common mercurial ointment, for bad cases, rub it down with three times its weight of lard—for ordinary cases, five times its weight of lard. Rub a little of this ointment into the head of the sheep. Part the wool so as to expose the skin in a line from the head to the tail, and then apply a little of the ointment with the finger the whole way. Make a similar furrow and application, on each side, four inches from the first, and so on over the whole body. The quantity of ointment (after being compounded with the lard) should not exceed two ounces; and considerably less will generally suffice. A lamb requires but one-third as much as a grown sheep. This will generally cure, but if the sheep should continue to rub itself, a lighter application of the same should be made in ten days.

No. 3.—Take of lard or palm oil 2 lbs., oil of tar ½ lb., sulphur 1 lb. Gradually mix the last two, then rub down the compound with the first. Apply in the same way as No. 2.

No. 4.—Take of corrosive sublimate ½ lb., white hellebore, powered, ¾ lb., whale or other oil 6 gallons, rosin 2 lbs., tallow 2 lbs. "The first two to be mixed with a little of the oil, and the rest being melted together, the whole to be gradually mixed." This is a powerful preparation and must not be applied too freely.

Mr. Spooner gives the preference to No. 1, as least troublesome; Mr. Youatt to No. 2; and the author of the Mountain Shepherd's Manual to No. 4. I should certainly prefer No. 3, if it is, as it is asserted to be, equally effectual, for the reason that it contains no poisonous or dangerous ingredients. But its perfect efficacy may be doubted.

Mr. Robert Smith, in his Prize Essay, several times cited, declares that scab "is never observed or known to arise spontaneously in a flock," in England. It is clearly and concededly not spontaneous in the United States. Mr. Smith adds:—"When first discovered, the whole flock should be carefully inspected and the diseased subjects removed to a separate field; it is best to give the whole flock a light dressing, as a preventive; no fear need to be entertained of dressing the inlambed ewes, as I have had occasion to practice it at different periods and have experienced no ill effects, observing not to dress the belly or points. The mercurial ointment in common use, prepared by all druggists, is found to be sufficiently good,* without resorting to other

---

* Mr. Smith undoubtedly means mercurial ointment prepared by druggists *for this especial object* — not mercurial ointment having the full strength of that prepared according to the London and New York pharmacopœia, which are the same, viz.,

recipes; when ordered the party should take care to name that it is required for the specific purpose of curing the disease, that attention may be specially paid to the grinding of the quicksilver. In mild cases one dressing by an experienced shepherd, at the rate of 3 lbs. to the score for full-grown sheep, and 2½ lbs. for younger ones, will prove sufficient,* plenty of shreds being the principal feature, and also observing to dress the points pretty freely; care should be taken to shut them up one or more nights, according to the case, and afterwards kept in a warmer situation, if possible, for a time, and given a good supply of food. In bad cases, it is proper to inspect them weekly, until the disease be entirely removed, and give opening medicines pretty freely. Should any die under the operation, the remainder should be washed immediately; if the disease do not then stop, they should be shorn, which is a certain remedy."

Tobacco has always been the favorite American remedy, but at the present time would be very expensive. If every farmer would, in a bed of his garden, raise a sufficient quantity of tobacco plants for this purpose and for dipping his lambs, it would cost him but a trifle.

Prof. Simonds, one of the most recent writers on the subject, recommends a liquid prepared as follows:

"Take two ounces arsenic and two ounces carbonate of potash, and boil in a quart of water till dissolved, and then add water enough to make a gallon of the solution. To this add a gallon of vegetable infusion, made by pouring a gallon of water over four ounces of fox-glove leaves, (*digitalis*,) and allowing the infusion to remain till cold, when it is poured off. 'These two gallons of liquid,' he says, 'constitute a safe agent, and one of the most potent remedies for scab. Half a pint of it (from a bottle with a quill in the cork,) on the skin at the back and sides of the sheep. Two or three dressings will be found sufficient to cure the most inveterate cases of scab in sheep.' The digitalis leaves can be obtained at any drug store."†

---

compounded of mercury 2 lbs., lard 23 oz., suet 1 oz. There is a mild mercurial ointment, prepared in London and sold under that name, which is compounded of strong mercurial ointment 1 lb., lard 2 lbs. The proper reduction of the strong mercurial ointment of the shops is given in No. 2 of the remedies mentioned in text.

* Mr. Smith writes of large English sheep. I should consider 1 oz. of the *reduced* ointment per head, quite enough for Merino sheep, and half that amount for lambs — in winter.

† I think I cut this from the American Stock Journal — but accidentally failed to mark it with its proper credit at the time.

ERYSIPELATOUS SCAB. — This is described by Mr. Stevenson (quoted by Mr. Youatt) as consisting of an "inflammation of the skin that raises it into blisters containing a thin, reddish and watery fluid. These continue for a short time, break and discharge their matter, and are followed by a blackish scab." Mr. Youatt says this disease is rare — that a little blood should be abstracted — and a purge of Epsom salts administered. External applications are not usually necessary, but if there are much burning and itching, sweet oil or camphorated oil will afford relief. I have never seen this disease.

WILD FIRE AND IGNIS SACER. — Mr. Youatt says: — "The wild fire, or more extensive vesication and torture, (than erysipelatous scab) and to a certain degree infectious, has occasionally existed as an epidemic [epizootic.] The *Ignis sacer* or violent cutaneous inflammation of the skin of the sheep, is occasionally mentioned in every history of the epidemics of sheep. As, however, a disease to be traced to any definite cause, and attacking solitary individuals of the flock, and thence communicated to others, it is unknown." I think these forms of cutaneous eruption are unknown in the United States.

OTHER CUTANEOUS ERUPTIONS.—I received numerous letters from Texas for a few months preceding the close of mail communication, by the present war, describing a cutaneous eruption of very general prevalence among the sheep of that State — and inquiring whether it was scab, and what was its proper remedy. The disease described by the different writers — twenty or thirty in all — appeared to be substantially the same. The sheep was uneasy and rubbed itself as if it itched more or less violently. Pustules did not appear on the skin, break and form sores as in scab — but the cuticle was thickened, rough, and sometimes rather red as if covered with a rash. I think the sheep did not usually tear off much of their wool. It seems to have spread rapidly from flock to flock as if contagious or epizootic.

I recommended dipping the sheep in tobacco-water strong enough to kill ticks. I had heard from several of these flocks before the mails were closed: and in every instance the remedy proved effectual.

Some forms of cutaneous disease, differing essentially from scab, have appeared occasionally, though very rarely, among sheep in New York, and, I think, in all the Northern States. Sometimes a flock in winter exhibit considerable itching about their under parts, and scratch them with their feet, pulling out the wool. This yields to an application of tobacco-water. I would suggest that sulphur ointment be tested as a remedy in these minor eruptive diseases. Number 3, among the scab remedies given above, would also seem to be a promising remedy for them. But if farmers will raise their own tobacco, it would probably leave nothing to desire, either on the score of efficacy or economy. Certain eruptions of the face, lips, &c., have been mentioned at pages 269, 271.

SMALL POX, OR VARIOLA OVINA.— When Messrs. Youatt and Spooner wrote their works on the sheep, this fearful malady had been long known on the Continent, but had never visited Great Britain. It however appeared in England, in 1847 I think, and committed desolating ravages. It has not been introduced into the United States, but as no one knows how soon it may be—by the same means by which that malady might be introduced among human subjects—its history and treatment deserve special attention.

La Clavelee, as it is termed in France, attacks sheep at all seasons of the year, and in all conditions—but lambs sooner than grown sheep. Half, and not unfrequently two-thirds of a flock used to perish by it. The sheep which recovers does not contract it the second time. It is communicated by contagion, and in every possible indirect way in which contagion is communicated among human beings, by substances which had been in contact with the subjects of the disease. A flock take it by being turned on a pasture which was occupied two or three months before by diseased animals—or by being driven over a road recently traveled by them. Mr. Youatt thus condenses and translates the statements of various French writers on the subject:

"In the regular *clavelee* there were four distinct periods; first, the symptoms which preceded the eruption, as dullness, loss of appetite and strength, and debility, marked by a peculiar staggering gait, the suspension of animation, and slight symptoms of fever. This continued during about four days, when commenced the second period, or that of eruption. Little spots of a violet color appeared in various parts, and from their center there sprung pustules accompanied by more or less inflammation, isolated or confluent, and with a white head; their base was well marked and distinct, they were

15*

surrounded by a red areola and their center was flattened. They were larger than an ordinary lentil. In some animals they were confined to a few spots, in others they spread over the whole body. They were scattered here and there, or disposed in the form of beads, or congregated together in a mass.

"When the disease was not of an acute character, and the eruption was not considerable, and the febrile symptoms were mitigated as soon as the pustule was developed, there was not much to fear. The eruption ran through its several stages, and no serious disorganization remained; but in too many cases the whole of the integument became reddened and inflamed, the flanks heaved, the pulse, whether strong or obscure, increased in frequency, the mouth was hot, the conjunctiva red, the breath fetid, the head swelled, the eyelids almost closed; rumination had ceased, the muscular power was exhausted, the pustules died away with little apparent fluid secretion, a fetid diarrhea ensued, and death speedily took place.

"The progress of the eruptive stage of the disease was frequently, however, a very unsatisfactory one. When the pustule had risen, and the suppuration had commenced, a new state of febrile excitement ensued, accompanied by more than usual debility. It lasted from three to four days, and during its continuance the pustules became whiter at their summit, and the fluid which they contained was of a serous character, yellow or red, transparent or viscid, and by degrees it thickened and became opaque, and then puriform; and at this period, when danger was to be apprehended, a defluxion from the nose ensued, and swellings about the head as already described.

"This was the contagious stage of the disease, and when it was too easily and fatally transmissible by accidental contact or by inoculation.

"Then came the last stage, that of desiccation, and about the twelfth day from the commencement of the disease. The pustules subsided, or the integument gave way, and the fluid which they contained escaped, and a scab was formed of greater or less size and density, yellow or black, and which detached itself bodily, or crumbled away in minute particles or powder. The contagion was now at an end, and the animal recovered his appetite and spirits and strength. This stage of desquamation frequently lasted three weeks or a month.

"A secondary eruption occasionally followed, of an erysipelatous character. There were no distinct suppurating pustules; but there was a more serous or watery secretion which soon died.

" This was the regular and the fortunate course of the disease; but too frequently there was a fatal irregularity about it. Almost at the commencement there was excessive fever, and prostration of strength and fetid breath, and detachment of large patches of the wool, and more rapid and bounding or inappreciable pulse, and strange swellings about the throat and head, and difficult deglutition. There was also a discharge of adhesive, spumy fluid from the mouth, and of ichorous or thick, and yellow, or bloody, and fetid discharge from the nostrils, often completely occupying and obstructing them. The respiration became not only laborious, but every act of it could be heard at a considerable distance — there was a distressing cough — the lips, the nostrils, the eyelids, the head, and every limb became swelled, the pustules ran together, and formed large masses over the face, and the articulations: diarrhea, that bade defiance to every medicine ensued, and the end was not far off."

The symptoms of the disease, after it appeared in England, are thus described by Mr. Thomas Wells, of Norwich, in the Norwich Mercury*:—" The leading symptoms of small pox are, a separation of the infected animal from the flock, a peculiar arching of the back, a drooping of the ears, a closing of the eyelids, amounting in some cases almost to blindness, and a pustular eruption, extending more or less over all parts of the body, but particularly those destitute of wool or covered with hair only; such for instance, as the cheeks, the skin inside the arms and thighs, the under surface of the tail, udder, etc."

The treatment of the malady, given by Messrs. Youatt and Spooner, (taken doubtless from Continental works on the subject,) is to separate out the diseased sheep from the flock, give them good food, protect them from wet and cold, open their bowels with Epsom salts during the febrile state, and afterward administer small doses of the salts with mild tonics, such as ginger and gentian. "Common salt was a favorite and very useful medicine, on account of its antiseptic and tonic properties."

The disease raged in Flanders, and I give the treatment

---

* I find it republished in the London Farmer's Magazine.

adopted in that country as more full, in some important particulars, than the preceding, and as describing, in detail, some of the minor manipulations and precautions necessary in treating the malady. In this light, it is a useful addition to the preceding prescriptions. If not adopted fully in this country — should the unfortunate occasion arise for our combating this malady — it at least furnishes useful hints. Professor A. Numann, of the Veterinary College of Utrecht, in his work on the diseases of animals, writes as follows:

"When the sheep eat freely and appear playful, while the pox comes out regularly, breaks, and dries up, no medicine is requisite; but should they lose their appetite, show an inclination to lie down, the heart beating quick and strong, and the pox not make its appearance on the third day, then nature requires assistance to drive the diseases outward; to this effect the following remedy is necessary:

"Take 2 oz. of juniper berries pulverized; a root of parsely cut, and split peas reduced to a powder, two handfuls each: boil all this in 4 lbs. of water; clear it off, mix in it $\frac{1}{4}$ oz. camphor, which has been previously dissolved in the yolk of an egg, and 1 oz. of good wine vinegar: this mixture to be divided in eight parts, one part to be administered night and morning till the pox is forced out. To obtain this point the following remedy will also be found efficacious:—Take flour of brimstone $\frac{3}{8}$ oz., the juniper berries, to be pulverized, the camphor mixed with the yolk of an egg, and the whole mixed with 4 oz. honey: to be divided in eight parts, one part to be given at night and morning.

"The stable in which the sheep are kept should be dry and airy, and not too warm; they ought to have fine, sweet hay, with barley straw cut very fine, which may be mixed with wheat bran moistened, bruised barley or flour of rice; a little salt to be mixed daily with it. When the pox is thrown out without containing any matter, the first given remedy is to be applied, and a seton to be set in the chest and each loin, which is to be effected in the following manner. Shear off the wool, to the size of a hand's breadth, from the part where you wish to place the seton; cut two small holes, the one above the other, through the hide, at a distance of three fingers; loosen the communication between one incision and the other by means of a flat stick; then draw through the opening a piece of linen half a finger's breadth, of which that part that goes under the hide must be besmeared on both sides with butter; the next day draw the band a little and besmear it afresh; take care that the band be long enough to enable you to tie it, to prevent its slipping out. On the fifth or sixth day, when the pock is charged with matter, the linen or band may be drawn, and the above remedy dispensed with.

"When the blood is not freed from pock matter, it often produces (when the pox is already cured) a swelling in one or other part of the body; as soon as such swelling is come to maturity, it ought to be opened, and the matter washed away quickly. If the eyes should be closed with a swelling, they must be often bathed with water, and when opened the matter carefully washed away. The following remedies may be applied in cases of malignant small pox. The pustules seldom burst without assistance, but the matter they contain spreading continually, they ought to be opened with a sharp-pointed knife as soon as they are in a state of maturity; and after squeezing out the matter, to be washed with a solution of salt and water until a cure is performed.

"As the small pox is very contagious, it is necessary to guard against it as much as possible, and when discovered to separate the sheep affected from the rest of the flock, and place them in another stable, which ought to be fumigated with juniper berries twice a day at least; the manure taken out, and fresh straw put in daily; besides, the stable must (after the complaint has subsided) be scoured with a solution of wood ashes, and then fumigated with chlorine, before it is made use of to receive sound sheep.

"In summer, sheep affected with the small pox may be driven in fine weather for a few hours, morning and evening, in the field, but care must be taken they do not go near the sound ones; the latter must not go into the field where the former have grazed: in general, all communication, of whatever nature it may be, between the sick and sound sheep must be avoided, and the shepherd who conducts and has care of the sick sheep should take care not to approach the sound sheep, lest he should communicate the contagion."

In 1760, healthy sheep were inoculated with the virus of the diseased ones, and the effects were found analogous to those of inoculation for small pox among human subjects. A disease having the same character was produced, but it was mild and rarely mortal. In a paper in the London Farmer's Magazine, September, 1848, Mr. O. Delafond states that to sum up the recorded cases of inoculations made in France, by Huard, Valois, Langlois, Guillaume, Buignot, D'Arboval, Gragnier, Girard, etc., between 1805 and 1848, the number of subjects which recovered was 28,248, and the number which died was 285 — or about one per cent. M. Gayot inoculated 10,000 in the departments of La Marne and La Haute Marne during the prevalence of the disease, when the mortality was twenty per cent. among those having it in the natural way; and he lost only one and one half per cent. of his patients. Messrs. Miquel and Thomieres inoculated between December, 1820, and January, 1822, 17,044 sheep, comprising eighty-four flocks, and forty-two of them infected ones. In some of the flocks not previously infected they did not lose a patient. In one, in which two-thirds of the sheep were already attacked by the disease, they lost about eight per cent. of the remaining number, many of which were doubtless in the incubatory state of the disease when inoculated. Out of 66,716 inoculated in Prussia, 65,042 recovered. Out of 8,000 sheep and 2,000 lambs inoculated in Austria, not one was lost. These examples might be indefinitely multiplied.

D'Arboval states that 7,697 sheep which had received the disease by inoculation and recovered, had been re-inoculated, made to cohabit with sheep laboring under the natural disease, &c., &c., and that in no instance did they again contract the malady.

Mr. Youatt declares that *variola ovina* is not identical with small-pox in the human being. He says there is an evident difference in the pustule — that of small-pox being "developed in the texture of the skin, and surrounded by a rose-colored areola, that of the *clavellee* evidently more deep-seated — reaching to the sub-cutaneous cellular tissue and surrounded by an areola of a far deeper color. The virus of small-pox was usually contained in a simple capsule which elevated the scarf skin — the virus of the sheep-pox seemed to be more diffused through the cutaneous and sub-cutaneous tissue, and there was abundantly more swelling and inflammation." He describes other differences in the appearance of the matter, scabs, &c.*

Vaccination followed the introduction of inoculation. To test their respective usefulness, 1,523 sheep were vaccinated in France, and the disease became fully developed in them. They were all subsequently inoculated with the virus of sheep-pox, and 308 took the disease, though in the mitigated form usual after inoculation. Other smaller experiments had a corresponding result; and, therefore, says Mr. Youatt — at

---

* I suppose that Mr. Youatt in expressing these opinions, expresses the opinions of the learned veterinarians of the Continent.

the period of writing his work on the sheep — "vaccine inoculation is now abandoned on the Continent, although it gives immunity to four-fifths of those that have been subjected to it, for inoculation with *le claveau*, or the virus of sheep-pox, will give immunity to all."

When this malady made its advent in England, it was by imported sheep — and the weight of testimony would seem to show that the disease was not apparent in them at the time, but was in its incubatory state. There can be no doubt that the period of incubation is long enough to allow infected sheep to be brought from Europe to America, in the swift-sailing steamers of the present day, before the disease would produce any appearances which those not practically familiar with sheep-pox would recognize as characteristic of the malady;* and the malady might progress much further without its nature being understood or suspected, in any region where it had not been previously known, and where its advent was totally unlooked for. And there is just as little doubt that it might be brought here at any time by wool, or pelts of diseased sheep, or any other substances infected by them, and under some disastrous combination of circumstances introduced, like fire to a train of powder, among the flocks of the American Continent. In England the flocks exposed to its ravages were larger than those of our Eastern States, and much nearer together than those in any part of our country — circumstances favorable to its more rapid propagation there: but there it was encountered with professional veterinary skill — cheap labor for attendance — and the determined efforts of a government and people which had vast interests at stake, and but a comparatively small home territory to watch over. Here, unless mitigated by climatic circumstances — a thing not to be anticipated from any analogy derivable from our experience with small-pox in human beings — it would advance more slowly, perhaps, but I apprehend with more destructive results. Our breeders, and the very intelligent and public-spirited breeders of Canada, who are constantly introducing sheep from Europe, are called upon, then, by every consideration of interest and propriety,

---

* Prof. J. B. Simonds' Lecturer on Cattle Medicine, etc., at the Royal Veternary College, England, and who was appointed government inspector of diseased sheep, when the sheep-pox appeared in England, states that it is about ten days from the time of the contact of a sound animal with a diseased one before the *first* symptoms appear. This is to be understood, doubtless, as the average period of incubation, and it might under various circumstances, or in different sheep, be extended several days longer.

to exercise a constant and watchful care on this subject — not only where sheep-pox is raging and is the subject of public attention in the foreign countries where they purchase sheep, but at *all times*, if it is known that the malady has *ever* visited those countries. The man who even *carelessly* brought this scourge to our shores, would deserve and receive the reprobation of a Continent.

I am not aware that any important discoveries have been made in the actual treatment of the disease in England; and owing to my failure to obtain certain expected English publications on the subject, before the completion of this volume, I cannot give any particular history of the disease in that country obtained from authoritative sources. The general tenor of my information on the subject is, that the *Variola Ovina*, in its natural form, is as destructive and contagious there as on the Continent; that the means relied on to counteract it are principally *preventive;* that the main modes of preventing it are by inoculation and vaccination. It seems that there are those who prefer the latter mode. I saw it stated in an article in the American Agriculturist, that an association of sheep breeders in Wiltshire, England, on trial, much preferred vaccination.*

The disease, after a lull of a few years, has recently, it would seem, re-appeared in England. I cut the following paragraphs on the subject from "Moore's Rural New-Yorker:"

"From Bell's Messenger we learn that the medicines employed in Mr. Parry's flock, where the disease was first apparent, are very simple, consisting chiefly of nitrate of potassa, mingled with the water which is placed in the troughs, until a subsidence of the fever takes place, after which sulphate of iron has been substituted. Where diarrhea has come on — as it not unfrequently does in the latter stage of the malady, more particularly if the pox becomes confluent — opium is recommended as a valuable agent to arrest the attack, which, if not quickly stopped, very soon carries off the sheep.

"Speaking of inoculation, the Messenger remarks:—

---

* It was stated in this article that the *inoculated* sheep "died off rapidly, and thus the proposed prevention only spread the infection." If this is a correct statement of the facts, it only shows, I imagine, that the flocks of Wiltshire were inoculated with improper virus, or that they were affected by exceptional and inauspicious circumstances. The alleged result is too much opposed to the well settled facts which attend inoculation, developed under upwards of a century and a half of observation — and to the combined experience of the Continental veterinarians — to be entitled to credit.

'Nearly three weeks have now elapsed since Mr. Parry's flock were inoculated; and it is worthy of remark that out of 446 ewes in which the disease was thus artificially, as it were, produced, he has lost only four; while of those which took the disease naturally, the losses have already been *sixty per cent.*, and there are numbers of other sheep of whose recovery there is little hope,—indeed, the total loss of those which have taken the disease in a natural way, Mr. Parry estimates will not be much short of 65 per cent. Putting this, therefore, in contrast with the results after inoculation — which, under the most favorable circumstances, are not expected to average a mortality of more than five per cent. — the desirableness of inoculation immediately upon the appearance of the disease in a flock is placed beyond doubt.'"

CHAPTER XXVIII.

DISEASES OF THE LOCOMOTIVE ORGANS.

FRACTURES — RHEUMATISM — DISEASE OF THE BIFLEX CANAL — GRAVEL — TRAVEL-SORE — LAMENESS FROM FROZEN MUD — FOULS — HOOF-ROT.

FRACTURES.— The most common fractures which occur in the brittle bones of the sheep are in the legs below the knees and hocks; and there is no difficulty in treating such cases. Any intelligent man is a sufficient surgeon for the occasion. The bones should be brought to their natural position and confined there with splints — or thin pieces of wood shaped to the leg and wound with strips of muslin, which confine a layer of cotton batting on the side next the leg. The splints are confined to the leg by winding twine around the whole when they are arranged in their places. I never had occasion to ease the limb on account of its swelling — or to administer purgatives in consequence of any ensuing fever in the sheep; though both might be called for. The limb is usually sound enough to remove the splints in the course of three or four weeks — though there is no occasion for haste in this particular. In default of other convenient materials, I have applied the bare splints over a wrapping of thick paper with cotton or wool laid evenly under it. Thick leather, shaped to the leg when wet, will support it without splints. If the fracture is of the arm or thigh and far above the knee or hock, it is not generally worth while to attempt any cure. Mr. Youatt says if the shoulder is fractured it can generally be successfully treated by removing the wool and applying a pitch plaster on the whole of the shoulder bone.

RHEUMATISM.— This has been sufficiently mentioned at page 155.

DISEASE OF THE BIFLEX CANAL.— We have owners of sheep who believe with Sir Anthony Fitzherbert, who flourished almost three centuries and a half ago, that, "There be some shape that hath a worme in his foote that maketh hym halte." The biflex canal, or "issue" as it is sometimes called, (in the front and upper part of the cleft between the toes,) gets some substance introduced into it which causes an irritation and swelling of the surrounding parts; and to cure this, believers in the "worme" actually, with a pocket knife, dissect out—or rather mangle out—the skin which surrounds the biflex canal! Such egregious ignorance and brutality, however, are now extremely rare. Two or three incisions in the swollen parts usually relieve them of the inflammation. The biflex canal and the other parts of the foot should be examined, of course, to see that no irritating foreign substances have become imbedded in them.

GRAVEL.— Gravel or dirt occasionally penetrates the foot of the sheep between its horny covering and the fleshy structures underneath. It ultimately produces an inflammation and swelling at the coronet, which at length breaks and expels the offending substances. As this process produces considerable pain and inconvenience to the animal, it is better, as soon as the lameness is observed, to remove enough of the horny covering of the foot to allow the escape of the gravel. It is well enough, then, to cover the parts with tar; but whether this is done or not, no injury will result from the removal of the necessary portion of the horn; and it will be very rapidly reproduced.

TRAVEL-SORE.— Sheep driven several hundred miles through mud and sand — say from Western Illinois to the banks of the Hudson — not only frequently become graveled, but the heels are sometimes worn so thin, and they and the skin between them become so tender, that the sheep proceed on their journey with pain, and fall off a good deal more than they otherwise would, in condition. The English sheep is much more subject to this than the Merino, both on account of its greater weight, and because the horny coverings of its feet are much thinner. The drover carries a phial of oil of vitriol in his pocket. The *bottom* of the heels are touched by a feather dipped in this, when the drove stops at night — and a little tar from the inn-keeper's bucket is smeared on the cauterized parts, on the backs of the heels, and between the toes. This gives great relief under any circumstances: and the sheep rapidly recovers if allowed a little rest. Butyr of antimony, acting much more as a purely superficial caustic, would be a better application than oil of vitriol.

LAMENESS FROM FROZEN MUD.—I have elsewhere mentioned that when sheep are kept in unlittered yards in winter, and especially when they are allowed to run over plowed ground, little pellets of mud often adhere to the hairs which hang down in the clefts of the feet, and a sudden and severe freeze converts these into pellets of the consistency of stone. Nay, the kneading operation of the toes on this lump of earth frequently gives it such consistency that, on becoming dry merely, it acts as a highly irritating body in the foot. I have seen half the sheep of a flock made lame enough in this way to give a strong suspicion of hoof-rot. On looking into the feet, the skin on each side of the little mud ball is found chafed and inflamed — sometimes worn through and matter formed in the wounds. I saw a purchase of a valuable flock of sheep broken off by this cause. The parties had agreed on the price, and both were anxious to complete the bargain. But there were a small number of lame sheep, and the purchaser demanded a guaranty against hoof-rot, which the seller refused to give, and consequently lost the sale of his sheep by his carelessness. The remedy, or rather the preventive, is too obvious to require mention.

FOULS.—Sheep are much less subject to this disease than cattle, but contract it if kept in wet, filthy yards, or on moist, poachy pastures. A wet season and tall grass sometimes produce it, even on dry uplands. The skin in the cleft of the foot first has a macerated or water-soaked appearance, which is followed by a degree of inflammation and lameness. It disappears when the sheep is removed to a dry yard or pasture — but more promptly if the parts have a solution of blue vitriol or turpentine applied to them, or are daubed with tar.

HOOF-ROT.—I mentioned in "Sheep Husbandry in the South," that the description of the early stages of this disease given in Mr. Youatt's justly popular work on Sheep, is almost wholly inapplicable to the malady in the United States, *and among Merino sheep.* I never have seen it among English sheep; and, in this country, they, like all our other coarse-wooled varieties, are notoriously less subject to it than Merinos, and far more readily cured when they contract it. Some of the reasons for this fact may probably be found in the different structure of their hoofs mentioned at page 168. As Mr. Youatt's works are received—and, as a general thing, justly received — as standard authority among the great mass of the reading farmers of our country, I feel called upon again to point out his errors on this subject. I may be allowed to speak with a degree of confidence in regard to a malady which has at four different periods attacked my flocks — embracing as many as five thousand sheep in its different visitations: and which has been, in every instance, fully and completely extirpated.

The following is Mr. Youatt's description of its first symptoms:— "The foot will be found hot and tender, the horn softer than usual, and there will be enlargement about the coronet, and a slight separation of the hoof from it, with portions of the horn worn away, and ulcers formed below and a discharge of thin, fetid matter. The ulcers, if neglected, continue to increase; they throw out fungous granulations, they separate the hoof more and more from the parts beneath until at length it drops off. All this is the consequence of soft and marshy pasture." Mr. Youatt attributes the disease not only to "infection by means of the virus," but to particles of earth or sand having forced their way through breaks in the hoof, and through "new pores," occasioned by the over-lapping portions of the horn breaking off. These particles "reaching the quick, an inflammation is set up, which, in its progress, alters or destroys the whole foot." He also attributes it to another cause. "The length to which the crust grows," he remarks, "changes completely the proper bearing of the foot, for, being extended forward, it takes the whole weight of the superincumbent parts. By the continual pressure on this lengthened part inflammation cannot fail to be set up." In describing the progress of the disease, he mentions the following circumstance:— "The whole of the inner surface of the pasterns is sore and raw." * * * "In some cases, as has appeared when the diseased state of this [the biflex] canal was examined, the malady commences here."

The hoof is not softened but rather hardened by the presence of hoof-rot, until its structure becomes to some extent disorganized. In not one case in a hundred is there any visible enlargement about the coronet in the early stages of the disease. The horn, so far from first separating from the coronet, generally adheres there to the last, even when the whole bottom of the hoof is gone, and nothing but a portion of its side shell remains. The hoof never "drops off," in the sense in which I here understand Mr. Youatt to mean — that is, entire; though it sometimes gradually becomes totally disorganized and thus disappears. The disease never commences between the quick of the foot and its horny shell, as it would do if caused by sand or other substances having penetrated through the hoof to those parts. The improper bearing of the foot occasioned by the extension of the fore part of the hoof forward, and of its side walls downward, frequently produces some degree of lameness; but it is that lameness of the ligaments, tendons and other tissues in and connected with the feet, which a man would incur by wearing a boot elevated three inches higher at the toe than at the heel, and then additionally tipped to one side or the other: but it bears not the slightest affinity to hoof-rot. And when the hoof is thus extended and thickened and elevated in front, Mr. Youatt is entirely mistaken in supposing that the "whole weight of the superincumbent parts" presses on this "lengthened part," or toe: it unquestionably presses mainly on the heel — as would a man's weight with his boot elevated in front, as already mentioned. I never saw a dozen sheep suffering under hoof-rot which had "the inner surface of their pasterns sore and raw." And, finally, genuine hoof-rot never "commences" at the biflex canal, any more than it does at the knee or nose of the animal. I could point out additional minor errors in Mr. Youatt's descriptions; but it is unnecessary. He must have written them without much personal observation of the disease, or he describes a different one — or hoof-rot presents essentially different early symptoms in Europe from what it does in the United States.*

The horny covering of the sheep's foot extends up, gradually thinning out, some way between the toes, or divisions of the hoof — and above these horny walls the cleft is lined with skin. Where the points of the toes are spread apart, this skin is shown in front covered with soft, short hair. The heels can be separated only to a little distance, and the skin that is in the cleft above them is naked. In a

---

* I can not, however, accept either of the latter explanations. Mr. Youatt too vividly and clearly describes the later characteristic lesions of hoof-rot — and which appertain to no other disease — to leave any chance for the supposition that he was describing a different malady; and I have no idea, after examining Continental and *other British* accounts of it, that there is any material difference in its diagnostics as between any part of Europe and America, — except that among certain breeds of sheep it is less virulent than among others. Mr. Youatt seems to me evidently to include among the initial symptoms of hoof-rot, those of gravel, inflammation of the biflex canal, and other scarcely named abnormal conditions of the foot, which come and go without any connection with hoof-rot — which never produced it — and which do not in one instance in a hundred, or five hundred, accompany it. Mr. Youatt, in tracing the origin and progress of this malady, seems to me to have leaned quite too heavily on the statements, or rather the *speculations* of Professor Dick, of Edinburgh — a writer of ability, but evidently a very unpractical one on the subject of hoof-rot.

healthy foot it is as firm, sound, smooth and dry as the skin between a man's fingers, which, indeed, it not a little resembles, on a mere superficial inspection. It is equally destitute of any appearance of redness, or of feverish heat.

The first symptom of hoof-rot, uniformly, in my experience, is a disappearance of this smooth, dry, colorless condition of the naked skin at the top of the cleft over the heels, and of its coolness. It is a little moist, a little red, and the skin has a slightly chafed or eroded appearance — sometimes being a very little corrugated, as if the parts had been subjected to the action of moisture. And on placing the fingers over the heels it will be found that the natural coolness of the parts has given place to a degree of heat. The inflammation thenceforth increases pretty rapidly. The part first attacked becomes sore. The moisture — the ichorous discharge — is increased. A raw ulcer of some extent is soon established. It is extended down to the upper portion of the inner walls of the hoof, giving them a whitened and ulcerous appearance. Those thin walls become disorganized, and the ulceration penetrates between the fleshy sole and the bottom of the hoof. On applying some force, or on shaving away the horn, it will be found that the connection between the horny and fleshy sole is severed, perhaps half way from the heel to the toe, and half way from the inner to the outer wall of the hoof. The hoof is thickened with great rapidity at the heel by an unnatural deposition of horn. The crack or cavity between it and the fleshy sole very soon exudes a highly fetid matter, which begins to have a purulent appearance. The extent of the separation increases by the disorganization of the surrounding structures; the ulceration penetrates throughout the entire extent of the sole; it begins to form sinuses in the body of the fleshy sole; the purulent discharge becomes more profuse; the horny sole is gradually disorganized, and finally the outer walls and points of the toes alone remain. The fleshy sole is now a black, swollen mass of corruption, of the texture of a sponge saturated with bloody pus, and every cavity is filled with crawling, squirming maggots. The horny toe disappears; the thin, shortened side walls merely adhere at the coronet; they yield to the disorganization; and nothing is left but a shapeless mass of spongy ulcer and maggots. Attempts to cure the disease, the state of the weather, and other incidental circumstances, cause some variations from the above line of symptoms. When the first attack occurs in hot weather, the progress of the malady is much more rapid and violent. The fly sometimes deposits

its eggs in the ulcer, and maggots appear almost before — sometimes actually before — there are any cavities formed, into which they can penetrate. The early appearance of maggots greatly accelerates the progress of disorganization in the structures.

The fore-feet are usually first attacked — sometimes both of them simultaneously, but more generally only one of them. The animal at first manifests but little constitutional disturbance. It eats as is its wont. When the disease has partly run its course in one foot, the other fore-foot is likely to be attacked, and presently the hind ones. When a foot becomes considerably disorganized, it is held up by the animal. When another one reaches the same state, the miserable sufferer seeks its food on its knees; and if forced to rise and walk, its strange, hobbling gait betrays the intense agony it endures on bringing its ulcerated feet in contact with the ground: There is a bare spot on the under side of the brisket of the size of the palm of a man's hand — but perhaps a little longer — which looks red and inflamed. There is a degree of general fever — and the appetite is dull. The animal rapidly loses condition, but retains considerable strength. No where else do sheep seem to me to exhibit such tenacity of life. After the disappearance of the bottom of the hoof, the maggot speedily closes the scene. Where the rotten foot is brought in contact with the side in lying down, the filthy, ulcerous matter adheres to and saturates the short wool of the shorn sheep: and maggots also are either carried there by the foot, or they are speedily generated by the fly. A black crust soon forms, and raises a little higher round the spot. It is the decomposition of the surrounding structures — wool, skin and muscle — and innumerable maggots are at work below, burrowing into the living tissues, and eating up the miserable animal alive. The black, festering mass rapidly extends, and the cavities of the body will soon be penetrated, if the poor sufferer is not sooner relieved of its tortures by death.

The offensive odor of the ulcerated feet, almost from the beginning of the disease, is so peculiar that it is strictly pathognomonic. I have always believed that I could by the sense of smell alone, in the most absolute darkness, decide on the presence of hoof-rot with unerring certainty. And I had about as lief trust my fingers as my eyes to establish the same point, from the hour of the first attack, if no other disease of the foot is present. But the heat, which invariably marks the earliest presence of hoof-rot, might arise from any other cause which produced a local inflammation of the same parts.

When the malady has been well kept under during the first summer of its attack, but not entirely eradicated, it will almost or entirely disappear as cold weather approaches, and not manifest itself again until the warm weather of the succeeding summer. It then assumes a mitigated form; the sheep are not rapidly and simultaneously attacked; there seems to be less inflammatory action in the diseased parts, and less constitutional disturbance; and the course of the disease is less malignant, more tardy, and it more readily yields to treatment. If well kept under the second summer, it is still milder the third. A sheep will occasionally be seen to limp, but its condition will scarcely be affected, and dangerous symptoms will rarely supervene. One or two applications of remedies made during the summer, will now suffice to keep the disease under, and a little vigor in the treatment will entirely extinguish it.

With all its fearful array of symptoms, can the hoof-rot be cured in its first attack on a flock? The worst case can be promptly cured, as I know by repeated experiments. Take a single sheep, put it by itself, and administer the remedies daily, after the English fashion, or as I shall presently prescribe, and there is not an ovine disease which more surely yields to treatment. But, as already remarked, in this country where sheep are so cheap and labor in the summer months so dear, it would be out of the question for an extensive flock-master to attempt to keep each sheep by itself, or to make a daily application of remedies. There is not a flock-master within my knowledge who has ever pretended to apply his remedies oftener than once a week, or regularly as often as that; and not one in ten makes any separation between the diseased and healthy sheep of a flock into which the malady has been once introduced. The consequence necessarily is that though a cure is effected of the sheep then diseased, it has infected or inoculated others — and these in turn scatter the contagion, before they are cured. There is not a particle of doubt—nay, I know, by repeated observation, that a sheep once entirely cured may again contract the disease, and thus the malady performs a perpetual circuit in the flock. Fortunately, however, the susceptibility to contract the disease diminishes, according to my observation, with every succeeding attack; and fortunately also, as already stated, succeeding attacks, other things being equal, become less and less virulent.

What course, then, shall be pursued? Shall the flockmaster sacrifice his sheep—shall he take the ordinary halfway course—or shall he expend more on the sheep than they are worth in attempting to cure them? Neither. The course I would advise him to pursue, will appear as I detail the experiments I have made.

Treatment.—The preparation of the foot is a subject of no dispute, but the labor can be prodigiously economized by attention to a few not very commonly observed particulars. Sheep should be yarded for the operation immediately after a rain, if practicable, as then the hoofs can be readily cut. In a dry time, and after a night which has left no dew on the grass, their hoofs are almost as tough as horn. They must be driven through no mud, or soft dung, on their way to the yard, which doubles the labor of cleaning their feet. The yard must be small, so they can be easily caught, and it must be kept well littered down, so they shall not fill their feet with their own manure. If the straw is wetted, their hoofs will not of course dry and harden as rapidly as in dry straw. Could the yard be built over a shallow, gravelly-bottomed brook,* it would be an admirable arrangement. The hoofs would be kept so soft that the greatest and most unpleasant part of the labor, as ordinarily performed, would in a great measure be saved; and they would be kept free from that dung which, by any other arrangement, will more or less get into their feet.

The principal operator or foreman seats himself in a chair—a couple of good sharp knives, (one at least a thin and narrow one,) a whetstone, the powerful toe-nippers (figured on page 169,) a bucket of water with a couple of linen rags in it, and such medicines as he chooses to employ, within his reach. The assistant catches a sheep and lays it partly on its back and rump, between the legs of the foreman, the head coming up about to his middle. The assistant then kneels on some straw, or seats himself on a low stool at the hinder extremity of the sheep. If the hoofs are long, and especially if they are dry and tough, the assistant presents each foot to the foreman, who shortens the hoof with the toe-nippers. If there is any filth between the toes, each man, after first using a stick, takes his rag from the bucket of water, draws it between the toes and rinses it, until the filth is removed. Each then seizes his knife, and the process of paring away

---

* A place might be prepared in any little brook by gravelling or by laying a floor of boards on the bottom.

the horn commences. *And on the effectual performance of this, all else depends.*

If the disease is in the first stage — i. e., if there is merely an erosion and ulceration of the cuticle and flesh in the cleft above the walls of the hoof, no paring is necessary. But if ulceration has established itself between the hoof and the fleshy sole, the ulcerated parts, be they more or less extensive, *must be entirely denuded of their horny covering, cost what it may of time and care.* It is better not to wound the sole so as to cause it to bleed freely, as the running blood will wash off the subsequent applications; but no fear of wounding the sole must prevent a full compliance with the rule above laid down. At worst, the blood can soon be staunched, however freely it flows, by a few touches of a caustic — say butyr of antimony.

If the foot is in the third stage—a mass of rottenness and filled with maggots — the maggots should first be killed by spirits of turpentine, or a solution of corrosive sublimate (see page 190) or other equally efficient application. It can be most conveniently used from a bottle having a quill through the cork. By continuing to remove the dead maggots with a stick, and to expose and kill the deeper lodged ones, all can be extirpated. Every particle of loose horn should then be removed, though it take the entire hoof, — and it frequently does take the entire hoof at an advanced stage of the disease. The foot should be cleansed if necessary with a solution of chloride of lime, in the proportion of a pound of the chloride to a gallon of water.* If this is not at hand, plunging the foot repeatedly in water, just short of scalding hot, will answer the purpose. And now comes the important question, what constitutes the best remedy?

The recipes for its cure are innumerable. One much used in New England at an early day, under the recommendation of "Consul Jarvis,"† was compounded as follows:

1. Roman or blue vitriol, pulverized very fine, three parts, with one part of white lead, mixed into a thin paste with linseed oil.

2. Another recipe, also much used in New England, is

---

* Mr. Youatt recommends this, and says it "will remove the fetor and tendency to sloughing and mortification which are the too frequent attendants on foot-rot." I never yet saw mortification (gangrene) of the foot result from this disease. Mr. Youatt's directions as to treatment are far more satisfactory than are his statements of the causes and symptoms of this malady.

† The Hon. William Jarvis is universally known under this appellation in New England.

as follows:—4 ounces blue vitriol, 2 oz. verdigris to a junk bottle of urine.

3. Spirits of turpentine, tar and verdigris, in equal parts.

4. The following recipe used to be hawked about the country at the price of $5, the purchaser having promised inviolable secrecy:—3 quarts alcohol, 1 pint spirits of turpentine, 1 pint of strong vinegar, 1 lb. of blue vitriol, 1 lb. of copperas, 1½ lbs. verdigris, 1 lb. alum, 1 lb. of saltpetre, pounded fine; mix in a close bottle, shake every day, and let it stand six or eight days before using: also mix 2 lbs. of honey and 2 quarts tar and apply it after the previous compound. "Two applications to entirely remove disease."

5. A saturated solution of blue vitriol applied through a quill in a cork—and finely pulverized vitriol dusted over the parts when wet. This was the favorite remedy of the farmers in the region where I reside, twenty-five years ago.

6. The most common and popular remedy now used in Central New York is:—1 lb. blue vitriol; ¼ lb. (with some ½ lb.) verdigris; 1 pint of linseed oil; 1 quart of tar. The vitriol and verdigris are pulverized very fine, and many persons before adding the tar, grind the mixture through a paint mill. Some use a decoction of tobacco boiled until thick, in the place of oil.

7. The remedy recommended by Mr. James Hogg, of Scotland, is turpentine 2 ounces, sulphuric acid 2 drachms—to be well mixed before it is used and applied freely to the diseased part.

8. Mr. Spooner thinks 1 oz. of olive oil and double the quantity of sulphuric acid, an improvement on the above. He says "the acid must be mixed carefully with turpentine, as considerable inflammation immediately takes place." He remarks that he has used all the powerful acids with success, and that he imagines it of but little consequence which caustic is employed, provided it be of sufficient strength.

9. Mr. Youatt recommends washing the foot in a strong solution of chloride of lime, and then resorting to "muriate or butyr of antimony." The foot to be dressed every day, and each new separation of horn removed, and every portion of fungus submitted to the action of the caustic, and a little clean tow to be wrapped round the foot and bound tightly down with tape, if the foot is principally stripped of its horn.

10. The following is the English "Halt Receipt." It is given in the prize essay of Mr. Robert Smith, already on numerous occasions cited; and Mr. Smith says he has found

this remedy "invaluable both in staying its progress, and curing the disease:"—1 oz. corrosive sublimate; 1 oz. blue vitriol; 1 oz. spirits of salts; 1 oz. verdigris; 1 oz. horse turpentine; 1 oz. oil of vitriol; ¾ oz. spirits of turpentine and 4 ozs. sheep ointment. (The last I presume means mercurial ointment.) "To be well mixed when prepared, and kept tied down [in a bottle] when not in use."

Mr. Smith also says:—"When the foot has become much diseased from neglect, it should be placed in an oil-cake poultice for twelve hours; then washed clean with warm water, and the poultice renewed again in twelve hours more; then to be again washed, and the diseased parts probed to the bottom and dressed; then to be tied up in common tar for twenty-four hours, and renewed when necessary, again applying the ointment. Opening medicine will materially assist in the cure of obstinate cases."

Any of these remedies, and fifty more that might be compounded, simply by combining caustics, stimulants, etc., in different forms and proportions, will prove sufficient for the extirpation of hoof-rot, *with proper preparatory and subsequent treatment.* On these last, beyond all question, principally depends the comparative success of the applications.

First. No external remedy can succeed in this malady unless it comes in contact with all the diseased parts of the foot—for if such part, however small, is unreached, the unhealthy and ulcerous action is perpetuated in it, and it gradually spreads over and again involves the surrounding tissues. Therefore every portion of the diseased flesh must be denuded of horn, filth, dead tissue, pus, and every other substance which can prevent the application from actually *touching* it, and producing its characteristic effects on it.

Second. The application must be kept in contact with the diseased surfaces long enough to exert its proper remedial influence. If removed, by any means, before this is accomplished, it must necessarily proportionably fail in its effects.

The preparation of the foot, then, requires no mean skill. The tools must be sharp, the movements of the operator careful and deliberate. As he shaves down near the quick, he must cut thinner and thinner, and with more and more care, or else he will either fail to remove the horn exactly far enough, or he will cut into the fleshy sole and cause a rapid flow of blood. I have already remarked that the blood can be staunched by caustics — but they coagulate it on the surface in a mass which requires removal before the application of remedies, and in the process of its removal the blood is very frequently set flowing again, and this sometimes several times follows the application of the caustic.* Cutting down to the crack between the horny and fleshy sole, is not enough. The operator must ascertain whether there is any ulceration between the outside horny walls and the fleshy part of the foot — or at the toe — or whether there is even a rudiment of an unreached sinus or cavity in any part of the foot where the ulceration has penetrated or is beginning to penetrate. The practiced eye decides these questions rapidly from the characteristic appearances, without the removal of unnecessary horn: but the new beginner must feel his way along cautiously, removing more horn where there is doubt, but so removing it that he will not unnecessarily cause an effusion of blood, or uncover the *healthy* quick, or disarrange the proper bearing of the foot. If the foot is in the third state, the removal of the maggots, the cleaning of the ulcers, the proper excision of the dead tissues, etc., require much time — sometimes more than half an hour to each foot. The most experienced operator cannot perform such processes in a hurry — the inexperienced one must perform them slowly, or all the time saved will be lost, twenty times over, in having to repeat them for an indefinite number of times.

English and Continental modes of treatment—the constant separation of the infected—daily dressings—poultices changed every twelve hours, with intermediate washings, probings, etc., as recommended by Mr. Smith—bandages—cloth boots —cloth boots with leather soles,† &c., &c., would cost more than the sheep would be worth after they were cured, in this country of high and scarce labor; and, in point of fact, in most regions, in the busy period of haying and harvesting when hoof-rot is at its height, the necessary labor could not be commanded at any cost. And yet if that labor could be obtained, the cost of it would not much exceed, in the long run, the aggregate cost of the labor actually devoted to the same end. The foreign practitioner promptly cures the malady. The American farmer, who has from one to two hundred infected sheep, during the first year when the disease is violent, drives them into a stable—a small one so they can be caught easily—once in ten days or a fortnight. An unoccu-

---
* I remember to have seen these recommended by some Continental Veterinarian whose name I can not now recall.
† The toe vein bleeds very freely, and it often requires some time and trouble to stanch it.

died afternoon—generally a rainy one, when nothing else can be done—is selected. Possibly a *little* straw is put on the floor of the stable at first—but in half an hour every foot is full of soft dung. The farmer, his hired man, and perhaps a boy or two, sit down to the task. Each is armed with a thick, broad-bladed, dull "jack-knife." The whole party then proceed literally to "cut and sear!" Some of the feet are cut down deeply into the quick, so that blood gushes from them in streams—others, which have too tough hoofs, have them left half an inch thick at the sole—covering up, very likely, rapidly developing ulcers. Blood and dung and "medicine" are left applied in about equal proportions to each foot. To crown the whole proceeding, the sheep are turned out as fast as "doctored" into the rain, or into wet grass, so that in ten minutes' time not a trace of the remedy applied remains on the foot. In such flocks the disease is barely kept at bay: it is never cured. The second and third year the "doctorings" become less and less frequent—but they must be resorted to occasionally, perpetually, or some of the sheep will get down on their knees. The farmer always finds bitter fault with the "medicine." He gets a new kind—with more ingredients*—and tries again. But he never finds the right one!

The above picture is, doubtless, an extreme one—but do we not all constantly witness more or less near practical approximations to it? The separation of the sheep, poulticing, inclosing of the foot, &c., I believe to be unnecessary—but the feet must be well prepared, and the sheep must be kept out of the rain, or grass wetted by rain or dew, for twenty-four or thirty-six hours afterward—the longer the better. *Without this the most careful preparation of the foot and the best remedies cannot be made effectual.* It is true that out-door moisture will not prevent the escharotic effects of powerful caustics, which do that portion of their work almost at once†—but these are not beneficially applied in ordinary cases; and when properly applied I am not prepared to say what would be the effects of the immediate and long continued

---

* In this particular Mr. Robert Smith's English "halt receipt," is a pattern! If such a jumble of ingredients, fortunately, do not chance to counteract each other, no well informed man ought for a moment to suppose there can be any utility in compounding so many articles together. This remedy I doubt not is a good one, as a whole,—but it does, in my opinion, contain some substances which neutralize each other's medicinal properties!

† I should consider that moisture highly *beneficial* in diluting those powerful and deeply corroding caustics which are sometimes *profusely* applied to the bottom and sides of the toes and in the cleft between them—such as aqua-fortis and oil of vitriol.

exposure of the unhealthy and cauterized surfaces to water. I venture to say that effect would not be auspicious. In the case of milder applications, which do not immediately cauterize—and from which different and less instantaneous effects are expected, as, for instance, from blue vitriol—the immediate and continued contact of water washes them off almost before they *begin* to exert their remedial effects. It is to prevent this that oil and tar are made portions of some of the preceding prescriptions. They will do some good in this way after they dry on the surface of the flesh; but they are wholly inadequate to the end, if the sheep is turned on wet grass immediately after their application. The best place to put sheep after applying the remedies to their feet, is on the naked floors of stables—scattering them over as much surface as practicable, so that there shall be as little accumulation of manure as possible under foot. Straw, especially if fresh littered down, absorbs or rubs off the moist substances which have been applied to their feet. The *bottoms* of the feet are soon thus cleaned off. A boy should go round with a shovel, until night, taking up the dung as fast as dropped. The sheep should be kept in the stables over the first night, and not let out the next day until the dew is off the grass: then they should be turned on the most closely cropped grass on the farm. It well pays for the trouble to put them in the stables the second night before the dew falls, and to keep them, as before, until it is dried off the next day.

I have never found that for moderate cases of hoof-rot— the worst ones which are allowed to occur in well managed flocks—that there is, in reality, any possible beneficial addition to mere blue vitriol, as a remedy, if it is applied in the most effective way. Twice I have cured a diseased *flock* by one application of it—and I never heard of it being done in any other way, or, indeed, on any other occasions. The following paragraph is from my "Sheep Husbandry in the South," published in 1848:

"I had a flock of sheep a few years since that were in the second season of the disease. They had been but little looked to during the summer, and as cold weather was setting in many of them were considerably lame—some of them quite so. The snow fell and they were brought into the yards, limping, and hobbling about deplorably. This sight, so digraceful to me as a farmer, roused me into activity. I bought a quantity of blue vitriol—made the necessary arrangements—and once more took the chair as principal operator. Never were the feet of a flock more thoroughly pared. Into a large washing tub, in which two sheep could stand conveniently, I poured a saturated solution of blue vitriol and water, *as hot as could be endured by the hand even for a moment.* The liquid was about four inches deep on the bottom of the tub, and was kept at about that depth by frequent additions of the *hot* solution. As soon as a sheep's feet were pared, it was placed in the tub and held there by the neck, by an assistant. A second one was prepared and placed beside it. When the third one was ready, the first was taken out, and so on. Two sheep were thus constantly in the tub, and each remained in it about ten* minutes. The cure was perfect. There was not a lame sheep in the flock during the winter or the next summer. The hot liquid penetrated to every cavity of the foot, and doubtless had a far more decisive effect even on the uncovered ulcers, than would have been produced by merely wetting them. Perhaps the lateness of the season was also favorable, as in cold weather the ulcers of ordinary virulence discharge no matter to innoculate the healthy feet; and thus at the time of applying the remedy there are no cases where there has been innoculation not yet followed by the actual disease. I think that the vitriol required for the above one hundred sheep was about twelve pounds, and that it cost me fifteen cents per pound. The account then would stand thus:—12 lbs. of vitriol at 15 cents, $1.80; labor of 3 men one day each, $2.25; total $4.05—or about *four cents* per sheep."

Many years after the above took place, I treated a flock of diseased lambs in the same way — except that they were put into a larger tub which would hold five of them, so that each stood in the hot fluid from twenty to twenty-five minutes: and again the cure was perfect. They too were handled just as winter was setting in; were wintered alone; and were turned early in the spring into a flock of about one hundred and fifty which had never had hoof-rot.

I believe the same remedy administered in the same way would be the cheapest† and most effectual one known for large flocks; and if I had to go through another war with the hoof-rot, I would construct a vat which would hold eight or ten sheep — perhaps with a grate over the fluid to prevent accidents‡ — and I would contrive some mode to *keep* the

---

* This, by a misprint, was published five, in Sheep Husbandry in the South.
† Blue vitriol costs more than formerly — but it is still a very cheap remedy when bought by the quantity.
‡ If ten sheep were put in a vat together, the attendant might have some difficulty

water hot, after it was poured into the vat. And if the disease was in its new and malignant stage, I might keep every sheep half an hour or upwards in the hot liquor. I do not aver that *one* application would cure all cases in that stage, but I judge from my experience that it would be more likely to do it than any other application I ever heard of.

When the disease is in what I have termed the third stage — when a decomposition of the tissues has taken place and a powerful escharotic is requisite to remove the dead structures and their ulcerous accompaniments — there is, as Mr. Youatt observes, no application comparable with chloride or buytr of antimony. I have used nitric, sulphuric and muriatic acid; and instead of finding, with Mr. Spooner, that it "is of little consequence which caustic is employed, provided it is of sufficient strength,"* I have found all the latter to possess too much "strength" — or, at least, to be too deeply corrosive when applied to flesh; while the butyr of antimony combines so readily with the fluids in the parts, that it very soon loses its caustic effects. It therefore, to a very considerable degree, possesses the admirable property of nitrate of silver (too expensive a material to be used in hoof-rot) of acting purely as a superficial caustic, so that an eschar is formed protecting the parts beneath, and promoting a new and healthy action. I much prefer muriatic acid to either nitric or sulphuric acid, when butyr of antimony is not to be obtained.

I have no space to discuss the questions whether hoof-rot is contagious — and whether it also originates without contagion. On the first point I will only say that I should esteem that man out of his senses who, after having very extended opportunities for observing its origin and for tracing its history in any particular region, should doubt its direct, decided and (after sufficient exposure of healthy animals to its virus,†) uniform contagiousness. Whether it is generated by

---

occasionally, in keeping them all on their feet. Lying down or falling down in the water would produce no catastrophe: it would not harm the sheep a particle — but it would dye it a light *blue*, and the liquor would be unnecessarily wasted.

* Spooner on Sheep, p. 441.

† I am inclined to believe that it is not communicated by effluvium — by infection — or even necessarily by contact between diseased and undiseased animals. I think it is chiefly, if not entirely communicated by a species of inoculation, if I may so term it — by the virus of a diseased foot being brought in contact with the inner portion of an undiseased foot. If this is a correct hypothesis, it would seem to follow that the malady would not be very likely to be communicated in all its stages; and that the rapidity of its transmission at all periods would depend somewhat upon chance. Sheep take it far most rapidly by being turned into pastures where diseased sheep have been some time running, and where a thousand blades of grass and other substances liable to come in contact with the inner portion of the foot, are charged with a quantity of the virus.

dirt, moisture, etc., without contagion, may perhaps be more doubtful. Some intelligent farmers take the affirmative of this question. In a letter I have just received from Mr. W. F. Greer, of Painesville, Ohio, he says:—"When in Lorain county (Ohio) in April last, I was told by many of the sheep men there that hoof-rot appears wherever sheep are kept in tall, rank feed — in the form of fouls at first. They were all of the opinion that it does appear spontaneously, and cited numerous instances to prove their theory."

Different countries and climates may probably be subject to the appearance of the disease under different circumstances. It is the prevailing view among English veterinary writers that hoof-rot frequently originates without contagion in Great Britain, though this opinion is not without its able dissentients. I have repeatedly known it to commence in that portion of the United States where I reside, without the owner of the sheep being able to trace it to any contagion. I have at page 165 mentioned two cases where my own sheep contracted it from flocks brought into contact with them accidentally, and for but a short period; and any one who reads the facts in those cases will readily see that the removal of the diseased sheep which communicated the malady, might readily have taken place without my ever being informed of the circumstances; and then I might have imagined that it was caused by "tall feed." But I never yet have heard of a case of fouls becoming hoof-rot, or of the latter disease occurring "spontaneously" in any new region where hoof-rot had never been previously introduced by diseased sheep, or where it was not at the time prevailing in a greater or less degree in some flock in the vicinity, or within a few miles. And we all know that there are very many regions where it has never been heard of among the sheep, though the grass is as tall, and all the other supposed exciting causes, except contagion, are fully equal.

CHAPTER XXIX.

OTHER DISEASES, WOUNDS, ETC.

THE ROT — SCROFULA — HEREDITARY DISEASES — CUTS — LACERATED AND CONTUSED WOUNDS — PUNCTURED WOUNDS — DOG BITES — POISONED WOUNDS — SPRAINS — BRUISES — ABSCESS.

THE ROT.— As already remarked, I have never witnessed an instance of the rot in the United States; although I have repeatedly seen sheep laboring under diseases called by that name. My opinion is that it has never appeared, at least, in our Northern States. Persons of much intelligence, residing in some of the Western and South-western States believe they have recognized it; but I do not remember to have seen any of their expressions to that effect properly supported by the published results of post-mortem examinations.

The symptoms of the disease are thus given by Mr. Spooner:— "The first symptoms attending this disease are by no means strongly marked; there is no loss of condition, but rather apparently the contrary; indeed, sheep intended for the butcher have been purposely cothed or rotted in order to increase their fattening properties for a few weeks, a practice which was adopted by the celebrated Bakewell. A want of liveliness and paleness of the membranes generally may be considered as the first symptoms of the disease, to which may be added a yellowness of the caruncle at the corner of the eye. Dr. Harrison observes, 'when in warm, sultry or rainy weather, sheep that are grazing on low and moist lands feed rapidly, and some of them die suddenly, there is fear that they have contracted the rot.' This suspicion will be further increased if, a few weeks afterward, the sheep begin to shrink and become flaccid about the loins. By pressure about the hips at this time a crackling is perceptible now or soon afterward, the countenance looks pale, and upon parting the fleece the skin is found to have changed its vermillion tint for a pale red, and the wool is easily separated from the pelt; and as the disorder advances the skin becomes dappled with yellow or black spots. To these symptoms succeed increased dullness, loss of condition, greater paleness of the mucous membranes, the eyelids becoming almost white and afterward yellow. This yellowness extends to other parts of the body, and a watery fluid appears under the skin, which becomes loose and flabby, the wool coming off readily. The symptoms of dropsy often extend over the body, and sometimes the sheep becomes chockered, as it is termed — a large swelling forms under the jaw, which, from the appearances of the fluid it contains, is in some places called the watery poke. The duration of the disease is uncertain; the animal occasionally dies shortly after becoming affected, but more frequently it extends to from three to six months, the sheep gradually losing flesh and pining away, particularly if, as is frequently the case, an obstinate purging supervenes."

Mr. Youatt thus describes the post-mortem appearances: " When a rotted sheep is examined after death, the whole cellular tissue is found to be infiltrated, and a yellow serous fluid everywhere follows the knife. The muscles are soft and flabby : they have the appearance of being macerated. The kidneys are pale, flaccid and infiltrated. The mesenteric glands enlarged, and engorged with yellow serous fluid. The belly is frequently filled with water or purulent matter; the peritoneum is everywhere thickened, and the bowels adhere together by means of an unnatural growth. The heart is enlarged and softened, and the lungs are filled with tubercles. The principal alterations of structure are in the liver. It is pale, livid, and broken down with the slightest pressure; and on being boiled it will almost dissolve away. When the liver is not pale, it is often curiously spotted. In some cases it is speckled like the back of a toad. Nevertheless, some parts of it are hard and schirrous; others are ulcerated, and the biliary ducts are filled with flukes. Here is the decided seat of disease, and it is here that the nature of the malady is to be learned. It is inflammation of the liver. * * * The liver attracts the principal attention of the examiner; it displays the evident effects of acute and destructive inflammation; and still more plainly the ravages of the parasite with which its ducts are crowded. Here is plainly the original seat of the disease—the center whence a destructive influence spreads on every side. * * * The Fluke—the *Fasciola* of Linnæus —the *Distoma hepaticum* of Rhodolphi—the *Planaria* of

Goese—is found in the biliary ducts of the sheep, the goat, the deer, the ox, the horse, the ass, the hog, the dog, the rabbit, the guinea-pig, and various other animals, and even in the human being. It is from three-quarters of an inch to an inch and a quarter in length, and from one-third to half an inch in greatest breadth.

"Figs. 1 and 3 represent this parasite of its usual size and appearance, and its resemblance to a minute sole, divested of its fins, is very striking. The head is of a pointed form, round above and flat beneath; and the mouth opens laterally instead

Fig. 1.   Fig. 2.   Fig. 3.

of vertically. There are no barbs or tentaculæ, as described by some authors. The eyes are placed on the most prominent part of the head, and are very singularly constructed (fig. 2). They have the bony ring of the bird. * * * The anastamoses of the blood vessels which ramify over the head are plainly seen through a tolerable microscope. The circulating and digestive organs are also evident, and are seated almost immediately below the head. The situation of the heart is seen in fig. 1, and the two main vessels evidently springing from it, and extending through almost the whole length of the fluke. Smaller blood-vessels, if so they may be called, ramify from them on either side. * * *

"In the belly, if so it may be called, are almost invariably a very great number of oval particles, hundreds of which, taken together, are not equal in bulk to a grain of sand. They are of a pale red color, and are supposed to be the spawn or eggs of the parasite. * * *

"There can be no doubt that the eggs are frequently received in the food. Having been discharged with the dung, they remain on the grass or damp spot on which they may fall, retaining their vital principle for an indefinite period of time. * * * They find not always, or they find not at all, a proper nidus in the places in which they are deposited; but taken up with the food, escaping the perils of rumination, and threading every vessel and duct until they arrive at the biliary canal, they burst from their shells, and grow, and probably multiply. * * * * *

"Leeuwenhoeck says that he has taken 870 flukes out of one

liver, exclusive of those that were cut to pieces or destroyed in opening the various ducts. In other cases, and where the sheep have died of the rot, there were not found more than ten or twelve. * * * Then, is the fluke worm the cause or the effect of rot? To a certain degree both. They aggravate the disease; they perpetuate a state of irritability and disorganization which must necessarily undermine the strength of any animal. * * * Notwithstanding all this, however, if the fluke follow the analogy of other entoza and parasites, it is the effect and not the cause of rot. * * *

"The rot in sheep is evidently connected with the soil or state of the pasture. It is confined to wet seasons, or to the feeding on ground moist and marshy at all seasons. It has reference to the evaporation of water, and to the presence and decomposition of moist vegetable matter. It is rarely or almost never seen on dry or sandy soils and in dry seasons; it is rarely wanting on boggy or poachy ground, except when that ground is dried by the heat of the summer's sun, or completely covered by the winter's rain. On the same farm there are certain fields on which no sheep can be turned with impunity. There are others that seldom or never give the rot. The soil of the second is found to be of a pervious nature, on which wet cannot long remain — the first takes a long time to dry, or is rarely or never so. * * *

"Some seasons are far more favorable to the development of the rot than others, and there is no manner of doubt as to the character of those seasons. After a rainy summer or a moist autumn, or during a wet winter, the rot destroys like a pestilence. A return and a continuance of dry weather materially arrests its murderous progress. Most of the sheep that had been already infected die; but the number of those that are lost soon begins to be materially diminished. It is, therefore sufficiently plain that the rot depends upon, or is caused by, the existence of moisture. A rainy season and a tenacious soil are fruitful or inevitable sources of it. * * * The mischief is effected with almost incredible rapidity."

Mr. Youatt here gives various instances to prove that rot is engendered in a few hours and even minutes.[*] He further says:—"It is an old observation that on all pasture that is suspected to be unsound, the sheep should be folded early in the evening, before the first dews begin to fall, and should not be released from the fold until the dew is partly evaporated. * * * Then the mode of prevention—that with which

---

[*] Youatt, p. 453.

the farmer will have most to do, for the sheep having become once decidedly rotten, neither medicine nor management will have much power in arresting the evil—consists in altering the character of as much of the dangerous ground as he can, and keeping his sheep from those pastures which defy all his attempts to improve them. * * * If all unnecessary moisture is removed from the soil, or if the access of air is cut off by the flooding of the pasture, no poisonous gas has existence, and the sheep continue sound. * * * The account of the treatment of rot must, to a considerable extent, be very unsatisfactory."

Mr. Youatt recommends the sale of sheep to the butcher after they are found to be rotted! To give what may be styled the butcher's autopsy, I copy his remarks:—"It is one of the characters of the rot to hasten, and that to a strange degree, the accumulation of flesh and fat. Let not the farmer, however, push this experiment too far. Let him carefully overlook every sheep daily, and dispose of those which cease to make progress, or which seem beginning to retrograde. It has already been stated that the meat of the rotted sheep, in the early stage of the disease, is not like that of the sound one; it is pale and not so firm; but it is not unwholesome, and it is coveted by certain epicures, *who, perhaps, are not altogether aware of the real state of the animal.* All this is a matter of calculation, and must be left to the owner of the sheep; except that, if the breed is not of very considerable value, and the disease has not proceeded to emaciation or other fearful symptoms, the first loss will probably be the least; and if the owners can get any thing like a tolerable price for them, the sooner they are sent to the butchers, or consumed at home, the better. Supposing, however, that their appearance is beginning to tell tales about them, and they are too far gone to be disposed of in the market or consumed at home, are they to be abandoned to their fate? No: far from it."

The above is a paragraph which I could most sincerely wish stricken from the writings of its accomplished author. It will astonish even those who are acquainted with the astute and calculating selfishness of Mr. Bakewell's character to learn that he *purposely* rotted sheep which were to be sold to the butcher, to avail himself of the superior fattening properties which the diseased animals temporarily possess! This remarkable fact is stated by both Mr. Youatt and Mr. Spooner.*

---

* The sale of the meat of diseased animals is regarded as infamous in all parts of

Of the treatment of rot Mr. Youatt continues:—"If it is suited to the convenience of the farmer, and such ground were at all within his reach, the sheep should be sent to a salt-marsh in preference to the best pasture on the best farm. There it will feed on the salt incrusted on the herbage, and pervading the pores of every blade of grass. A healthy salt-marsh permits not the sheep to become rotten which graze upon it; and if the disease is not considerably advanced, it cures those which are sent upon it with the rot. * * * Are there any indications of fever, heated mouth, heaving flanks, or failing appetite. Is the general inflammation beginning to have a determination to that part on which the disease usually expends its chiefest virulence? Is there yellowness of of the lips and of the mouth, of the eyes and of the skin? At the same time are there no indications of weakness and decay? Nothing to show that the constitution is fatally undermined? Bleed—abstract, according to the circumstances of the case, eight, ten, or twelve ounces of blood. There is no disease of an inflammatory character at its commencement, which is not benefited by early bleeding. To this let a dose of physic succeed—two or three ounces of Epsom salts, administered in the cautious manner so fre-

---

the United States and in almost the only case where I ever knew it to occur in my vicinity, the seller—a very low man—was punished in exemplary damages by a jury and received the soubriquet of "Stinking Meat," which followed him through life. The meat sold by him had no offensive odor; its condition was not discovered until it was partly consumed; nor was it then discovered by the taste or appearance of the meat —but by the information of a person privy to the facts. But it was proved that the animal was affected by a disease usually mortal, and expected to prove mortal in the present case. I have forgotten what the disease was. There might have been some color of proof that the meat proved unwholesome. Selling "unwholesome provisions" is a misdemeanor at common law, and therefore indictable. And every contract for provisions implies that they shall be wholesome. (See 3 Blackstone's Com., 165.) The vendor is bound to know they are sound and wholesome at his peril. American courts and juries have given an extensive construction to the term wholesome. In this connection. In fact, if it can be proved that a person has sold meat knowing that the animal, when killed, was laboring under any constitutional or serious malady, or even that it was killed to anticipate death from unnatural and accidental causes—if there is any good reason to suppose those causes placed the meat in the situation of that of a *diseased* animal—I say courts and juries under these circumstances demand only colorable proof that the meat is *unwholesome* to assess damages on the vendor. Thus in Fonda *vs.* Van Bracklin, the plaintiff recovered five dollars damages of the defendant, for selling him a quarter of beef from a cow that "had eaten *shortly before she was killed*, a very large quantity of peas and oats, and that was slaughtered for fear she would die in consequence of her having eaten them." It was proved that those who ate the beef "were generally made very sick and that one of Fonda's servants was sick two weeks from eating it." The case was carried up to the Supreme Court, &c., of the State of New York, and the judgment affirmed. I know nothing about the parties, but infer from the amount of the verdict that they were persons of low character, and that the proof of the subsequent illness of the persons who eat the meat, was not much credited. In other words, I infer that the verdict was, more than anything else, an expression that no man has a right to sell the meat of a diseased animal. The jury could not have believed that the meat produced the alleged effects. Such a verdict could not have been founded on that hypothesis.

quently recommended; and to these means let a change of diet be immediately added—good hay in the field, and hay, straw, or chaff, in the straw-yard.

"The physic having operated, or an additional dose, perchance, having been administered in order to quicken the action of the first, the farmer will look out for further means and appliances.   *   *   *   Two or three grains of calomel may be given daily, but mixed with half the quantity of opium, in order to secure its beneficial and ward off its injurious effects on the ruminant. To this should be added—a simple and cheap medicine, but that which is the sheet anchor of the practitioner here—common salt.   *   *
In the first place, it is a purgative inferior to few, when given in a full dose; and it is a tonic as well as a purgative.   *   *
A mild tonic, as well as an aperient, is plainly indicated soon after the commencement of rot. The dose should be from two to three drachms, repeated morning and night. When the inflammatory stage is clearly passed, stronger tonics may be added to the salt, and there are none superior to the gentian and ginger roots; from one to two drachms of each, finely powdered, may be added to each dose of the salt.   *
 *   *   The sheep having a little recovered from the disease, should still continue on the best and driest pasture on the farm, and should always have salt within their reach.
 *   *   *   The rot is not infectious."

SCROFULA.—I have never witnessed an instance of this malady in our country. Mr. Spooner says of it:—"Sheep are liable to a scrofulous disease which is almost uniformly fatal. It is called the *evil* in some places, and elsewhere receives other denominations. A hard swelling of the glands under the jaws is first observed; after a time small pustules appear about the head and neck, which break, discharging a white matter, then heal, and are followed by others more numerous. This gradually robs the animal of flesh, and slowly pining away, it becomes at length quite useless, and in this state is destroyed. It seldom attacks great numbers at a time, but selects generally a few individuals from a flock.

"The writer, though he cannot say that he has perfectly succeeded in effecting a cure, has done so to a certain extent, so that the tumors disappeared and the animals improved in flesh and health, but afterwards relapsed. This he has accomplished by administering four or five grains of hydriodate of potash daily in gruel, and rubbing the parts likewise with ointment of iodide of mercury. As soon as the animal is considerably better, it should be sent to the butcher."

HEREDITARY DISEASES. — Mr. Finlay Dun, Lecturer on Materia Medica and Dietetics at the Edinburgh Veterinary College, after giving much attention to the subject of hereditary diseases in domestic animals, summed up his conclusions as follows:

1. "They are transmitted by the male as well as by the female parent, and are doubly severe in the offspring of parents, both of which are affected by them.

2. They develop themselves not only in the immediate progeny of one affected by them, but also in many subsequent generations.

3. They do not, however, always appear in each generation in the same form; one disease is sometimes substituted for another, analogous to it, and this again, after some generations, becomes changed into that to which the breed was originally liable, as phthisis (consumption) and dysentery. Thus, a stock of cattle previously subject to phthisis, sometimes become affected for several generations with dysentery, to the exclusion of phthisis, but by and by, dysentery disappears to give place to phthisis.

4. Hereditary diseases occur to a certain extent independently of external circumstances, appearing under all sorts of management, and being little affected by changes of locality, separation from diseased stock, or such causes as modify the production of non-hereditary diseases.

5. They are, however, most certainly and speedily developed in circumstances inimical to general good health, and often occur at certain, so called, critical periods of life, when unusual demands on the vital powers take place.

6. They show a striking tendency to modify and absorb into themselves all extraneous diseases; for example, in an animal of consumptive constitution, pneumonia seldom runs its ordinary course, and when arrested, often passes into consumption.

7. Hereditary diseases are less effectually treated by ordinary remedies than other diseases. Thus, although an attack of phthisis, rheumatism, or opthalmia may be subdued, and the patient put out of pain and danger, the tendency to the disease will still remain and be greatly aggravated by each attack.

"In horses and neat cattle, hereditary diseases do not

usually show themselves at birth, and sometimes the tendency remains latent for many years, perhaps through one or two generations, and afterwards breaks out with all its former severity."

Mr. Dun's omission of sheep from the list of animals above named, as subjects of hereditary disease, is merely accidental, for in a paper on the "Hereditary Diseases of Sheep and Pigs," published in the Journal of the Royal Agricultural Society, 1856, he mentions, as among the hereditary diseases of the former, epilepsy or fits, hydatids in the brain (or turnsick or sturdy), chronic cough, chronic diseases of the respiratory organs generally, diseases of the digestive organs, which produce diarrhea and dysentery, rheumatism, scrofula, tabes mesenterica (a variety of scrofulous disease,) and consumption. Mr. Dun says :—" When a scrofulous constitution presents itself prominently in an adult sheep, it is generally in the form of pulmonary consumption, or, as it is technically termed, *phthisis pulmonalis*." He subsequently adds :

"But these are not the only evils which assail sheep of a scrofulous constitution. They are occasionally affected by chronic swellings about the neck and throat, at first hard, but afterwards softening, bursting externally, and discharging an unhealthy pus. These swellings are analogous to clyers in cattle, and like them are most apt to occur in scrofulous subjects living in localities exposed to east winds. Scrofulous sheep are likewise subject to intractable swellings of the joints, to foot-rot in its most tedious and aggravated form; and to rickets, a disease of the bones, occurring in early youth, from perverted nutrition, and consisting in a softening of the osseous tissue. They are further of such a weak and depraved constitution as to fall easy and early victims to any ordinary or prevailing diseases which, moreover, are in them developed with unusual severity."

CUTS.— When a sheep has received a simple, clean cut, the edges of the wound should be brought accurately together and the skin confined by stitches. A bandage, if the situation of the wound admits of its use, will keep the separate parts better in place, and prevent the stitches in the skin from tearing out, as they are apt to do when the cut is *across* the muscular fibers, so that their retraction has a tendency to pull the wound open. A little turpentine applied to the wool near the parts, or to the bandage, will prevent the attack of the fly.

LACERATED AND CONTUSED WOUNDS.—If the wound is torn, and contused, the parts if not too much injured should be placed as near as practicable in their natural position; if too much lacerated or crushed, the loose and disorganized parts should be carefully cut away. If the situation of the wound admits of it, a warm poultice may be applied, and changed twice a day, until healthy suppuration ensues. Afterward it will only require to be kept covered with an oiled or greased cloth, sufficiently touched with turpentine to repel the fly. If the situation of the wound does not admit of poulticing, it should be fomented until clean, and some mild stimulant applied. Mr. Spooner recommends tincture of myrrh, and an astringent powder compounded as follows: powdered chalk 4 oz., Armenian bole 1 oz., powdered charcoal 1 oz., powdered alum $\frac{1}{2}$ oz., sulphate of zinc $\frac{1}{2}$ oz.

PUNCTURED WOUNDS.—These are made by pointed instruments, splinters of wood, etc. Fomentations are generally made use of, and the Mountain Shepherd's Manual states that if these are made with a decoction of chamomile flowers, their good effects will be increased. "The method of applying [fomentations] is to dip a piece of woolen cloth into the decoction when hot, then to wring it, and apply it to the parts, dipping the cloth again when the heat has abated." If the wound heals on the outside before it does within, and matter forms in it, it must again be opened.

DOG BITES.—From their torn and lacerated character, dog bites are generally very fatal,—and the more so, from the fact that the skin is generally stripped from large surfaces. When the latter has occurred in hot weather, and particularly when the skin is removed from the *body*, gangrene generally ensues speedily. I have attempted to procure the re-adhesion of the skin by carefully cleansing it, restoring the remaining portions to their natural position, and stitching the edges together: but in hot weather, and when the flies are abundant I have never succeeded when the denuded surfaces were at all extensive. In the latter cases, fatal gangrene has usually even anticipated the attacks of the fly. Wounds from bites are to be treated like other wounds exhibiting the same kind of injuries.

POISONED WOUNDS.—In the Mountain Shepherd's Manual it is stated that sheep in Europe are not unfrequently bitten by venomous snakes, and that when this occurs, "a spoonful of rape or olive oil should be given several times a day, or the same quantity of the solution of an ounce of volatile salt in two quarts of water." In France snakes have been thought to suck the milk of ewes, inflicting wounds in the teat which cause them to dry up permanently.* I have never seen or heard of a case of either kind in the United States; and I attach no credit to the supposed snake-sucking in any part of the world.

SPRAINS.—The mode of treating sprains recommended in the Mountain Shepherd's Manual is the only one I have ever heard of which was attended with any observable success. The limb is immediately immersed in hot water for half an hour, and this repeated several times a day. The cure is often rapid.

BRUISES AND STRAINS,—Are treated on the sheep, if at all, with hot fomentations, and the application of camphor.

ABSCESS.—I have never seen a case of this. Mr. Spooner says:—"Abscess, which is a collection of pus or matter under the skin, may be produced by a bruise, or by some constitutional cause. Whilst collecting, the surface of the skin is usually very tender, and sometimes there is also much constitutional irritation present. A collection of matter may be known by the heat, swelling and pain of the part. On pressing it the contained fluid is felt to fluctuate; and the pressure being removed, the part immediately assumes its former shape, whilst a watery or dropsical swelling, on being pressed, leaves for some time the marks of the fingers. After some time the abscess points; that is, the matter can be more distinctly felt at one particular part, at which, if permitted, the abscess would soon burst. This, however, should not be permitted; but at this stage the abscess should be opened at the lowest part, or that which would admit most readily of its discharging itself. The opening should be large, and no dressing will be required except the continuance of the fomentation, which should previously be used. It should be observed that, if the abscess is languid and slow in forming, a stimulant, such as hartshorn and oil, rubbed in occasionally, will be useful."

---

* Mountain Shepherd's Manual, p. 8.

CHAPTER XXX.

MEDICINES USED IN THE TREATMENT OF THE DISEASES OF SHEEP.

THE list of medicines below, comprises the principal ones employed in the treatment of the diseases of sheep.

EXPLANATION OF MEDICAL TERMS USED IN THIS CHAPTER.

Absorbent.— A medicine used for absorbing acidity in the stomach and bowels.

Anodyne.— A medicine which relieves pain.

Antacid.— A remedy which removes sourness in the stomach.

Anthelmintic.— A remedy which destroys or expels worms.

Antiseptic.— A preventive of putrefaction.

Antispasmodic.— A preventive of spasms.

Aperient.— A medicine which opens the bowels.

Astringent.— A medicine externally that has the property of contracting organic textures, and internally of diminishing evacuation or dunging.

Carminative.— A remedy which allays pain and causes the expulsion of flatus, or wind, from the body.

Cathartic.— A medicine which causes an increase of evacuation or dunging.

Caustic.—A body which has the power of burning or consuming flesh and other animal substances.

Cordial.— A tonic medicine which excites the system and increases the rapidity of the circulation of the blood.

Diaphoretic.— A medicine which promotes perspiration or sweating, and a carrying off of the humors of the body through the pores of the skin.

Disinfectant.— An agent which removes the causes of infection.

Diuretic.— A medicine which increases the discharge of urine.

Emetic.— A substance capable of producing vomiting or puking.

Emollient.—A substance which allays irritation and softens and relaxes parts that are inflamed, and swollen hard.

Febrifuge.— A medicine which abates or drives away fever.

Febrile.— Feverish.

Laxative.— A medicine which loosens or opens the bowels.

Lubricant.— A substance which makes the body to which it is applied soft and slippery.

Narcotic.— A medicine which, by acting on the brain, relieves pain, allays morbid susceptibility, and produces sleep. In too large doses narcotics produce stupor and death.

Purgative.— A medicine which operates more powerfully in opening the bowels than a laxative.

Rubefacient.— An application which produces redness or irritation of the skin.

Sedative.— A medicine employed to depress unnaturally increased action of the vital forces, and thus quiet the system.

Stimulant.— A medicine which has the power of exciting the action of the organs and the discharge of the functions of the animal system.

Stomachic.— A medicine which strengthens the stomach and gives more activity to its functions.

Sudorific.— A medicine that causes sweating.

Tonic.— A medicine that gives increased strength and vigor to the action of the system.

LIST OF MEDICINES.

ALCOHOL, (Spirit of Wine).—Added in small quantities as a stimulant to purgatives, in low forms of disease.

ALE.— Administered to sinking ewes before and after parturition, usually in doses of about a gill; after long and exhausting parturition, it is mixed with two to four drachms of laudanum; the dose repeated at intervals of three or four hours. Ale is sometimes given to chilled lambs.

ALOES.— Occasionally used as a purgative; administered by some good shepherds in combination with oil for colic or stretches. But there are better cathartics for sheep. The tincture of aloes, says Mr. Youatt, "is a very useful, stimulating and healing external application to wounds. Two ounces of powdered aloes and a quarter of an ounce of powdered myrrh, should be macerated [soaked] in a pint of rectified spirit and diluted with an equal quantity of water."

ALTERATIVES.— Ethiop's mineral, (black sulphuret of mercury,) nitre and sulphur, compounded in the proportions of one, two and four, is useful in the cutaneous diseases of sheep. Average dose two drachms, administered daily.

ALUM — Is used in some astringent medicines, but is inferior to various other articles for this purpose. It is sometimes employed in solution to bathe an inverted womb or a prolapsed rectum. (See page 145.) Burnt alum is used as a stimulant and caustic on wounds, but there are better ones.

AMMONIA — In the form of hartshorn, enters into various liniments, and some other medicines. It is an excellent external stimulant, and rubefacient. Internally it is an antacid and sudorific, but is not often thus used in sheep diseases.

ANODYNES.— Opium is chiefly employed for this purpose. The modes and times of its exhibition have been pointed out throughout this volume.

ANTIMONY, (the Chloride of Butyr of?)—By far the best caustic in advanced stages of hoof-rot. For the causes of its superiority, and for its mode of application, see page 370.

AQUA-FORTIS.— See Nitric Acid.

ARSENIC.— Used in solution to kill ticks and cure scab. See pages 188, 341–343.

ANTISPASMODICS.— Opium is employed in the case of tetanus or locked-jaw, colic, etc.

ASTRINGENTS.— Opium (acting as an anodyne,) is chiefly relied on internally. Catechu frequently forms a part of medicines intended to have a slightly astringent effect. Alum is used in the form of alum whey— 2 drachms of pulverized alum dissolved in a pint of hot milk. The external astringents most used in sheep diseases are a solution of alum, or a decoction of white oak bark, burnt alum, bole Arminian, etc.

BALLS.— Medicine should never be administered in the form of balls to sheep. For the reasons see page 299.

BLISTERS.— Not often resorted to in sheep practice. When used the hair or wool must be closely shaved, and the parts well rubbed with an ointment composed of one part of powdered cantharides, four of lard, and one of resin.

BOLE ARMINIAN.—An argillaceous earth used occasionally as a mild external astringent on wounds. Internally an astringent and absorbent.

BONESET OR THOROUGHWORT.—(*Eupatorium perfoliatum*.) A good tonic and diaphoretic; in large doses emetic and aperient. A tea made of this is administered with excellent effect to lambs, under the circumstances stated at page 150.

CALAMINE.—See Zinc.

CALOMEL.—See Mercury.

CAMPHOR.—Used internally as a narcotic, diaphoretic and sedative. It is used externally, dissolved in alcohol, to reduce the swelling of the thyroid glands in the necks of lambs, (see page 153.) Dissolved in the same way, or in oil, it is applied to the udders of sheep having garget, (see page 331.) It is also used with good effect as an external stimulant in rheumatism, strains, bruises, swellings, swelling of the joints, &c.

CANTHARIDES.—The principal ingredient in blistering ointments.

CARRAWAY SEED.—Given in doses of two or three drachms as a stomachic, with other medicines.

CASTOR OIL.—A safe and excellent purgative, but not used as much as saline purgatives.

CATECHU.—A valuable astringent, in doses of half a drachm. It is one of the ingredients of the celebrated "sheep's cordial."

CAUSTICS.—Butyr of antimony, muriatic acid, sulphuric acid, nitric acid, blue vitriol, burnt alum, &c., &c. Blue vitriol is immeasurably the best application and mild caustic in the early stages of hoof-rot, as butyr of antimony is in later ones. (See Hoof-Rot.)

CHALK, (*Prepared*)—By its alkaline properties, neutralizes the acidity of the stomach, and thus checks diarrhea. It is a very valuable remedy in doses from half an ounce to an ounce, given as directed under the head of "diarrhea." Mr. Spooner also recommends it as a useful external application to wounds and sores.

CHAMOMILE.—A mild tonic and febrifuge. Used externally in fomentations for wounds, ulcers, &c.

CHARGES,—Thick adhesive plasters spread over parts that

require support, and the application of a constant and moderate stimulus. They are placed on the shoulder of the sheep when the bones underneath are fractured.

CLYSTERS.—See Injections.

COPPER.—See Verdigris and Vitriol.

COPPERAS.—See Sulphate of Iron.

CORDIAL.—Sheep's cordial. An excellent remedy for Diarrhea. For mode of preparing it see page 307.

CORROSIVE SUBLIMATE (Bi-chloride of Mercury)—The most convenient form in which mercury can be administered internally. The proto-chloride, or calomel, from its great gravity, could not, with any certainty, be made to reach the fourth stomach. It would seem that mercury should be a useful remedy in several of the diseases of sheep. Corrosive sublimate dissolved in alcohol is an excellent application to old and ill-conditioned wounds or ulcers. It is very effectual in destroying maggots in wounds and in repelling the attacks of the fly.

CROTON SEEDS OR OIL—A very powerful and rapid purgative rarely resorted to in the diseases of sheep. Dose, says Mr. Spooner, from 5 to 15 drops of the oil. It is sometimes applied externally, sufficiently reduced, in glandular and and other indolent swellings.

DIGITALIS (Fox-glove)—A sedative, and it lowers the frequency of the pulse. It enters into most of the fever medicines of the English veterinarians. Dose, 1 scruple.

ELDER—The ointment of, has been once or twice mentioned in this volume. It is made by boiling elder leaves in lard, and forms one of the most softening and soothing applications known for inflamed and irritated surfaces.

EPSOM SALTS (Sulphate of Magnesia)—In doses from half an ounce to one, and in some few cases two ounces, the best purgative which can, in almost every disease, be administered to sheep.

FOMENTATIONS—To reduce swellings, lessen inflammation and relieve pain in inflamed udder, garget, and various other cases, these are invaluable. They are applied by dipping a woolen cloth constantly in hot water—as hot as can possibly be endured by the hand—and laying or gently pressing it on the parts. To be effectual, fomentations must be continued a long time; and the part should be left covered if practicable, particularly in cold weather, or cold may be taken in it, and the original difficulty only aggravated.

GENTIAN.—The best vegetable tonic in use. Dose from 1 to 2 drachms.

GIN.—Given in doses of half a teaspoonful to a teaspoonful in warm milk to chilled lambs, with admirable effect. I omitted to mention, when speaking of the mode of procuring the adoption of a foster lamb, that gin rubbed on the nose of the ewe and sprinkled over the lamb, promotes that object.

GINGER.—A highly useful cordial and stomachic, given with most aperient medicines to prevent griping. Dose from half a drachm to two drachms.

HARTSHORN.—See Ammonia.

INJECTIONS.—These are of the utmost importance to relieve constipation in lambs. For their composition, and the mode of administering them, see page 150.

IODINE.—The hydriodate of potash in the proportion of one part to seven or eight parts, by weight, of lard, constitutes an ointment which is a powerful stimulant to the absorbing vessels, and therefore is an excellent application to glandular swellings, or to indurated tumors. It is a good application to swelled udder, (garget,) or to enlarged thyroid glands. (See Goitre and Garget.)

LARD.—A gentle purgative in doses of two ounces. The basis of most ointments, and applied externally in almost every case as an emollient and lubricant in the place of oils.

LAUDANUM.—See Opium.

LEAD (Acetate or Sugar of Lead).—Mixed with other ingredients to form caustic applications in hoof-rot.

LEAD (White).—Is used in cooling and drying ointment.

LIME (Carbonate of).—See Chalk.

LIME (Chloride of).—Is a powerful disinfectant and antiseptic. It is used to disinfect and purify stables, &c., in which contagious diseases have occurred, and to clean the foot and remove stench in the worst stages of hoof-rot. It is administered internally for hoove. (See Hoove.)

LINSEED.—Or flax-seed, is invaluable as an emollient poultice. It forms an excellent gruel for animals during illness.

LINSEED OIL.— In doses of two ounces is a safe purgative where the intestines are irritated. Ordinarily it is not so efficient as Epsom salts.

MAGNESIA (Sulphate of) — See Epsom salts.

MERCURY.— Mercurial ointment is used as an application for scab, and also to kill ticks. For the mode of compounding, see pages 189, 342.

MERCURY (Proto-Chloride of, or Calomel)—It is not much used in diseases of sheep. (See pages 322, 323, 378.)

MERCURY (Bi-Chloride of.) — See Corrosive Sublimate.

MURIATIC ACID (Spirit of Salt) — Next to chloride of antimony, the best caustic in the worst stages of hoof-rot.

MYRRH.— A stimulant tonic, used in applications to wounds. (See under Aloes.)

NITRATE OF SILVER, (Lunar Caustic) —Is the best superficial caustic, but is far too expensive to be used on ordinary occasions. In one place, however, it is indispensable—in the bites of rabid animals, or where their saliva is brought in contact with surfaces denuded of skin. It has not yet been shown that the saliva of the rabid sheep will communicate that dreadful disease, but the opposite fact is not positively established. Some of the sheep, an account of which is given at pages 283–290, had been handled by laborers, after becoming rabid, and the saliva might have come in contact with their hands; and one of the Messrs. Freer, whose sheep were bitten at the same time, not suspecting the nature of the malady, sponged out the mouth of an animal in the last stages of rabies! On examining his hands the skin was found slightly fractured in several places. Whatever might have been the result, it relieved some painful anxieties to see his hands, and those of my son's laborers, well and thoroughly cauterized with lunar caustic, so as to form a thick black eschar or scab wherever there was the least break in the skin. Mr. Youatt, who probably had more experience with rabies than any man who ever lived, considered immediate and thorough cauterization, by nitrate of silver, of the wounded or exposed parts —to the very bottom of the the tooth holes in case of bites— a perfect protection against the virus of rabies.

NITRE, OR SALT-PETRE, (Nitrate of Potash)—In doses of a drachm or two, enters into every fever medicine. It is cooling, diuretic, and diaphoretic. Externally it is a powerful antiseptic.

NITRIC ACID (Aqua-fortis)—Sometimes used as a substitute for chloride of antimony, or muriatic acid, as a caustic in hoof-rot. Used by drovers also to harden the soles of sheep's feet which have become thin and tender by traveling. It is touched over the soles with a feather.

OLIVE OIL (Sweet Oil)—Is used in many external applications, and sometimes internally as a laxative; but for the last purpose is inferior to the other oils given as cathartics, and to Epsom salts.

OPIUM—As an antispasmodic, sedative and astringent it stands unrivaled. Mr. Youatt remarks:—"A colic drink would have little effect without it; and if opium were omitted in the medicines for diarrhea and dysentery, every other drug would be given in vain." In the form of gum the dose is about 10 grains; in the form of laudanum, from 1 to 2 drachms.

PEPPER (Black)—Given pulverized, in doses of half a teaspoonful in warm milk, to chilled lambs. It is a warm carminative stimulant, and is capable of producing general arterial excitement.

PIMENTO (Allspice)—A substitute for ginger in the same doses, but not as valuable.

PUMPKIN SEEDS—A tea of, is an excellent diuretic for very young lambs, when their urine does not pass with sufficient freedom.

RHUBARB—Unites the properties of a cathartic and subsequent astringent. In small doses, it is a tonic and stomachic, invigorating the digestion. When the bowels are relaxed and torpid, and the stomach in a feeble state, it would seem the most appropriate purgative, when a purgative is required.

RYE, (Ergot of)—A powerful stimulant to the womb—resorted to in England in very protracted lambing. Mr. Spooner says the dose is one scruple infused in hot water, and repeated if required in the course of one or two hours.

SPIRITS.—Brandy, rum, whisky, etc., may be made a substitute for gin for chilled lambs. See Gin.

SALT, (Muriate of Soda.)—An ounce constitutes a light purgative; in small quantities a tonic and stomachic. The necessity of keeping sheep freely supplied with salt has been referred to under Summer and Winter Management. For its great efficacy in Rot, see page 378.

SALT-PETRE.—See Nitre.

SETONS.—The mode of inserting these is pointed out at page 348.

SPIRIT OF NITROUS ETHER, (Sweet Spirit of Nitre.)—In doses of two drachms; a valuable diaphoretic, diuretic and anti-spasmodic. It is much used in fibrile affections.

SULPHATE OF COPPER.—See Blue Vitriol.

SULPHUR.—Internally an aperient in doses from one to two ounces. Externally it forms the basis of ointments used in various cutaneous diseases.

SULPHURIC ETHER.—A powerful stimulant and anti-spasmodic. Dose one drachm.

SULPHURIC ACID.—A powerful caustic, used alone, or in combination with other ingredients, in advanced stages of hoof-rot.

SULPHATE OF IRON, (Copperas or Green Vitriol.)—Used in hoof-rot remedies, but much less valuable for that purpose than blue vitriol. Internally a tonic.

SPIRIT OF TAR—Destroys maggots, and prevents the fly from depositing its eggs in ulcers or wounds.

TAR—Is an impure turpentine, but it contains several distinct principles, of which creosote is one. Internally it is stimulant, diuretic, anthelmintic, and in large doses is laxative. Externally it is a stimulant, produces a good effect on foul or indolent ulcers, and repels attacks of the fly. It is also resorted to as a mechanical coating for the feet, &c., when denuded of their natural coverings, in order to retain other applications underneath, keep out water, &c

TOBACCO.—A decoction of it kills the acarus of scab, and thus cures that disorder. It also kills ticks, lice, &c. An injection of it, or the smoke of it blown into the nostrils, causes the larvæ of the the Gad-fly to be dislodged from the cavities of the head. Altogether it is a most valuable sheep medicine, and every sheep farmer should cultivate it in his garden for that purpose. Tobacco ointment, made by boiling an ounce of fresh tobacco leaves cut fine, in a pound of lard over a gentle fire until it becomes friable, would be an admirable application on irritable ulcers of the foot or other parts.

TURPENTINE (Spirits of).— Has about the same internal and external effect with tar; but it lacks the creosote, which may render it a little less effective on old ulcers. (See Tar.)

VERDIGRIS (Acetate of Copper) — Often used in hoof-rot in combination with blue vitriol. Its medicinal properties are very similar, and I doubt whether it forms any useful addition to the former in such cases.

ZINC (Carbonate of) — Mixed with lard, constitutes a valuable emollient and healing ointment. It is mixed in the proportion of one part of the carbonate, by weight, to eight of lead.

CHAPTER XXXI.

THE DOG IN ITS CONNECTION WITH SHEEP.

THE INJURIES INFLICTED BY DOGS ON SHEEP — THE SHEEP DOG — THE SPANISH SHEEP DOG — THE HUNGARIAN SHEEP DOG — THE FRENCH SHEEP DOG — THE MEXICAN SHEEP DOG — THE SOUTH AMERICAN SHEEP DOG — OTHER LARGE RACES OF SHEEP DOGS — THE ENGLISH SHEEP DOG — THE SCOTCH SHEEP DOG, OR COLLEY — ACCUSTOMING SHEEP TO DOGS.

THE DOG is justly a favorite animal with man, and I cannot deny that I have written some prose heroics in his praise.* But on the whole, on summing up the advantages and disadvantages which he produces to mankind — and especially to sheep growers — there can be no doubt that the balance is enormously against him.

THE INJURIES INFLICTED BY DOGS ON SHEEP. — I had purposed to collect some statistics of the annual injuries inflicted on sheep by dogs in the State of my residence, (New York,) but I found them already prepared to my hand in reference to the State of Ohio — which will answer equally well for the purposes of illustration — over the signature of the able and efficient Corresponding Secretary of the Ohio State Board of Agriculture, John H. Klippart, Esq. I need not say that his name is an ample guaranty of the accuracy of his statements. I cut the paper from a recent number of the Ohio Farmer. Mr. Klippart says:

"In 1858, the Legislature of Ohio enacted a law making it the duty of the Township Assessors to return a list of the sheep killed and wounded by dogs, in every township in the State. From the annual returns we are now obtaining data from which to estimate the amount of damage done by dogs

---

* In Sheep Husbandry in the South.

to the wool-growing interest. Last winter the Legislature enacted a law requiring the Assessors to return a list of all the dogs in their respective townships or wards. Up to the present time I have returns from eighty counties — eight counties having failed to make returns; but the returns from the eighty counties furnish sufficient data 'to do some figuring.'

"Eighty counties return 162,933 dogs, or nearly 2,037 dogs per county; if the remaining eight counties, viz.: Allen, Ashland, Fulton, Licking, Mahoning, Montgomery, Noble and Putnam, maintain the same average, the complete returns will then show 179,256 dogs in Ohio! This will give $4\frac{1}{2}$ dogs to every square mile in the State, 1 dog to every 13 inhabitants in the State. We have a population of 58 inhabitants to the square mile.

"In 1860 the Legislature of Rhode Island appropriated the requisite sum of money to enable the United States Marshal to collect some special statistics of that State, among which was dogs; the number returned by the Marshal was 6,854 dogs. Rhode Island has 1,306 square miles of territory, 173,869 inhabitants. This gives $5\frac{1}{4}$ dogs to every square mile, or more than 1 dog to every 25 inhabitants, whilst there are 133 inhabitants to the square mile.

"The probabilities are that not more than half the dogs in Ohio have been returned to the Assessors. Many instances have come to my knowledge, where parties preferred killing the dogs to paying taxes on them, and the dogs were accordingly destroyed. In the city of Columbus, one ward returns three dogs only; but private information assures me of more than forty in the same ward. Franklin county returns 2,167 dogs, whilst well informed parties assert that there are more than that number in the city of Columbus. It is safe to assume that there are at least 200,000 dogs in the State.

"What does it cost to keep (feed) these dogs? In towns and cities it will be no exaggeration to value the food consumed by dogs at fifty cents each per week, or twenty-six dollars per annum; it is worth just as much in the country or on the farm to keep a dog; but their food can be procured cheaper there, and is worth at the lowest estimate ten dollars per annum. If we estimate the cost of keeping the dogs in the State at the *town* rate, the figures show that the cost of keeping them is five millions of dollars, but if we take the country rate it will amount to two millions of dollars—these are the two extremes, the truth lies in the middle, because there are fully as many dogs in cities, towns and villages, as there are in the country. Therefore the amount expended for food for dogs in Ohio, is worth annually the sum of three and a half millions of dollars, or more than *three-fourths the total amount of State taxes* for the years 1861 or 1862, and just the amount of State taxes for 1860! Reflect for a moment on this fact, that if the amount of food consumed by the dogs in the State in a single year were properly disposed of, the sum obtained for it would pay the State taxes! How desperately some people complain at the amount of taxes, yet none complain of the cost of keeping a dog. Aside from the expense of keeping dogs, they have killed and injured sheep in

1858, to the amount of		$146,758
1859, " "		102,398
1860, " "		86,795
1861, " "		87,092
1862, " "	about	85,000
Total in five years,		508,043
Annual average,		101,608

"There are then $100,000 worth of sheep killed and injured every year by dogs; and this has been going on ever since sheep were in the State. In 1846, sheep were first enumerated and valued for taxation; in that year the number in the State was 3,141,946. In 1862, the number was 4,448,227, an increase of 1,306,281, or 41½ per cent. in 16 years. In this same period of time, the number of swine has more than doubled, cattle have just doubled, and horses not quite doubled. Were it not for the destruction of sheep by dogs, Ohio would to-day have ten million head of sheep; but when sheep growers are compelled to pay an annual tax of $100,000 to $150,000, according to the caprice of worthless dogs, aside from the regular township and county tax, it is no wonder that they become discouraged, invest their surplus capital in Western lands, and thus let the productive interests of the State suffer. There is no kind of doubt that the dogs have annually destroyed $100,000 worth of sheep from 1846 to the present time, or an aggregate of $1,700,000, and to what purpose? Who has been benefited by this destruction of sheep? NOBODY! When the lightning strikes down one of the 'monarchs of the forest,' or destroys a house and kills some of the inmates, the benefits in health and continuation of life to those remaining is still of greater benefit, than the loss incurred is a damage. The explosion of the electric fluid purifies the atmosphere, and is a guarantee for the continuance of health; whereas if we had no electrical phenomenon, the air would become very impure, and epidemics or other diseases engendered by the impurity, would destroy vastly more lives than the lightning does. But, in the destruction of sheep by dogs, there is no benefit or advantage of any kind arising to anybody.

"Finally, are dogs of as much benefit to the State in the aggregate as they cost? What this cost is I have endeavored to show, and if any person will show me that they are worth what they cost, I will be much obliged to him for his pains.

"It is no argument to say that the food would have been lost at all events—and that it costs nothing to keep a dog; a hog will eat all the refuse from the kitchen, and drink the swill besides, and pays for its keeping in good fat pork and lard, or if taken to market commands cash. In fact I know several instances where poor men grow rich by keeping hogs, and other instances where men, comparatively well off, grew poorer by keeping dogs."

And now *per contra!*

THE SHEEP DOG.—Buffon thus eloquently describes the sheep-dog, and compares his sagacity and value to man, with other races:—"This animal, faithful to man, will always preserve a portion of his empire and a degree of superiority over other beings. He reigns at the head of his flock, and makes himself better understood than the voice of the shepherd. Safety, order and discipline are the fruits of his vigilance and instinct. They are a people submitted to his management, whom he conducts and protects, and against whom he never applies force but for the preservation of good order. * *
If we consider that this animal, notwithstanding his ugliness, and his wild and melancholy look, is superior in instinct to all others; that he has a decided character in which education has comparatively little share; that he is the only animal born perfectly trained for the service of others; that, guided by natural powers alone, he applies himself to the care of our flocks, a duty which he executes with singular assiduity, vigilance, and fidelity; that he conducts them with an admirable intelligence, which is a part and portion of himself; that his sagacity astonishes at the same time that it gives repose to his master, while it requires great time and trouble to instruct other dogs for the purposes to which they are destined; if we reflect on these facts, we shall be confirmed

in the opinion that the shepherd's dog is the true dog of Nature, the stock and model of the whole species."

I shall call attention to but a few of the most distinguished varieties of the sheep dog.

ARROGANTE—A SPANISH SHEEP DOG.

THE SPANISH SHEEP DOG.—The cut above affords a faithful representation of a thorough-bred Spanish Sheep Dog imported with a flock of Merino sheep a number of years since into England.

Soon after Arrogante's arrival in England, a ewe under his charge chanced to get cast in a ditch, during the temporary absence of the Spanish shepherd, who had accompanied the flock and dog at their importation. An English shepherd, in a spirit of vaunting, insisted on relieving the fallen sheep, in preference to having the absent shepherd called, though warned by his companions to desist. The stern stranger dog met him at the gate and also warned him with sullen growls, growing more menacing as he approached the sheep. The shepherd was a powerful and bold man, and felt that it was too late now to retract with credit. On reaching the sheep, he bent carefully forward, with his eyes on the dog, which instantly made a spring at his throat. A quick forward movement of his arm saved his throat, but the arm was so dreadfully lacerated that immediate amputation became necessary. To save the dog, which had but done his duty, as he had been taught it, from the popular excitement, he was shipped in a vessel which sailed that very afternoon, from Bristol for America. He was sent to Francis Rotch, Esq., then a resident of New-Bedford, Massachusetts.

Fifteen or sixteen years ago, when I was writing "Sheep Husbandry in the South," Mr. Rotch wrote to me as follows:

"I have, as you desired, made you a sketch of the Spanish sheep dog Arrogante, and a villainous looking rascal he is. A worse countenance I hardly ever saw on a dog. His small, blood-shot eyes, set close together, give him that sinister, wolfish look, which is most unattractive; but his countenance is indicative of his character. There was nothing affectionate or joyous about him. He never forgave an injury or an insult; offend him, and it was for life. I have often been struck with his resemblance to his nation. He was proud and reserved in the extreme, but not quarrelsome. Every little cur would fly out at him, as at some strange animal; and I have seen them fasten for a moment on his heavy, bushy tail, and yet he would stride on, never breaking his long, 'loping,' shambling trot. Once I saw him turn, and the retribution was awful! It was upon a large, powerful mastiff we kept as a night-guard in the Bank. He then put forth his strength, which proved tremendous! His coat hung about him in thick, loose, matted folds, dirty and uncared-for —so that I presume a dog never got hold of anything about him deeper than his thick, tough skin, which was twice too large to fit him anywhere, and especially around the neck and shoulders. The only other evidence of his uncommon strength which I had observed, was the perfect ease with which he threw himself over a high wall or paling, which often drew my attention, because he seemed to me wanting in that particular physical development which we are accustomed to consider as necessary to muscular power. He was flat-chested, and flat-sided, with a somewhat long back and narrow loin. (My drawing foreshortens his length.) His neck, forearm and thigh certainly indicated strength. If the Spanish wolf and the dog ever cohabit, he most assuredly had in him such a cross; the very effluvia of the animal betrayed it. In all in which he differed from the beautiful Spanish shepherd dog, he was wolfish, both in form and habits. But, though no parlor beauty, Arrogante was unquestionably a dog of immense value to the mountain shepherd. Several times he had met the large wolf of the Appenines, and without aid slain his antagonist. The shepherds who bred him said it was an affair of no doubtful issue, when he encountered a wolf single-handed. His history, after reaching England, you know."

I have been unable to procure any new portrait, known to be authentic, of a dog of this breed. The American editor of Mr. Youatt's work on the Dog, (Dr. Lewis,) states the Spanish sheep dog "is of the same breed" as the great Alpine Spaniel or "Bernardine dog" which is employed by the monks of St. Bernard in rescuing travelers among the storms and avalanches of the Alps. I have seen several of these, and Arrogante resembles them as nearly as can a spare, attenuated, ugly man resemble one of massive proportions and noble countenance — the height, length, contour, loose hide, etc., are the same.* But while I strongly incline to credit Dr. Lewis' assertion of the identity of the breeds, I have not felt authorized to give a portrait of a Swiss dog as characteristic of a race of Spanish dogs.

Arrogante proved himself an animal of immense value. Dull, almost stupid, and apparently sleeping much of the day, nothing, however, escaped his observation, or was subsequently erased from his memory. If led round a building, or inclosure, or even an open space, at night-fall, in a manner to evince particular design, during the entire night like a sentinel he traversed some part of the guarded ring, permitting neither man nor beast to pass in or out from it. When miserable curs intruded on his charge, they were slain in an instant. He possessed almost human intelligence in protecting property of every kind belonging to his master. But, though never the aggressor, the terrible vindictiveness of his temper, when injured, finally cost him his life.

Mr. Trimmer, in his work on the Merinos, thus describes the mode of employing the Spanish Sheep Dog:— "There is no driving of the flocks; that is a practice entirely unknown; but the shepherd, when he wishes to remove his sheep, calls to him a tame wether accustomed to feed from his hands.

---

* The cut of the Bernardine dog, in Mr. Youatt's work, represents a magnificent animal—but the kind of resemblance I have named between it and Arrogante plainly exists.

The favorite, however distant, obeys his calls and the rest follow. One or more of the dogs, with large collars armed with spikes, in order to protect them from the wolves, precede the flock, others skirt it on each side, and some bring up the rear. If a sheep be ill or lame, or lag behind unobserved by the shepherds, they stay with it and defend it until some one returns in search of it. With us, dogs are often used for other and worse purposes. In open, uninclosed districts they are indispensable, but in others, I wish them, I confess, either managed or encouraged less. If a sheep commits a fault in the sight of an intemperate shepherd, or accidentally offends him, it is *dogged* into obedience, the signal is given, the dog obeys the mandate, and the poor sheep flies round the field to escape from the fangs of him who should be his protector, until it becomes half dead with fright and exhaustion, while the trembling flock crowd together dreading the same fate, and the churl exults in this cowardly victory over a weak and defenceless animal."

Mr. John Hare Powell, in the Memoirs of the Pennsylvania Agricultural Society, describes some Spanish dogs, imported with the early Merinos into this country, and then owned by himself, as possessing "all the valuable characteristics of the English shepherd's dog, with sagacity, fidelity and strength peculiar to themselves." He adds:—"Their ferocity when aroused by any intruder, their attachment to their own flock, and devotion to their master, would, in the uncultivated parts of America, make them an acquisition of infinite value, by affording a defence against wolves, which they ready kill, and vagrant cur dogs, by which our flocks are often destroyed. The force of their instinctive attachment to sheep, and their resolution in attacking every dog which passes near to their charge, have been forcibly evinced upon my farm."*

THE HUNGARIAN SHEEP DOG.—The following description of the Hungarian Sheep Dog, occurs in Paget's "Hungary and Transylvania."†—"It would be unjust to quit the subject of the Puszta Shepherd without making due and honorable mention of his constant companion and friend, the juhasz-hutya—the Hungarian shepherd dog. The shepherd dog is

---

* Mr. Powell's paper is copied into Memoirs of the Board of Agriculture of the State of New York, Vol. 3, 1826. With it is an illustration of a Spanish Sheep Dog, which looks like a cross between a cur and a bull-dog. But it is so completely out of drawing that I am led to infer that it was drawn by a wholly incompetent artist and that it bears no resemblance to the original.

† Hungary and Transylvania, by John Paget, Esq., Vol. 2, p. 12, *et. seq.*

commonly white, sometimes inclined to a reddish brown, and about the size of our Newfoundland dog. His sharp nose, short erect ears, shaggy coat, and bushy tail give him much the appearance of a wolf; indeed, so great is the resemblance, that I have known a Hungarian gentleman mistake a wolf for one of his own dogs. Except to their masters, they are so savage that it is unsafe for a stranger to enter the court-yard of a Hungarian cottage without arms. I speak from experience; for as I was walking through the yard of a post-house, where some of these dogs were lying about, apparently asleep, one of them crept after me, and inflicted a severe wound on my leg, of which I still bear the marks. Before I could turn round, the dog was already far off; for, like the wolf, they bite by snapping, but never hang to the object like the bull-dog or mastiff. Their sagacity in driving and guarding the sheep and cattle, and their courage in protecting them from wolves and robbers, are highly praised; and the shepherd is so well aware of the value of a good one, that it is difficult to induce him to part with it."

I have little doubt that the Hungarian dogs above described are the descendants of the Spanish ones, introduced into Hungary with the Merino sheep, though possibly they may be somewhat crossed by inter-breeding with the dogs of the country.

FRENCH SHEEP DOG.—Professor Grognier gives the following account of this breed:—"The Shepherd's Dog, the least removed from the natural type of the dog, is of a middle size; his ears short and straight; the hair long, principally on the tail, and of a dark color; the tail is carried horizontally or a little elevated. He is very indifferent to caresses, possessed of much intelligence and activity to discharge the duties was designed. In one or other of its varieties it is found in every part of France. Sometimes there is but a single breed, in others there are several varieties. It lives and maintains its proper characteristics, while other races often degenerate. Everywhere it preserves its proper distinguishing type. It is the servant of man, while other breeds vary with a thousand circumstances. It has one appropriate mission, and that it discharges in the most admirable way: there is evidently a kind and wise design in this."

THE MEXICAN SHEEP DOG.—The following account of these noble dogs appears as a communication from Mr. J. H.

Lyman, in the third volume of the American Agriculturist: "Although Mr. Kendall and some other writers have described this wonderful animal as a cross of the Newfoundland dog, such, I think, cannot be the fact: on the contrary, I have no doubt he is a genuine descendant of the Alpine mastiff, or more properly Spanish shepherd dog, introduced by them at the time of the Conquest. He is only to be found in the sheep-raising districts of New Mexico. The other Mexican dogs, which number more than a thousand to one of these noble animals, are the results of a cross of everything under the sun having any affinity to the canine race, and even of a still nobler class of animals if Mexican stories are to be credited. It is believed in Mexico, that the countless mongrels of that country owe their origin to the assistance of the various kinds of wolves, mountain cats, lynxes, and to almost if not every class of four-footed carnivorous animals. Be this as it may, those who have not seen them can believe as much as they like; but eye-witnesses can assert, that there never was a country *blessed* with a greater and more abundant variety of miserable, snarling, cowardly packs, than the mongrel dogs of Mexico. That country of a surety would be the plague-spot of this beautiful world, were it not for the redeeming character of the truly noble shepherd dog, endowed as it is with almost human intellect. I have often thought, when observing the sagacity of this animal, that if very many of the human race possessed one-half of the power of inductive reasoning which *seems* to be the gift of this animal, that it would be far better for themselves and for their fellow creatures.

"The peculiar education of these dogs is one of the most important and interesting steps pursued by the shepherd. His method is to select from a multitude of pups a few of the healthiest and finest-looking, and to put them to a sucking ewe, first depriving her of her own lamb. By force, as well as from natural desire she has to be relived of the contents of her udder, she soon learns to look upon the little interlopers with all the affection she would manifest for her own natural offspring. For the first few days the pups are kept in the hut, the ewe suckling them morning and evening only; but gradually, as she becomes accustomed to their sight, she is allowed to run in a small inclosure with them until she becomes so perfectly familiar with their appearance as to take the entire charge of them. After this they are folded with the whole flock for a fortnight or so, they then run about during the day with the flock, which after a while becomes so accustomed to them as to be able to distinguish them from other dogs — even from those of the same litter which have not been nursed among them. The shepherds usually allow the slut to keep one of a litter for her own particular benefit; the balance are generally destroyed.

"After the pups are weaned, they never leave the particular drove among which they have been reared. Not even the voice of their master can entice them beyond sight of the flock; neither hunger nor thirst can do it. I have been credibly informed of an instance where a single dog having charge of a small flock of sheep was allowed to wander with them about the mountains, while the shepherd returned to his village for a few days, having perfect confidence in the ability of his dog to look after the flock during his absence, but with a strange want of foresight as to the provision of the dog for his food. Upon his return to the flock, he found it several miles from where left, but *but on the road leading to the village*, and the poor, faithful animal in the agonies of death, dying of *starvation*, even in the midst *of plenty;* yet the flock had not been harmed by him. A reciprocal affection exists between them which may put to blush many of the human family. The poor dog recognized them only as brothers and dearly loved friends; he was ready at all times to lay down his life for them; to attack not only wolves and mountain cats, with the confidence of victory, but even the bear, when there could be no hope. Of late years, when the shepherds of New Mexico have suffered so much from Indian marauders, instances have frequently occurred where the dog has not hesitated to attack his human foes, and although transfixed with arrows, his indomitable courage and faithfulness have been such as to compel his assailants to pin him to the earth with spears, and hold him there until dispatched with stones.

"In the above instance the starving dog could have helped himself to one of his *little brother* lambs, or could have deserted the sheep, and very soon have reached the settlements where there was food for him. But faithful even unto death, he would neither leave nor molest them, but followed the promptings of his instinct to lead into the settlement; their unconsciousness of his wants and slow motions in traveling were too much for his exhausting strength.

"These shepherds are very nomadic in character. They are constantly moving about, their camp equipage consisting merely of a kettle and a bag of meal; their lodges are made in a few minutes, of branches, &c., thrown against cross-sticks. They very seldom go out in the day-time with their flocks, intrusting them entirely with their dogs, which faithfully return them at night, never permitting any stragglers behind or lost. Sometimes different flocks are brought into the same neighborhood owing to scarcity of grass, when the wonderful instincts of the shepherds' dogs are most beautifully displayed; and to my astonishment, who have been an eye-witness of such scenes, if two flocks approach within a few yards of each other, their respective proprietors will place themselves in the space between them, and as is very naturally the case, if any adventurous sheep should endeavor to cross over to visit her neighbors, her dog protector kindly but firmly leads her back, and it sometimes happens, if many make a rush and succeed in joining the other flock, the dogs under whose charge they are, go over and bring them all out, but, strange to say, under such circumstances they *are never opposed by the other dogs.* They approach the strange sheep only to prevent their own from leaving the flock, though they offer no assistance in expelling the other sheep. But they *never permit* sheep not under canine protection, nor dogs not in charge of sheep, to approach them. Even the same dogs which are so freely permitted to enter their flocks in search of their own, are driven away with ignominy if they presume to approach them without that laudable object in view.

"Many anecdotes could be related of the wonderful instinct of these dogs. I very much doubt if there are shepherd dogs in any other part of the world except Spain, equal to those of New Mexico in value. The famed Scotch and English dogs sink into insignificance by the side of them. Their superiority may be owing to the peculiar mode of rearing them, but they are certainly very noble animals, naturally of large size, and highly deserving to be introduced into the United States. A pair of them will easily kill a wolf, and flocks under their care need not fear any common enemy to be found in our country."

Mr. Kendall speaks of "meeting, on the Grand Prairie, a flock numbering seventeen thousand, which immense herd was guarded by a very few men, assisted by a large number of noble dogs, which appeared gifted with the faculty of keeping them together. There was no running about, no barking or biting in their system of tactics; on the contrary, they were continually walking up and down, like faithful sentinels, on the other side of the flock, and should any sheep chance to stray from its fellows, the dog on duty at that particular post, would walk gently up, take him carefully by the ear and lead him back to the flock. Not the least fear did the sheep manifest at the approach of these dogs, and there was no occasion for it."

Capt. Allison Nelson, of Bosque county, Texas, visited me in 1860. He had started to bring me a pair of these Mexican dogs, but unfortunately permitted himself "to be laughed out of it"—his friends being under the impression that it would be carrying coals to New Castle to take sheep dogs to a region where the Scotch colley was to be found in abundance. Capt. Nelson confirmed Mr. Lyman's statement in regard to their sagacity and courage. His sheep were herded in the Mexican way, around fires and not in folds. He said that after night-fall the dogs separated themselves from the sheep and formed a cordon of sentries and pickets around them,—and woe to the wolf that approached too near the guarded circle! The dogs crouched silently until he was within striking distance, and then sprang forward like arrows from so many bows. Some made straight for the wolf and some took a direction to cut off his retreat to forest or chaparral. When overtaken his shrift was a short one.

Such dogs would be invaluable on the broad prairies of the North-western States, to save the labor, trouble, and sometimes injury of folding flocks each night in a stationary and distant fold.

SOUTH AMERICAN SHEEP DOG.—Similar to the preceding in character and habits, are the sheep dogs to be found in various parts of South America. They, too, are undoubtedly an offshoot from the Spanish stem. The following interesting account of them is from Darwin's Journal:

"While staying at this estancia (in Banda Oriental,) I was amused with what I saw and heard of the shepherd dogs of the country. When riding it is a common thing to meet a large flock of sheep guarded by one or two dogs, at the distance of some miles from any house or man. I often wondered how so firm a friendship had been established. The method of education consists in separating the puppy, when very young, from the bitch, and in accustoming it to its future companions. A ewe is held three or four times a day for the little thing to suck, and a nest of wool is made for it in the sheep-pen. At no time is it allowed to associate with other dogs, or with the children of the family. The puppy, moreover, is generally castrated: so that when grown up, it can scarcely have any feelings in common with the rest of its kind. From this education it has no wish to leave the flock, and just as another dog will defend its master, man, so will these the sheep. It is amusing to observe, when approaching a flock, how the dog immediately advances barking—and the sheep all close in his rear as if around the oldest ram. These dogs are also easily taught to bring home the flock at a certain time in the evening. Their most troublesome fault when young is their desire of playing with the sheep, for in their play, they sometimes gallop their poor subjects most unmercifully. The shepherd dog comes to the house every day for some meat, and immediately it is given to him he skulks away as if ashamed of himself. On these occasions the house dogs are very tyrannical, and the least of them will attack and pursue the stranger. The minute, however, the latter has reached the flock, he turns round and begins to bark, and then all the house dogs take very quickly to their heels. In a similar manner a whole pack of hungry wild dogs will scarcely ever (and I was told by some, never,) venture to attack a flock guarded even by one of these faithful shepherds. The whole account appears to me a curious instance of the pliability of the affections of the dog race; and yet, whether wild, or however educated, with a mutual feeling of respect and fear for those that are fulfilling their instinct of association. For we can understand on no principle the wild dogs being driven away by the single one with its flock, except that they consider, from some confused notion, that the one thus associated gains power, as if in company with its own kind. F. Cuvier has observed that all animals which enter into domestication consider man as a member of their society, and thus they fulfill their instinct of association. In the above case the shepherd dogs rank the sheep as their brethren; and the wild dogs, though knowing that the individual sheep are not dogs, but are good to eat, yet partly consent to this view, when seeing them in a flock, with a shepherd dog at their head."

OTHER LARGE RACES OF SHEEP DOGS.—There are one or two fine species in France, as those of Brie and Auvergne. In a letter from G. W. Lafayette, to John S. Skinner, Esq., the latter are pronounced equal to Spanish dogs.* Large,

---

* See Farmers' Library, Vol. I, p. 405.

powerful races, possessing the same general characteristics, are to be found in almost every country excepting our own, where the fine-wooled breeds of sheep have been extensively introduced. With a commerce extending to all the maritime nations of the world, it is singular that so little pains have been taken to introduce them.

THE ENGLISH SHEEP DOG.— The following cut presents an accurate portrait of an animal of this breed, imported by Mr. B. Gates, of Gap Grove, Lee county, Illinois. It is taken from The Farmer's Library:

DROVER'S DOG.

The Drover's Dog, or English Sheep Dog, or Butcher's Dog— for by all these different names is he known — is thus described by Mr. Theodore C. Peters, of Darien, New York, in third volume of the American Agriculturist, 1844:

"I purchased a bitch of the *tailless species*, known as the English drover dog, in Smithfield market, some two years ago. That species is much used upon the downs, and is a larger and fleeter dog than the Colley. We raised two litters from her, got by Jack, [a Colley,] and I think the cross will make a very valuable dog for all the purposes of the farmer. They learn easily, are very active, and so far they fully answer our expectations.

"A neighbor to whom we gave a bitch of the first litter, would tell her to go into such a lot and see if there were any stray cattle there; and if there were any there, detect them and drive them down to the house. He kept his cattle in the lot, and it was full eighty rods from the house. The dog was not then a year old. We had one of the same litter, which we learned to go after cows so well, that we had only to tell him it was time to bring the cows, and he would set off for them from any part of the farm, and bring them into the yard as well as a boy. I think they would be invaluable to a farmer on the prairies. After raising two litters, we sent the bitch to Illinois. I hope farmers will take more pains in getting the shepherd dog. There is no difficulty in training. Our old one we obtained when a pup, and trained him without any trouble, and without the help of another dog. Any man who has patience, and any *dog knowledge* at all, can train one of this breed to do all that he can desire of a dog."

THE SCOTCH SHEEP DOG OR COLLEY.—The light, active, sagacious Colley admits of no superior — scarcely an equal — where it is his business merely to manage his flock, and not to defend them from beasts larger than himself. Mr. Hogg says that a "single shepherd and his dog will accomplish more in gathering a flock of sheep from a Highland farm than twenty shepherds could do without dogs. Neither hunger, fatigue, nor the worst treatment will drive him from his master's side, and he will follow him through every hardship without murmur or repining."

The same well known writer, in a letter in Blackwood's Magazine, gives a most glowing description of the qualities of his Colley, "Sirrah." One night a flock of lambs, under his care, frightened at something, made what we call in America a regular *stampede*, scattering over the hills in several different bodies. "Sirrah," exclaimed Hogg in despair, "they're a' awa!" The dog dashed off through the darkness. After spending with his assistants the whole night in a fruitless search after the fugitives, Mr. Hogg commenced his return to his master's house. Coming to a deep ravine, they found Sirrah in charge, as they first supposed, of one of the scattered divisions, but what was their joyful surprise to find that not a lamb of the whole flock was missing!

THE COLLEY.

Mr. Peters, in the same paper from which we have just quoted, thus speaks of the Colley:—"I think the shepherd dog the most valuable of his species, certainly for the farmer. Our dog Jack, a thorough-bred Scotch Colley, has been worth $100 a year in managing our small flock of sheep, usually about seven hundred in number. He has saved us more than that in time in running after them. After sheep have been once broken in by, and become used to the dog, it is but little trouble to manage them; one man and the dog will do more than five men in driving, yarding, &c. Let any man once possess a good dog, he will never do without one again.

"The sagacity of the shepherd's dog is wonderful; and if I had not seen so much myself, I could hardly credit all we read about them. It is but a few days since I was reading in a Scotch paper a wonderful performance of one of these

Colley dogs. It seems the master of the bitch purchased at a fair some eighty sheep, and having occasion to stay a day longer, sent them forward and directed his faithful Colley to drive them home, a distance of about seventeen miles. The poor bitch when a few miles on the road dropped two whelps; but faithful to her charge, she drove the sheep a mile or two farther — then allowing them to stop, she returned for her pups, which she carried some two miles in advance of the sheep, and thus she continued to do, alternately carrying her own young ones, and taking charge of the flock, till she reached home. The manner of her acting on this occasion was gathered by the shepherd from various persons who had observed her on the road. On reaching and delivering her charge, it was found the two pups were dead. In this extremity the instinct of the poor brute was yet more remarkable; for, going immediately to a rabbit brae in the vicinity, she dug out of the earth two young rabbits, which she deposited on some straw in a barn, and continued to suckle them for some time, until they were unluckily killed by one of the farm tenants. It should be mentioned that the next day she set off to the place where she left her master, whom she met returning when about thirteen miles from home."

I have to make a sad draw-back on these statements. It is well known in the region of New York where I reside, and where the Colley dog is quite common, that it is sometimes — under the instruction of vicious associates perhaps — taught in its youth to kill sheep: and when this occurs, it is proverbial that the sheep has no other so fell and destructive canine enemy. Its extreme activity, and the keenness of its bite, causes a wholesale slaughter. Two dogs of this kind killed eight Merino ewes for me this year, and had they not fortunately been detected at the outset of their attack, they would soon probably have added fifty to the number of their victims. When first seen they were darting about, biting one sheep after another — a single touch of their teeth being apparently sufficient to strip off half the skin — as if they were committing the havoc solely for their amusement, and were prompted neither by hunger nor thirst. Indeed, I ascertained from their owners that they had both been well fed within an hour of the time of their entering the flock. They were moreover habitually well fed dogs, and were in excellent case. I think the mongrel Colley learns to kill sheep as readily as a cur; but whether this is true of the pure blood dog, I am not prepared to say.

ACCUSTOMING THE SHEEP TO THE DOG. — It is a mistake to suppose that a trained sheep dog will manage any strange flock, however wild and unaccustomed to such company. The sheep must be gradually made acquainted with, and accustomed to, the dog. They must know — and they will readily learn it — that he is their friend, their guardian and protector, instead of that hereditary enemy which their instinct teaches them to fly from. A want of knowledge of this fact has frequently led to disappointment and disgust, to a giving up of the valuable dog which it has cost pains and money to procure. My friend, the late Col. John S. Skinner, related to me a ludicrous accident which befel President Jefferson, or rather his sheep dogs, when he undertook to show off some newly imported ones, *a la philosopher*, without being apprized of the above-mentioned fact. The tale is told in my Sheep Husbandry in the South. The comedy turns on the fact that the great political sage took out some admiring visitors to witness the wonderful exploits of his dogs: "let" them "slip" on some raw ovine subjects, whereupon the latter dashed themselves over precipices, &c.: and the "valuable dog which it had cost pains and money to procure," was so mortified at the proceeding that he ran the other way, was never again heard of, and is supposed by some to be running to this day!

As in the case of so many "good stories," there was not a word of truth in it! Some years after my publication of it, I chanced to be in conversation with Mr. Jefferson's family on this very subject and learned that the dogs were sent to him from France — that they were admirably broken and possessed almost human intelligence — that neither of them ever brought man or beast to grief, except that the bitch, who took it upon herself to herd the *hens* every night, insisted on doing it about half an hour before the latter wished to retire for the night — and *they* sometimes made *loud* complaints on the subject!

APPENDICES.

APPENDIX A — (page 122.)

ORIGIN OF THE IMPROVED INFANTADOS.

To gratify the curiosity of some thorough-paced Merino sheep breeders, as well as to illustrate the rapid "march of improvement," when the right animals are bred together, I will present a few facts culled from a large body of notes in my possession, giving full descriptions of the leading animals named in the pedigrees on page 121, and in the remainder of Mr. Hammond's flock.

"Old Black," was bought of S. Atwood, by Mr. Sanford, of Orwell, and was owned and used by him and Mr. Hammond together. He weighed about 135 lbs., and yielded about 14 lbs. of wool. (Unless otherwise stated, all fleeces named here will be understood to be unwashed.) He was long, tall, flat-ribbed, rather long in the neck and head, strong-boned, a little roach-backed, deep-chested, moderately wrinkled: his wool was about 1¼ inches long, of medium thickness, extremely yolky, and dark colored externally: face a little bare, and not much wool on shanks. He did not possess a very strong constitution. He proved an admirable sire of ewes, but was not so good for rams.

"Old Matchless" run well into the blood of Mr. Atwood's lighter colored sub-family, though he himself was darkish colored. (Mr. Atwood had either found two such sub-families in the Humphreys' sheep, or he had gradually created and established them in his flock to attain certain breeding objects: I think the latter was the case.) He weighed about 150 lbs., was a sheep of excellent form, commanding appearance, and strong constitution. He yielded 10¼ lbs. of wool when a lamb, but his usual fleece afterwards was only 12 or 13 lbs. His fleece was about two inches long, coarsish, of medium thickness, pretty yolky — but thin and short on the belly. He was not well covered on the head, and was bare on the shanks. He got large, strong, but not very well covered lambs. He was not as good a stock ram as Old Black. He died early.

"Wooster" weighed about 100 lbs. He was well shaped and compact, with short legs, a short, thick head, and neck of medium length and thickness. He was very heavily wrinkled under the neck, and also at the elbow and tail. His wool was nearly two inches long, quite thick, very dark and yolky. He was well covered on belly and foretop, and middling well on the face. He yielded 19¼ lbs. of wool at two years old. He was an excellent stock getter, and bred extremely well with the light colored ewes. He sold a lamb for $300, but Mr. H. continued to use him. (See page 113.)

"Old Greasy" weighed about 110 lbs. He was light boned and rather long and thin in every part, though the rib was tolerably full. He was but little wrinkled, having simply the cross on the brisket, the convolution of skin under the chops called by many "the double," and a narrow dewlap between them. He was exceedingly yolky, and his wool very long and thick for so yolky a sheep. The wool was about $2\frac{1}{4}$ inches long, was fine and even, covered belly and foretop fairly, but not the shanks, and the fleece weighed 22 lbs. His constitution was medium, and he was an excellent stock getter, so far as fleece was concerned. He was used to darken the produce of the light colored ewes.

"Old Wrinkly" weighed from 125 to 130 lbs., and was a strong-boned, low, compact sheep, with round carcass and short legs, short thick head and neck, but was a little too light in the hips. He was very heavily wrinkled over and under neck, and also about elbow, tail, thigh and flank. His flank was deep and tail broad. His fleece was thick, about two inches long, of medium quality, not entirely even, and showed a little jar on the neck wrinkles. He was well covered on head and belly, and wooled to the foot. His fleece weighed 23 lbs. It was rather light colored, though very yolky. His yolk was yellow. The wool opened well. He had a strong constitution, and was a good sire ram. He was sold for $300.

"Little Wrinkly" weighed about 110 lbs. He had bones of good size, was about medium in respect to compactness, and was round in the rib. He was much less wrinkly than Old Wrinkly, and was inferior to him in general appearance. His fleece was very fine and even, and possessed a good deal of style. It was of medium length, (two inches long,) thick, and coated with dark external gum. He was as yolky as Old Greasy, and his yolk white. His fleece weighed about $19\frac{1}{4}$ lbs., a good deal of weight considering its quality. He would not have been used had Long Wool or Old Greasy been alive; yet he proved a good stock ram, in some cases, getting Sweepstakes and two large, very heavy fleeced ewes. He got them when a lamb. He died at three.

"Sweepstakes" weighs about 140 lbs. Taken all in all he is about as perfect a formed Merino ram as was ever seen, and defective in no essential particular. His wool is $2\frac{1}{4}$ inches long, fine, extremely even, and does not contain a particle of jar. His belly, head, etc., are admirably covered, and he is wooled profusely to the feet all round. He has no external gum, is medium in point of color, but possesses abundance of thin, yellowish yolk. His wool opens brilliantly and with a beautiful style. He has produced a single year's fleece of 27 lbs. His constitution is powerful. He impresses his own characteristics unusually strongly on his get. He took the first premium of the Vermont State Agricultural Society as a lamb, as a yearling, and as a grown ram. In 1861 he met several of the best rams of the State (the best of his competitors were got by himself) in a sweepstakes, and was victorious. Mr. Hammond has been several times offered $2,500 for him.

"California," the next named ram in the pedigree published at page 121, was the property of Henry Hammond, as is his dam Beauty 1st. (His stock is the same with that of his uncle, Edwin Hammond, being half of the same common flock.) California was sold for $1,000, and I think was less than a year old when sold. I have no description of him. His dam brings to him and to Gold Drop the blood (individual blood) of several very celebrated animals which do not appear in the pedigree of Sweepstakes, viz., Young Matchless, the Lawrence Ewe, Long Wool and Old Queen.

"Young Matchless" was in the light colored line. He weighed about 150 lbs. He was a model of strength, compactness, symmetry and showiness. He had immense constitution. He was well wrinkled under the neck, at the elbow, thigh and tail. His fleece was about 2½ inches long, extremely thick, of medium quality, of good style, even and had no jar. It covered him well on belly, head, legs, etc. He was particularly well wooled over the eye. He was rather light colored. He was less yolky than any ram heretofore described, and his yolk was white. His fleece weighed 23 lbs., and is believed to have contained more *pure* wool than that of any other ram Mr. Hammond ever owned except Sweepstakes. He gave his get great length and thickness of wool, and the *great round carcass* so conspicuous in the flock. He took the first State premium, &c. A half interest in him and Greasy was sold to Wm. R. Sanford for $500.

"The Lawrence Ewe" combined the size, beauty, constitutional vigor and wooliness of both her sire and dam. She weighed about 110 lbs., and did not lack a single property of excellence or showiness. She was dark externally, yolk yellowish, and had some external gum. Her fleece was of good quality, and weighed 14 lbs. She was sold for $600, which was esteemed a remarkable price at that day. She was the dam of two very famous rams, viz., Long Wool and the Lawrence Ram.

"Long Wool" took something of his form from his sire, and accordingly was not as low, compact and round as his immediate maternal ancestors, but he was considerably better formed than Old Greasy. He weighed from 125 to 130 lbs. His wrinkles, &c., resembled his sire's, but he had more of them, and some small ones about elbow and tail. His fleece was about 2½ inches long, very thick, yolk white and brilliant, style excellent. He was wooled to the feet all round, well wooled on the belly and head. He was not quite as well wooled over eye as Young Matchless or the Queen family — but did more to improve this point among the Queens than any other ram. His fleece was dark colored. No memorandum is preserved of its weight: it was over 20 lbs. He was an admirable sire for ewe lambs — the best, perhaps, Mr. H. ever had. They were long and thick wooled, dark externally, and particularly well covered. He improved the flock, especially in wool over the eye. His lambs were also low, round, thick and of strong constitution. Mr. H. declined $500 for him when two years old. He was killed early, in fighting.

"The Lawrence Ram" is not named in the pedigrees, published on page 121, but has been one of the most celebrated rams of the flock. He was got by Old Wrinkly, dam, the Lawrence ewe. He weighed about 130 lbs. He was a short, stout, heavy-boned, low sheep, with a remarkably short and heavy neck, and a broad loin and rump. He had a powerful constitution. He was heavily wrinkled in front, with folds at elbow, tail, thigh and flank. He was dark colored and yolky. His wool was of medium length, (two inches,) very thick, of medium quality, even, and the yolk yellowish. He was well covered on face, belly, &c. His fleece weighed 24 lbs. He was a capital sire for both ram and ewe lambs. The heaviest fleeced ewes now in Mr. H.'s flock were got by him. He was sold in his old age for $200.

"Old Queen" is but two removes from the "First Choice of Old Ewes," and is considered by her owner the mother of more valuable sheep than any other ewe ever owned by him.

"First Choice of Old Ewes" was of the medium size of Atwood ewes of that day, weighing about 80 lbs. She was fine in the bone, of about medium length, with a short, wide head. Her general form was compact, and good, with the exception of a slight flatness in the ribs. She was but little wrinkled, having only the cross and double with a dewlap between. Her wool was hardly two inches long, but was fine, even, thick, dark, and well filled with white yolk. It covered her well on belly, but she was bare on the forehead compared with the sheep of the present flock, and had not much below the knees. Her washed fleece weighed about five pounds. She proved an extraordinary breeder, and her line — the "dark or Queen line" — has always been carefully preserved.

The "Light Colored Ewe" weighed 85 or 90 lbs. She was shortish, very square built, with a short, thick head and neck, medium length of leg, and rounder ribs than most of the Atwood sheep. She was high headed, had the cross and double with dewlap between and under the chops. Her wool was about 2¼ inches long, very thick, and covered her well on the face and belly. She was wooled to the foot. Her fleece was even but not very fine. It was light colored and rather destitute of yolk. Her fleece weighed about six lbs. washed. She was an excellent breeder, but not regarded as equal to the First Choice of Old Ewes, in this particular. She was the origin of the "light colored line," always preserved in the flock to interbreed with the "dark or Queen line."

"First Choice of Ewe Lambs," at maturity, weighed from 90 lbs. to 95 lbs. She was strong boned, low, short, and thick in every part except the neck, which was slightly too long and thin. Her ribs were well arched. She had the cross on the brisket, but no double or dewlap, and was smooth under the chops. She was regarded, however, as the best formed sheep, on the whole, bought of Mr. Atwood, and also the best covered one. She was well wooled on the belly, head and shanks. Her fleece was about two inches long, dark externally, and well filled with white yolk. Crossed in the Queen line, she produced Wooster: crossed in the light colored line, she produced the Lawrence ewe. She died early. Her blood was lost to the flock by the sale of Wooster and the Lawrence ewe — but brought back by Mr. Hammond's putting ewes to the Wooster ram, and by his subsequently re-purchasing the Lawrence ram.

I have not space here to follow out the course of breeding between the three lines which has led to such extraordinary improvement. The best sheep of the flock have always been produced by interbreeding between them. The mode in which Sweepstakes unites the three strains will be seen from his pedigree at page 121. "21 per Cent.," so often named in this work, unites them through some of the most celebrated animals of each line. He was got by the Lawrence ram; dam, Old Tulip, an own sister of Old Queen. The "Thousand Dollar Ram" now owned by Mr. Asahel F. Wilcox, of Fayetteville, New York, was got by Sweepstakes out of Old Queen's dam. "Wrinkly 3d," now owned by Capt. Davis Cossit, Onondaga, New York, was got by Sweepstakes, dam, Countess, by Little Wrinkly—Countess' dam in the light colored line, &c., &c.

The first *great* change in Mr. Hammond's weight of fleeces was made by Young Matchless; and he equally improved the form, size and constitution. His only deficiency was in yolk, and consequently in dark color, and his get resembled him in that particular.

"Old Greasy" and "Long Wool," and particularly the latter, made a marked improvement in the fleece. They added materially to its yolkiness, and consequently to its dark, external color, without either shortening it or rendering it thinner; and they also added to its fineness and style. They both gave better forms to their progeny than their own, but Old Greasy's get were sometimes deficient in this particular. Long Wool did not deteriorate the form, particularly in his female get. Old Greasy gave a good, and Long Wool an excellent, constitution to his descendants.

"Old Wrinkly" rendered the flock more stocky, and wrinkly, and shorter in the legs, head, &c.

The "Lawrence ram" got large, strong, round carcassed, and well-formed offspring — possessing a remarkable constitution. His get on ewes by Greasy and Long Wool were as dark colored as their dams, and had heavier fleeces.

"Sweepstakes" has done much to harmonize the different strains of blood and give uniformity to the flock — improving defects where they existed. In the external color of their wool, he, 21 per Cent., and the Thousand Dollar Ram, are about midway between the light and dark colored lines — the point where weight of fleece and bodily development are best combined.

APPENDIX B — (page 128.)

ORIGIN OF THE IMPROVED PAULARS.

THE following is a full, and it is believed, accurate account of the crosses of blood contained in some of the principal improved Paular stocks of the present day, with such notices as I could obtain of the leading animals in the establishment of the crossed family:

In 1844, Judge M. W. C. Wright, of Shoreham, Vermont, bought a ram bred by Mr. Stephen Atwood, and brought by him to the New York State Fair, held that year at Poughkeepsie. Mr. Hammond, of Vermont, and myself, were present at the purchase. My recollections of him entirely coincide with those of Judge Wright, and his subsequent owners, Messrs. Elithorp and Remelee. He did not weigh, with his fleece off, to exceed 100 lbs. "He was," Mr. Elithorp writes me, "a low, short-legged, square-built sheep, short-bodied, short and rather heavy-necked, with a few moderate-sized folds about the neck, and a brace or fold [of pendulous skin] extending from his hind-leg to his flank. He was flat on the back, had a deep chest, and possessed a good constitution." His fleece was fine, glossy, even, highly crimped, thick and "long for an Atwood sheep in those days." It covered his head and belly unusually well, and extended to his hoofs, "making his legs look short and heavy." His yolk was abundant, entirely fluid, and white in color; and his external color was very dark for a sheep unhoused in summer.

His fleece in 1845, of two years' growth, weighed 22 lbs. unwashed; his subsequent fleeces ranged from 13 lbs. to 15 lbs., and averaged about 14 lbs. He was an admirable sire ram with ewes of all descriptions, stamping his individual characteristics strongly on his progeny.

On his return with this sheep to Vermont, Judge Wright sold him to Prosper Elithorp, of Bridport, and Loyal C. Remelee, of Shoreham, after reserving to himself the use of him for a certain period that fall; and he also used him in part for two succeeding years. He was thenceforth called the "Atwood ram." He got the "Elithorp ram" out of a ewe bred by Mr. Remelee, and sold by him to Mr. Elithorp. This ewe was got by Judge Wright's "Black Hawk," out of a pure Jarvis ewe purchased by Mr. Remelee of Mr. Jarvis. The dam and grand-dam of the Elithorp ram, writes Mr. Elithorp, "were essentially Jarvis sheep in their appearance, except that they carried darker coats on the outside, and their wool was thicker set. It was long, fine, splendid wool. They were good shaped and hardy for Jarvis stock." The Elithorp ram "weighed from 130 to 140 lbs., in good condition: was formed considerably like his sire except that he was more leggy; his wool was long and fine, resembling the Jarvis wool, except in its mode of opening, which was not in ringlets, but in flakes up and down." It "covered him well, was not yolky to excess, was heavy for those days, but its precise weight is not remembered." He was also an excellent stock ram. Judge Wright's Black Hawk was got by "Fortune" out of a pure Jarvis ewe purchased by Judge W. of Mr. Jarvis. "Fortune" was bred by Tyler Stickney, of Shoreham, and got by "Consul" out of a pure Paular (Rich) ewe. Consul was a pure Jarvis ram purchased by Mr. Stickney of Mr. Jarvis. Black Hawk, Fortune, (for a long time owned by S. W. Jewett,) and Consul, were all highly celebrated animals in their day, the two first especially. Fortune was sold for a higher price than any ram of his day. His dam was an exceedingly choice animal.

Mr. Elithorp sold the Elithorp ram, then a lamb, in the fall of 1845, to Erastus Robinson, of Shoreham, Vermont. While owned by Mr. Robinson, he got the "Old Robinson ram" out of a ewe bred by Mr. Elithorp, and sold by him with 29 others to Mr. Robinson in the spring of 1848. This ewe was got by the Atwood ram, above mentioned, out of a pure Paular (Rich,) ewe bred by Mr. Robinson and sold by him to Mr. Elithorp in the fall of 1843. She was the second choice of Mr. R.'s flock. "She (the grand dam of the Old Robinson ram,) was a model in every particular that constitutes a good sheep, except size, which was below medium, and she had quite short legs." Her daughter (the dam of the Old Robinson ram) was a counterpart of her, except that she was a good size larger." Both "were heavy shearers, yielding from 8 lbs. to 9 lbs. each of white, glossy wool. They were peculiar for heavy caps on their foreheads, short, bull-dog noses, thick ears, and very short necks. They had no short wool on their noses or ears, but were coated on these parts with white glossy hair." The Old Robinson ram "partook of the strong characteristic points of his dam" in carcass, "while his fleece was more of the Atwood stamp. His legs, like those of his dam and grand dam, were *very* short." Judge Wright describes him as "a small ram, (weighing about 100 lbs.,) low in the leg, with a heavy neck and a large and deep chest, covered with large folds or corrugations from his head to his tail. His wool was of medium length, compact, almost too fine, and covered him to the hoofs. He partook of many of the qualities of his

18*

grand-sire, the Atwood ram: he had a large amount of yolk; it was creamy, and of course his fleece partook of that color in the inside. On the outside it was quite dark." When five years old, says David Cutting, who sheared him that year, he yielded 11 lbs. 11 oz. of wool. Mr. Stickney, who purchased him of his brother-in-law, Mr. Robinson, in about 1855, and who was familiar with him all his life, informed Judge Wright "that he was very uniform in his weight of fleece, and that its average weight was about 14 lbs." (unwashed.)

This ram, in the hands of Mr. Robinson and Mr. Stickney, got an immense number of lambs, which were very strongly marked with his own characteristics. They were generally small, short, and exceedingly compact, with fine, yolky, and for those times, heavy fleeces. They became great favorites, and sold far and near under the name of the "Robinson stock." This was an obvious misnomer, as Mr. Robinson, (a valuable man and intelligent breeder,) was not the founder of either of the three American families which constituted the new family, or the originator of the cross that produced it. Messrs. Robinson and Stickney commenced their original flocks with prime Rich sheep, purchased from a member of that family. In 1845, Mr. Robinson bred 20, and in 1846, 23 of his ewes to the Atwood ram, owned by Mr. Elithorp. In the spring of 1848 he bought 30 ewes of Mr. Elithorp, "a majority of which were Atwood and a cross of Atwood and Rich — with some Jarvis blood in a small number of them." These are believed to be nearly as many as the other ewes then owned by him; and he thenceforth bred the flocks together, using first the Elithorp ram, and the Old Robinson ram when he became old enough, with them. The flock at Mr. Robinson's death contained about an equal amount of Paular (Rich) and Infantado (Atwood) blood, and it was very celebrated for its excellence. The Stickney branch of the family contained a larger proportion of the Paular blood. The old Rich flock proper was crossed somewhat with the Atwood blood, as I have mentioned while describing them.

Mr. Elithorp, from whom I have derived most of the above account of his own and Mr. Robinson's flocks, is, by the common voice of his fellow-citizens, a judicious breeder and excellent judge of sheep. And his candor and integrity are wholly above suspicion.

APPENDIX C — (page 242.)

ENGLISH EXPERIMENTS IN FEEDING SHEEP.

The following accounts of further experiments in feeding sheep are selected from Mr. T. E. Pawlett's already cited Essay on the Management of Sheep, which received the commendation of the Royal Agricultural Society of England. Mr. Pawlett says:

"The following experiments were all made with sheep of the Leicester breed; and before I proceed further (that I may not be misunderstood, as some of my statements may appear surprising to those unaccustomed to make experiments and weigh sheep,) I shall state the weight of Swedes, &c., &c., sheep and lambs will daily consume; also the live weight they will generally gain in four weeks, according to their age and season of the year. I am enabled to do this without much fear of contradiction, as I have been in the habit of regularly weighing my sheep and lambs nearly every month for more than twenty years.

An ewe lamb-hog in the month of February will eat of cut
Swedes in twenty-four hours, about................... 18 lbs.
A wether lamb-hog. 20
A ram lamb-hog.. 22
A shearling wether-.. 22
A feeding or breeding ewe................................. 24
A sucking ewe... 28
A ram above two years old................................. 30

— no other food but cut Swedes being given to them: but if the weather is mild or warm they will not eat so much as I have stated by about one-fourth. If corn or oil-cake, or any other dry food is used, they will consume less turnips in proportion to the quantity given. I have found that by giving sheep one pint of beans each per day, they will not require so many Swedes by about five lbs. or six lbs. each.

"Lambs and sheep managed and fed well, if in small lots, will gain in live weight each on the average per month:

Young lambs in the month of April	9 lbs.	Lambs in the month of October	12 lbs.
May	16	November	8
June	18	December	6
July	15	January	5
August	12	February	7
September	12	March	10

— being about 130 lbs., in twelve months, of live weight, or about 84 lbs. of mutton. Some lambs will, however, greatly exceed in gain the weights that I have stated. * * * * *

"In carrying out these experiments, I was obliged, for the most part, to keep them in small yards, a system which I am generally opposed to (for any length of time;) believing that sheep and all other animals ought, as far as regards situation, to be kept in a state as near approaching to that which nature assigned for them as possible, provided always that their lair be kept clean and dry, and shelter allowed them from the cold winds. When yards must be used for sheep, they ought always to be kept tolerably free from manure, well littered, and to have plenty of fresh air.

"*Experiment No.* 1.— In the month of March, 1845, I selected twelve couples from the flock; the lambs being then about a fortnight old. These were divided into two lots, as equally as well could be with respect to the condition of the young lambs, and put into two separate small yards. Six of them were fed on clover-hay chaff entirely; the other six couples had 140 lbs. of cut Swedes, and half a peck of beans daily; both lots having water. At the end of the trial of about a month, the lambs of each were carefully compared; and those certainly looked the best and most thriving whose dams had been fed only on clover-hay chaff.

The six ewes fed on cut Swedes, ate 140 lbs., or 1¼ cwt. every day,   *d.*
   at 6d. per cwt., cost per week each........................8¼d.
Also, half a peck of beans daily for the six ewes, at 4s. per
   bushel, cost each per week, 7d; making the cost of keeping
   each ewe per week 15¼
The six ewes fed on clover-hay chaff only, ate daily 21 lbs., or 3½
   lbs. each, at 6d. per stone, or £4 per tun, cost per head per
   week.. 10½

*   *   *   *   *   *   *   *   *

"*Experiment No. 2.*— Being desirous to prove further the value of clover-hay chaff for ewes and lambs, I again selected twelve couples from the flock, and divided them equally into two lots; they were also put into separate small yards. On the 3d of April, 1845, the lambs being weighed alive:

Six couples were fed on 9 lbs. of bran daily, which cost per head   *d.*
   for each ewe per week, 5¼d.; they had also 15 lbs. of clover-
   hay chaff daily at 6d. per stone, cost each ewe per week 7¼d.,
   making the total cost of keeping each couple per week...... 13
The other six couples were fed on clover-hay chaff only, and ate
   3¼ lbs. each ewe per day, at 6d. per stone, cost............. 10¼

All the lambs were weighed again on the 17th of April, and the result was as follows:

Six lambs, whose dams were fed on clover-hay chaff and bran,   lbs.
   gained each on the average in 14 days....................... 6
Six lambs, whose dams were fed on clover-hay chaff only, gained
   in the same time.. 4¾

A difference is here shown of 1¼ lb. of live weight per lamb in favor of the use of bran, but when the cost of it is taken into consideration there does not appear to be much advantage in the use of it.

"*Experiment No. 3.— Mangel Wurzel against Swedes.*—March 11th, 1846, I drew 12 couples from the flock, the lambs being about a month old; these were divided fairly into two lots, and put into separate yards; six of them were fed on mangel wurzel cut and put into troughs, with a little hay-chaff; the other six couples were fed on cut Swedes, with a little hay-chaff also; they were all weighed alive when put in on the 11th of March, and again on the 2d of April, when I found the following result:

	Lambs gained each on the average in twenty-two days.	Ewes lost in the same time.
	lbs.	lbs.
Couples fed on yellow globe mangel wurzel and chaff.............	8¼	8
Couples fed on cut Swedes and chaff	9¼	3½

"This experiment does not speak much in favor of mangel wurzel for couples early in the spring, but my ewes did not appear to like them, and would not take to them well only as they were fresh cut. I found, upon weighing the food of both sorts, that the ewes ate of mangel about 14 lbs. each per day, and those that had Swedes 22 lbs. each, which was a great difference in the consumption of food; mangel being of a softer nature than Swedes, they ought to have eaten the most of them, but as the contrary was the fact, I suppose made the great difference stated in the loss in weight of those ewes fed on the mangel over those that had Swedes, whilst the gain in weight of their lambs was much about the same. Hence, I conclude that if ewes are fed with mangel wurzel, they should have them thrown whole to them, either on grass land or in the yard, with plenty of good clover-hay chaff, or they will not do well; but this more particularly applies to their use in the early spring months, when they are in a very succulent state; they, however, lose much of this by keeping toward the summer, when their value becomes apparent, as I shall endeavor to show hereafter. * * * *

"When young lambs are about three weeks old they will begin to eat, and should have some food given them apart from the ewes, or run upon some green food, such as clover, tares, or grass. I generally make a yard or fold with common hurdles (kept very airy and well littered) on my land intended the following summer for turnips, into which I put my ewes when their lambs begin to eat, and let the lambs run through a hurdle set up endwise upon a piece of tares or vetches sown for the purpose the preceding autumn. The couples are kept in this way until the pastures intended for them have grown sufficiently high to carry the number required until the lambs are weaned. Although some extra expense is incurred by this system in the use of dry food, a good return is obtained by the outlay, as the clovers and grass, by not being stocked early, carry a much greater number of couples during the summer. The usual method is to turn the ewes and lambs upon the clovers and grass as soon as the turnip season is over, allowing them to range indiscriminately over the whole field, which is decidedly a bad practice. I would strongly recommend that part of the field should be fenced off for the lambs to feed upon apart from their dams, which may be done by setting upright some common hurdles.

"*Method of Keeping Couples during the Early Summer Months.*—In the year 1845 I had a field of land, one side of which was sown with white clover and trefoil, the other side with tares, and a piece of red or broad clover was sown between each. The white clover and the tares were fed off with ewes and lambs in the usual way, the ewes on either part being kept asunder, but the lambs from each lot ran together through the hurdles upon the red clover, which was a good pasture; they had also a few split beans every day. To ascertain the value of tares against clover and trefoil, for this purpose, I made

"*Experiment No. 4.*—From each of the above lots I took a few lambs and weighed them alive twice during the month of May, and found their increase in live weight per month to be as follows:

Average gain, in weight, of seven lambs, whose dams were fed lbs.
   on clover and trefoil...................................... 20
The like, whose dams were fed duing the same period on tares.. 16¼
— being a difference of 3¾ lbs. each lamb in favor of the clover and trefoil. * * * * * * * *

"In the spring of 1846, having a considerable quantity of the yellow globe mangel wurzel left on hand, I determined on making a further trial of them as a summer food for sucking ewes, conceiving that they would, when bereft of much of their succulent qualities through keeping, feed sheep better than I found to be the case, as related in experiment No. 3. I, therefore, selected from the flock a few couples in the middle of May; one part of them were folded in the clover field, and fed with plenty of cut mangel wurzel and a little hay-chaff; their lambs ran through the hurdles on a good pasture of red clover. The other lot of ewes were left at large in the common way on white clover and trefoil; their lambs, also, had a good piece of red clover to feed upon: both lots of lambs had a small quantity of peas.

"*Experiment No. 6.*—On the 25th of May the lambs from each of the lots above described were weighed alive, and again on the 22d of June, when the result was as follows:

Those lambs belonging to the ewes fed on mangel wurzel, gained each, on the average, in 28 days........................... lbs. 21

Lambs from ewes fed in the usual way on clover and trefoil, gained each, in the same time........................... 18

Difference each lamb in favor of mangel wurzel................. 3

"This statement, as well as others preceding, of lambs gaining in live weight of about 20 lbs. each in 28 days, may appear startling to those unaccustomed to weigh them alive, but this is no uncommon weight for lambs to gain, if well fed and attended to in the early summer months. Those ewes fed on mangel ate about $22\frac{1}{4}$ lbs. each per day, care being taken that their lambs had none of it on those days that the food was weighed, and, unlike those ewes fed on it in March, (see Experiment No. 3,) I found them to thrive and do well with it. It should, however, be remembered that the summer of 1846 was very favorable for the use of mangel, the weather being very dry the whole of the period the above trial was carried on, and, consequently, more unfavorable for those ewes fed on the clovers, which, toward the end of the time, were nearly dried up. From this trial it appears that mangel wurzel is of great use as a summer food for sheep, and as it will keep a long time, if properly stored the preceding autumn, must be very useful in a dry season for any kind of stock. * * * *

"Having proved by many experiments the advantage of putting young lambs, after weaning, upon old keeping—namely, pastures that have been stocked from the commencement of the spring—over eddishes or pastures that have been previously mown the same season, I will state one experiment as a sample of the rest. In the year 1834, I put a lot of lambs on some old sainfoin, having a few tares carried to them, and another lot of lambs were put on young sainfoin, or an eddish which had grown to a pasture; these, also, had some tares. Each lot was weighed at the commencement, and again at the end of the trial:

"*Experiment No. 7.*—Gain in weight of a lot of lambs fed on old sainfoin, from July 10 to August 10, each on the average,.. lbs. $14\frac{1}{4}$

Lambs fed on sainfoin eddish, gained each in the same time,. $8\frac{1}{4}$

Difference,.. 6

* * * * * * * *

"*Experiment No. 8.*— June 10, 1844, ten lambs were weaned, and weighed alive, and put on red clover, with some tares and beans given; on the same day, ten lambs were weighed alive, remaining with their dams on white clover and trefoil, but allowed to run through hurdles upon good red clover. Each lot was weighed again on July 5th, when it was found that they had increased in weight as follows, each lamb on the average:

Lambs not weaned gained each, in thirty-three days,........ 17 lbs.
Lambs, weaned, gained in the same time,................... 16¼ "

"*Experiment No. 9.*—June 4, 1845, twelve lambs were weaned and put upon red clover, tares, and a few beans, twelve other lambs lying with their dams on white clover, but run through hurdles upon good red clover. Both lots were weighed when put to trial, and again at the end of a month.

Gain in weight of lambs not weaned,...................... 21 lbs.
Gain in weight of lambs that were weaned during the same time.. 20¾ "

"These experiments are nearly equal; but I must remark, that many of those lambs that were weaned early wintered the best."

* * * * * * *

"*Experiment No. 10.*—In the month of October, I selected two lots of lambs, and weighed them alive. To one lot was given, in troughs, cut Swedes; and to the other was given, in troughs, the common white turnip, also sliced. At the expiration of a month they were weighed again, and gained each, on the average, as follows:

The lambs fed on common white turnips cut gained each,.... 10 lbs.
Those fed on cut Swedes, gained in the same time each,..... 4½ "

In favor of the white turnip,......................... 5½

To show that the white turnip loses much of its value as the winter approaches, agreeably to what I have stated, I will just show the result of another experiment.

"*Experiment No. 11.*—On the 8th of November two lots of lambs were weighed alive. One lot was fed on cut Swedes only, and the other lot had only cut white turnips. They were weighed again December 6, and gained each as follows, on the average:

Lambs fed on white turnips gained each, in a month,....... 6¼ lbs.
The lambs fed on Swedes gained, in same time,............. 5 "

"The same lambs were continued to be fed as before for three weeks longer, when I found, upon weighing them again, that the white turnips quite gave place to the Swedes.

"*Experiment No. 12,* (dry food, with Swedes, against Swedes only.—In 1833 I weighed two lots of lambs on the 19th of November. To one was given cut Swedes, with clover-hay chaff and maltcoom mixed; the other lot had only cut Swedes. They were all weighed again on the 16th of January, and gained in weight as follows:

Lot of lambs fed on cut Swedes, with clover-chaff and maltcoom, lbs.
 gained each, in two months,............................. 14¼
Lot of lambs fed on Swedes only, gained each, in the same time, 8

In favor of dry food,.................................. 6¼

"*Experiment No. 13.*—Being again desirous of testing the use of dry food for lambs at turnips, I took sixteen lambs from my flock on February the 18th, 1846, and weighed them; eight of them were penned and fed with cut Swedes only. The other eight lambs had cut Swedes, with 2 lbs. of clover-hay chaff and 2 lbs. of bran mixed together for the eight per day, or half-a-pound each. They were weighed again on the 17th of March, when the result was as follows:

	lbs.
Gain in weight of lambs, on the average, fed on Swedes, bran, and clover-chaff, in a month,............................	7¼
Gain in weight of lambs fed only on Swedes, during same time,	3¼
Difference in favor of dry food,........................	3½

The cost of dry food was

	s.	d.
2 lbs. of bran per day amongst eight lambs for 28 days, or 4 stone at 5s. per cwt. cost,...........................	2	6
2 lbs. of clover per day for eight lambs, during 28 days, gives 4 stone at 4s. per cwt...........................	2	0
	8)4	6
Cost of dry food for each lamb, per month,...........		6¾

* * * * * * * * *

"*Experiment No. 14.*—Having used linseed for some years with success in the feeding of cattle, I determined to try whether it would answer equally as well for sheep. I therefore gave a lot of eight lambs, feeding on cabbages with white turnips, half a pint of linseed each per day. To another lot of eight lambs, also upon cabbages with white turnips, clover-chaff was given, as much as they would eat. They were all weighed on the 27th of October, and again at the end of the trial.

Lambs fed on cabbage and linseed gained each per month,... 16 lbs.
Lambs fed on cabbage and clover-hay chaff gained each, in same time,.. 16 "

"*Experiment No. 15.*—Having determined some years ago to have nothing more to do with feeding sheep in yards, I was, however, last season induced, through the favorable representations of some persons, to give it a further trial. I took some of my best lambs, that I intended to show for premiums, and put them into a warm, well-sheltered yard, with a lofty hovel to feed under, being kept well littered with dry, fresh straw; and their quarters appeared so comfortable, that I thought they must go on well. They were fed with Swedes and corn in the usual way. I weighed them alive when put into the yard, December 4th, 1845, against some other lambs fed on the same food, but in the field, kept in the ordinary way. Both lots were weighed again on February 3d, 1846.

	lbs.
Those fed in the turnip-field gained each, on the average, in eight weeks,...	13
Those lambs fed in the yard gained each, on the average, in the same time,..	3
Against the yard-feeding system,.....................	10

"These lambs did not appear to like the confinement of being in a yard, and would take every opportunity of getting out if they could. This system is not natural for sheep, and cannot answer for long-wools, or be depended on. * * * * *

"*Experiment No. 16.—* lbs.
On grass land, lambs fed with Swedes and chaff gained each, on the average, from December 10 to March 5,............ 18
On turnip land, lambs fed in the same way gained each, in the same time,.. 17

"*Experiment No. 17.—*
On grass land, lot of lambs fed with cut Swedes and chaff, in Dec., Jan., and Feb., 1835, gained each, on the average, in three months,... 21
On turnip land, lot of lambs, fed in the same manner, gained each, in the same time,................................. 19

Being only a gain of 2 lbs. each during three months.

"*Experiment No. 18.—*
On grass land, lambs fed on carrots, Swedes, and chaff, gained each, on the average, from Jan. 27, 1836, to March 2,....... 7
On turnip land, lambs fed in the same manner gained each, in the same time,.................................. $2\frac{1}{4}$

"The difference here is greatly in favor of feeding on grass land, but not for carrots, (see other experiments.)

"*Experiment No. 19.—*
On grass land, lambs fed on Swedes, carrots, and chaff gained each, on the average, from Nov. 16, 1837, to Feb. 10,....... 16
On turnip land, lot of lambs, fed in the same manner, gained each, in the same time,................................ $18\frac{1}{4}$

"This experiment differs much from the last; but it is the result of three or four experiments that must be looked to, for I well know that no single experiment can be depended on.

APPENDIX D — (page 248.)

SHEEP AND PRODUCT OF WOOL IN UNITED STATES.

The following statistics are from the United States Census of 1860. Under the extraordinary demand for wool which has existed for the last two years, the number of sheep has probably increased far more since 1860 than it did for the ten preceding years.

APPENDIX D.

STATES.	WOOL.		SHEEP.
	1850.	1860.	1860.
	Pounds.	*Pounds.*	*Number.*
Alabama,	657,118	681,404	12,404
Arkansas,	182,595	410,285	6,481
California,	5,520	2,681,922	23,414
Connecticut,	497,454	335,986	2,700
Delaware,	57,768	50,201	559
Florida,	23,247	58,594	1,675
Georgia,	990,019	946,229	120,596
Illinois,	2,150,113	2,477,563	33,822
Indiana,	2,610,287	2,466,264	32,012
Iowa,	373,898	653,036	22,267
Kansas,		22,593	1,145
Kentucky,	2,297,433	2,325,124	67,161
Louisiana,	109,897	296,187	21,643
Maine,	1,364,034	1,495,063	61,926
Maryland,	477,438	491,511	1,135
Massachusetts,	585,136	377,267	8,616
Michigan,	2,043,283	4,062,858	47,916
Minnesota,	85	22,740	2,473
Mississippi,	559,619	637,729	1,062
Missouri,	1,627,164	2,069,778	96,005
New Hampshire,	1,108,476	1,160,212	6,191
New Jersey,	375,396	349,250	12,093
New York,	10,071,301	9,454,473	3,065
North Carolina,	970,738	883,473	77,296
Ohio,	10,196,371	10,648,161	132,653
Oregon,	29,686	208,943	10,788
Pennsylvania,	4,481,570	4,752,523	53,225
Rhode Island,	129,692	90,699	5,458
South Carolina,	487,233	427,102	
Tennessee,	1,364,378	1,400,508	29,854
Texas,	131,917	1,497,748	320,926
Vermont,	3,400,717	2,975,544	18,015
Virginia,	2,860,765	2,509,443	112,591
Wisconsin,	253,963	1,011,915	11,885
Total States,	52,474,311	59,932,328	
TERRITORIES.			
Columbia, District of	525	100	62
Dakota,			
Nebraska,		3,312	52
New Mexico,	32,901	479,245	142,110
Utah,	9,222	75,638	4,325
Washington,		20,720	212
Total Territories,	42,648	579,015	
Aggregate,	52,516,959	60,511,343	1,505,810

I give these figures for what they are worth. It will be seen that the number of sheep reported in 1860 bears no correspondence whatever with the product of wool the same year. It assuredly required over twelve millions of sheep, taken as they average, to produce sixty million pounds of wool; and then the lambs of the year, not sheared, would at least equal six millions more. I have no doubt there were twenty millions of sheep in the United States in 1860, and probably the present number equals twenty-five millions.

APPENDIX E — (Page 250.)
STARTING A SHEEP ESTABLISHMENT IN THE NEW WESTERN STATES.

The following letter is from an intelligent gentleman residing in Essex County, New York, whom I knew a few years since as a highly respectable member of the New York Legislature:

CHICAGO, ILLINOIS, May 1, 1863.

Hon. H. S. RANDALL—*Dear Sir:* Yours dated April 20th came duly to hand. I should have replied at once, but have not had a spare moment for the last four weeks, as my sheep have required my undivided attention. I am here on business for a day, and will take time to give you a few facts as far as my experience is concerned.

About the 20th of last July I started from Calhoun County, Michigan, with two droves of sheep, about 1,700 in each drove. My destination was Southern Minnesota. In consequence of the Indian outbreak in that section of country, I changed my plan and stopped in Northern Iowa, about twenty miles west of McGregor, on the old military road to Forts Crawford and Atkinson. My sheep stood driving remarkably well, and arrived at that point about the 10th of September. I found good feed, and by the time winter set in my sheep were in fine order. I sold about 300 in the autumn, thinking I would winter the remainder. I then set about preparing winter quarters for 3,000 sheep. I did not erect my sheds at one place, (on account of the inconvenience of hauling the feed I had purchased to one place,) but about two miles apart, where water was convenient. I succeeded in getting a grove, at each place, and built my sheds fronting the grove and parallel with each other, about 500 feet long. I built them of poles and posts from the groves, and covered them with straw. The front posts were about six feet above ground and the back ones about four. I employed Irishmen that were in the habit of using the spade and covered the back side with dirt, and then covered this smoothly with sod, which made them very warm — being left open in front, this was important. I then cut the sheds up with board fences about 22 feet apart, commencing under the shed and running out about 50 feet in front, making yard and shelter for about 50 sheep. I forgot to mention the width of the sheds, which was 13 feet. I then sorted my sheep, putting heavy wethers by themselves, heavy ewes by themselves, &c.; in short, I went through the flock grading them according to strength and sex. I started with prepared winter quarters for 3,000, but continued to sell some through the early part of winter. By the 1st of January I had reduced my flock to 2,200. After that I declined selling more.

I will now give you a brief account of my feeding, its quantity, quality, &c. I procured what hay I conveniently could, about half of which was nice timothy. I expected to buy from time to time during the winter, which I have been able to do at fair rates, say from $3 to $4 per ton. I would quite as soon have good upland prairie hay as timothy, provided it is cut early. The sheep will eat it better. I also bought what corn I could in the field, paying from $4 to $7 per acre.

This I cut while the fodder was green, before frost, shocking it in the field and drawing in after the ground froze. This I found excellent feed. I fed it once a day, usually at noon. After that was used up I fed corn in the ear to all except my yearling lambs. The latter I fed a mixture of shelled corn, oats and shorts from the mill, mixing it as follows:—½ corn, ¼ oats, ¼ shorts. I gave a pen of 50 lambs one-half bushel once a day (at 11 o'clock.) This, with what hay they could eat, made them prosper finely. I fed hay to all my sheep twice a day; but the lambs generally got it three times.

My sheep have been remarkably healthy. Of course one dies occasionally, but I have got them well through the winter. I have just finished tagging. On coming to handle them, we find them very heavy. A large number are good mutton. Since putting up my sheep last fall, I have lost less than one per cent. of 630 lambs that I went into winter with. Only one has died. I think the feed I have used for lambs can't be bettered. My sheep are about two-thirds ewes. I can't give any definite idea of how many lambs I shall have, as I did not put my bucks in with my ewes until the first of December. I was unfortunate enough in the autumn to have a native buck get in with my flock once in a while, and the result has been that I have had about ninety lambs during the winter, scattered along. I had from the ninety ewes eighty-four good healthy lambs. I should, however, have had but very few of the lambs living, coming as they did, had it not been for the care of my yard-master. A lamb will chill in one hour in cold weather if not taken to the fire to dry, which is found necessary in most cases.

I am satisfied that Iowa and Southern Minnesota are especially adapted to wool growing. The country where I am keeping my sheep is somewhat uneven and rolling, and a good farming country. The country seems prosperous. Improved farms are selling from $15 to $20 per acre, and unimproved lands from $3 to $10 per acre.

I am sorry that I am obliged to give you such a hurried statement of my experience with sheep in the West. Any farther inquiries you may be pleased to make, I shall be happy to answer.

Yours truly, R. A. LOVELAND.

APPENDIX F — (page 257.)

CLIMATE OF TEXAS.

THE following account of some of the peculiarities of the climate of Texas, of the seasons and crops and their vicissitudes, I extract from articles on the Climatology of that State, contributed to the Texas Almanacs of 1860 and 1861, by Professor Caleb G. Forshey, Superintendent of the Military Institute, in Fayette County:

TEXAS NORTHERS.

Number and Duration.—1. During seven or eight months of every year, Texas is liable to a class of storms, or winds, styled "*northers,*" from the direction from which they come.

2. In the year 1857, there were twenty-six northers experienced at the Texas Military Institute, in Fayette county. Of these some two or three were gentle or baffled northers. They occupied fifty-seven days, having an average of two and one-fifth days in length. The latest in spring, was May 16, and earliest in autumn, was Nov. 7.

3. In the year 1858, there were thirty-seven northers, about thirty-three of which might be classed as *well marked,* the others being either gentle or baffled northers. These occupied seventy-eight days. The latest in spring, was May 9, and the earliest in autumn, was Oct. 7.

4. In the first half of 1859, there have been twenty-four northers, of which four may be described as gentle or baffled northers. They have occupied forty-seven days in their transit, and the latest was May 24.

5. It is proper to remark that nearly all the northers of May and October are mild, and rarely do much damage, or produce so low a temperature as to be severely felt. All the other months, November to April inclusive, are liable to northers of considerable severity.

6. It appears then, that in thirty months last past, of which eighteen months are liable to distinct northers, we have experienced eighty northers, not including the feeble ones of May and October. The same period has seventy-seven weeks, very nearly affirming the hypothesis of *weekly returns* of the norther. An inspection of the table shows a large number of punctual weekly recurrences of this meteor.

7. At this place of observation their duration varies from one to four days.

Area and Boundaries of Norther.—8. The region over which this peculiar storm has its sweep, is not very great, though its precise limits can not be defined. By diligent inquiry from persons of great experience, we submit the following limits:

9. On the north, by the valley of Red river, in the Indian Territory; on the east, by the second tier of counties from the east boundary of Texas, near meridian 95°, south to the Trinity and thence south-east to the mouth of the Sabine. On the south they are felt across the Gulf, to the coast of South-Mexico and Yucatan. On the west they are bounded by the Sierra Madre, up to the mouth of the Pecos, and thence by about the 101st meridian to the sources of Red river.

10. Within this area, there are various degrees of violence, having their axis of intensity between meridians 97 and 98, and increasing in force and duration, the further south. At Red river, on this line, they are usually limited to a day or two; whereas at Corpus Christi and Matamoras, one norther often continues till the next supersedes it; and at Vera Cruz, a twenty-days norther is not remarkable.

West of Fort Belknap, to the Pecos, the northers grow feebler and rarer. North of Red river, on the route from Fort Washita to Fort Smith, they are rarely felt.

On the east margin they are much modified by the forests of the timbered region. At all points, an open prairie increases their vigor.

Forces and other Phenomena.—11. The norther usually commences with a violence nearly equal to its greatest force, if its initial point be near the observer. If it has traveled some distance, it will be warmed up, and moderated in its violence, at first attack. Its greatest force might be marked five, in a scale between a gentle breeze, at one, and a hurricane, at ten. The writer has measured one traveling at about thirty-two miles per hour—but many others at twelve to eighteen miles. The mean progress seems to be about fifteen miles per hour.

12. Just before a norther, two to six hours, the south wind lulls, and the still air becomes very oppressive. A low black cloud rolls up from the north, and when it comes near the zenith, the wind strikes with vigor. Sometimes we have a sudden dash of rain; but generally northers are intensely dry, and soon drink up all the moisture of the surface earth, and of the objects upon it, capable of yielding their humidity.

Great thirst of man, and all other animals, is experienced; an itching sensation over the skin; a highly electric condition of the skin of horses and cats; a wilting and withering of vegetation, even when the temperature would not account for it; a reduction of temperature, usually very sudden, sometimes, though rarely, a degree per minute, for twenty minutes; and in winter commonly a reduction from 70° or 75°, to 30° or 40°.

This fall of temperature is the more severely felt from the drying power of the north wind—evaporation from the surface of the skin increasing the severity of the temperature.

13. Nervous, rheumatic, and gouty persons suffer more severely than others. To invalids suffering from other maladies, it has not been found unhealthy; and for persons of weak lungs, if not too much exposed to its direct fury, it is found. to be more salubrious than the humid south winds. *Consumptions do not originate over the area of the norther.* On the contrary, many persons afflicted with weak or diseased lungs, resort to this region, and find relief. The western and northern portions of this area are most salubrious, and best adapted to weak lungs.

* * * * * * * *

Phenomena not readily explicable.—When a dry norther commences, the whole air, in an hour or two, curdles, and becomes smoky, or rather whitish, and has a distinct smell. Its odor sometimes resembles that which is developed by a flash of lightning, though, at other times, it reminds one of fine straw smoke, in its odor.

It is highly probable that this turbidness and odor, are due to the ozone set free, by the high electrical excitation, in a dry norther. Experiments instituted to test the matter, last April, were too late in the season.

Sirocco.—When the norther has a little westing, it is observed to be more intensely dry, and to be destructive to vegetation, even before the frost which usually follows it. Corn, beans, young foliage, and the grass and weeds of the prairie, bow and wither before it.* A few of these I have called *Siroccos.* They occur as well in summer as in spring or autumn, and differ, in several respects, from the true norther.

---

* The citizens of Galveston, and the southern portions of Texas, will remember the violent north-wester in 1856, which preceded and attended the storm which wrecked the Nautilus. It was, in my judgment, a true Sirocco. In like manner the north-west wind, that withered the corn-fields in Lamar, Fannin, and Grayson, and the counties south of these, on the 17th day of August, 1858, deserves a like name.

APPENDIX F.

SEASONS AND CROPS: THEIR VICISSITUDES.

1857.	1858.	1859.	1860.
January.—No rain.	January. — No severe cold; abundant rain.	January—Some severe weather. Rain 2¼ inches.	January. — Moderately cold. Rain, 1.5 inch.
February 6.—Prairies getting green.—10th. Corn, peas, lettuce, and radishes coming up. Rain 1 inch.	February 3.—Violent storm. 1st. Brazos overflows. 22d. Peaches killed by frost, 25 deg. 27th. Growing weather.	February 15—Grass covers woods and prairies; corn-planting begins. 24th.—Woods gray. Rain 1 inch.	February 1, 2, 3, 24, 25, 26.—Frost. 17th. Rain copious, East-Texas. Whole rain of month, 5 inches.
March 7.—Corn six inches high; prairies one month forward. 12th. Terrible frost kills every thing—fruit and crops. Rain 1 inch.	March 2.—Freeze, 24 deg. 20th. Woods greenish; grasshoppers hatching, west. 27th. Make havoc and migrate. 17th. Corn planted. 25th. Squirrels migrate on Trinity.	March 6. — Woods half-green; rye heading; dogwoods bloom; corn coming up generally. 20th. Good stand; post oaks naked, blackjacks green. 23d. Wild geese leave, and doves coo. Rain—7.87.	March 5.— Prairies green; corn-planting; woods gray. Frost, 28-9 cuts off cotton and some corn, and gardens. 14th. Radishes and lettuce.—Whole rain, 1.5 in. 28th. Geese migrate; good prospects of crop.
April 5.—All green again; new crops up and vigorous. 6th. Norther, hail, and freeze; all crops, fruit, and mast, killed. 11-12th. Sleet, snow, and freeze, again. 24th. Frost in valleys.—Rain, ½ inch.	April 1—Grasshoppers bad in Guadaloupe; May 20, country eaten up by them west of 97° 10'.	April 1.—Radishes and lettuce. 23d.—Frost kills corn and cotton in low grounds. Rain, 0.69 in.	April 1. — Whip-poor-wills. 5th.—Woods quite green. 14th. Ground cracking from drouth.—21st. Dewberries ripe. 19th-27th, good rains; total, 3.8 inches.
May 30.—Rain two inches—not 12 inches in a year.	May 1 to 9.—Rain 5½ inches; wheat, oats, rye and millet die of rust. 10-15th. Rivers overflow. 25-30th. Corn tasseling; beans, peas and potatoes in use from 10th.	May 7.—Fair rains start the re-planted crops; not one grasshopper in the land. 22d. Crops look well; wheat harvest begins. 28th. Wheat harvest closes; early corn tassels. Total rain, 6.76 inches.	May 1.—Crops very promising; no grasshoppers. 15th. Crops wilt for want of rain. 25th. Corn tasseling; very dry. 21st. Rye ripe. 25th. Oats cut. 30th. Wheat ripe and cutting. Rain, 0.35 in.
June 11. — Wheat reaped; good crop; man and beast suffering for water. 20th. Grass all dead.	June. — Showery weather. 11th. Great rain. Rain in June, 6¼ inches. 6th. Roasting ears.	June 3.— Roasting ears. 11th. Rain saves corn; total, 0.50 in.	June.—No rain this month. Corn perishes, gardens die, creeks and springs dry up. Much corn cut up west of Colorado. Fayette and Washington make half-crops corn; wheat, oats, rye, and barley good. Greatest drouth over United States ever remembered.
July. — No rain! August, no rain!	July.—Rain 1 inch. Good corn crops over most of the State.—Rust kills all small grain.	July.—Very dry.—Total rain, 0.90. 30th. Cattle suffer for water	July 1.—Cattle suffer for water; ponds and creeks all dry; continues to July 18th, when this report closes.

1857.	1858.	1859.	1860.
	August and September.—Dry; only 1 inch rain.	August—Rain, 0.50; west of 97° no rain; all summer corn and cotton dead. August gave showers in Guadaloupe, etc.	
September 7.—Oaks drying from drouth, except live oak. First good rain this year, 2 inches.		Sept,—Good rains; 5.85 inches.	
October.—Rain, 3½ inches. The prairies green.	October. — Good rains, 3.7 inches.	October. — Good rains, 6.60 inches.	
November.—Grasshoppers, west. Reasonable rains; good fall gardens. 26-27th. Hard storms very extensive; Nebraska wrecked at Galveston. Rain, 2½ inches.	November. — Some rain—2¼ inches.	November.—Warm and pleasant month; no rain.	
December—Lowest temperature, 30°.	December.—Rains copious, 4.4 inches. No severe cold.	December 1 to 8.—Terrible winter weather; snow, sleet, rain and freeze; kills cattle, horses and sheep in vast numbers. Hardest December ever known.	

NORTHERS, WINTER OF 1859-60.

First genuine norther,..............Sept. 30
Last genuine norther,..............April 23
Number of weeks' time,..................28
Number of northers,......................28
Number of days occupied,..............101
Average duration, hours,..............89
Lowest day's temperature, Dec. 6th,...16°
Lowest 3 days' norther, Dec. 6th,......20.3

TEMPERATURE AND HYGROMETRY OF 1859 AND PART OF 1860.

	1859.						1860.				
	TEMPERATURE.				WET BULB	RAIN:	TEMPERATURE.				WET BULB	RAIN:
	SUNR.	2 P.M.	9 P.M.	MEAN	MEAN	IN.	SUNR.	2 P.M.	9 P.M.	MEAN	MEAN	IN.
January,..	41.00	63.58	47.19	50.57	48.00	4.75	45.11	60.00	50.00	52.03	48.73	1.40
February,.	55.19	73.32	58.82	62.44	50.50	0.80	46.04	67.20	53.04	55.40	50.17	4.85
March,....	53.71	71.50	59.00	61.50	54.56	1.56	53.16	73.06	58.17	61.13	55.24	1.35
April,.....	59.44	72.60	63.60	65.31	59.33	0.75	63.60	78.63	66.20	69.47	64.20	3.80
May,......	71.48	84.22	71.13	75.61	69.33	1.75	73.40	85.30	72.52	76.22	69.04	0.35
June,.....	72.23	88.38	80.07	81.56	76.45	2.50	81.21	94.21	81.27	85.58	5.18	0.00
July,......	82.06	89.77	82.10	84.76	77.40	2.30						
August,...	79.01	93.02	82.04	84.90	79.36	0.40						
September,	75.30	85.00	78.00	79.42	75.40	5.85						
October,...	59.80	75.20	63.86	66.29	64.53	6.60						
November,	55.16	74.43	61.16	63.92	61.80	0.10						
December,.	35.00	54.00	40.00	43.00	42.25	3.00						
Annual,.				68.04	63.62	30.36	½ yr.	60.42	76.51	63.03	66.67	60.44	11.75

APPENDIX G.

PROPORTION OF WOOL TO MEAT IN SHEEP OF DIFFERENT AGES, SEXES AND SIZES.

THE following was not received until this work was nearly through the press, and too late to refer to it except in this place:

POMPEY, Onon. Co., N. Y., Aug. 22, 1863.

HON. HENRY S. RANDALL — *Dear Sir:* Agreeable to your request, I herewith send you my investigations and observations upon the comparative weight of wool and bodies of sheep. I hope they will be of benefit to the sheep breeder, as well as the wool grower; and that I shall have the satisfaction of knowing that I have in part repaid to the world much that I owe for the investigations of those who have gone before me. With high hopes, but no higher ambition than to be called a "good farmer," I remain your obedient servant,

HOMER D. L. SWEET.

COMPARATIVE WEIGHT OF WOOL AND BODIES OF SHEEP.

BY H. D. L. SWEET.

The Hon. Robert R. Livingston, the first President of the first Agricultural Society of the State of New York, in his justly celebrated essay on Fine-Wooled Sheep, uses the following language:

"The inferiority in the size of the Merino to some other breeds, which some make as an objection, is, in my opinion, an important advantage, not only in sheep but in every other stock not designed for the draft; because they will fatten in pastures in which larger cattle would suffer from the fatigue they must undergo, in order to procure the food that is necessary for their support.

"This meaning applies more strongly to sheep than to any other stock. They are generally kept upon high and dry pastures, that are frequently parched in summer, when fatigue is most irksome to them. To which we may add that the fleece is not proportioned, as the food is to the *bulk* of the animal, but to his *surface*, and a small sheep having more surface in proportion to his bulk, must also have wool in the same proportion. That is, a sheep whose live weight shall be 60 lbs., and who, of course, will require but one-quarter of the food of a sheep that weighs 240 lbs. will, notwithstanding, have *half* as much wool (if the fleeces are equally thick,) as his gigantic brother." [*]

---

[*] Transactions of the Society for the Promotion of Useful Arts, Vol. II, p. 86.

In proof of the first proposition, that sheep do consume in proportion to their bulk, Mr. Livingston submits, in an appendix to his essay, the record of many experiments which show conclusively that such is the fact; but of his second proposition, that they shear in proportion to their surface, he gives no facts, and I suppose it to be mere theory. The attention of the writer was called to this subject by the Hon. George Geddes, some four years since, and at his request the trial was made, and the result has been given to the world by yourself. Experiments of the same character on the same flock have been conducted for three successive years, and their results are recorded in the following tables.

In one or two points they are not as perfect as I could wish, but they are the best that could be done with so small a flock. Had there been from forty to fifty in each class and every year, the natural law in relation to them might be nearer in accordance with the facts noted; for as there are exceptions to all *rules*, I may be giving the exception and not the rule. This can be true only in regard to five and six year old ewes, and five year old wethers. In all other cases, taking the three years collectively, I am confident that facts of value have been obtained.

The base of the flock a few years since was Saxon; they are now classed from one-half to seven-eighths Spanish Merino—a portion of the largest, in 1861, was one-quarter French Merino. In 1861 the ewes raised 35 lambs; in 1862, 30, and in 1863, 70. In the fall of '61 the oldest and largest were sold and replaced by 60 lambs purchased. In the fall of '62, 70 wether lambs were purchased, part of the smallest of them were sold, some three-year old ewes purchased; and some older ones sold. Other discrepancies that may be noted are attributable to death. They were all brook-washed about two weeks before shearing. The flocks at the time of shearing were in good condition — some of the ewes thin, of course. The four rams in the flock are included with the wethers, to save space, figures and calculation. The first table is the same as published in 1862, in Mr. Randall's Essay, in the Transactions of the N. Y. S. Agricultural Society, except that I have subdivided the sexes. The fifth table is the same as the second one then published, except that I have added the last three classes, and called them one. They were sheared the 26th and 27th of June, 1861; 27th, 28th and 30th of June, 1862, and 25th, 26th and 27th of June, 1863. Every sheep and fleece were weighed separately and recorded on the spot.

[The tables referred to in the preceding paragraphs are given on the two following pages.]

SWEET BROTHER'S FLOCK, POMPEY, N. Y.

TABLE 1. 1861.— Classified by Age and Sex.

No. in Class.	Ages.	SEXES. Ewes.	SEXES. Weth's.	Gross Weight.	W't. of Bodies.	Wt. of Wool.	Aver. of Bodies.	Aver. of Fleeces.	Lbs. of Body to 1 of Wl.	Per Ct. of W. to Gr. Wt.
19	1	E		1,193.72	1,097	96.72	52.47	5.09	10.44	8.10
13	1		W	965.23	894	71.23	68.77	5.48	12.55	7.37
15	2	E		1,124.37	1'048	76.37	69.86	5.09	13.72	6.88
15	2		W	1,383.92	1,299	84.92	86.66	5.66	15.29	'6.53
9	3	E		759.14	710	49.14	78.88	5.45	14.45	6.46
42	3		W	4,155.11	3,891	264.11	92.64	6.28	14.73	6.83
41	4	E		3,738.	3,557	181.	86.75	4.41	19.65	4.84
26	4		W	2,921.13	2,736	185.13	105.11	7.12	14.76	6.33
180	1 to 4	84	96	16,341.	15,331	1.010	85.17	5.38	15.17	6.18

TABLE 2. 1862.— Classified by Age and Sex.

42	1	E		2,378.57	2,189	189.57	52.11	4.51	11.60	7.96
52	1		W	3,224.51	2,985	239.51	57.40	4.60	12.46	7.42
19	2	E		1,387.16	1,292	95.16	68.	5.	13.57	6.86
13	2		W	1,225.16	1,147	78.16	88.23	6.	14.66	6.46
14	3	E		1,026.31	960	66.31	68.57	4.70	14.47	6.46
13	3		W	1,297.36	1,215	82.36	93.40	6.33	14.75	6.35
9	4	E		726.59	679	47.59	77.44	5.28	14.20	6.54
27	4		W	2,693.06	2,505	188.06	92.77	6.96	13.32	6.98
15	5	E		1,178.15	1,111	67.15	74.	4.47	16.54	5.77
11	5		W	1,153.40	1,075	78.40	97.72	7.12	13.71	7.00
215	1 to 5	99	116	16,290.27	15,158	1,132.27	70.50	5.26	13.30	6.95

TABLE 3. 1863.— Classified by Age and Sex.

14	1	E		955.78	877	78.78	62.64	5.62	11.00	8.24
78	1		W	5,623.84	5,201	422.84	66.67	5.42	12.30	7.71
42	2	E		2,861.64	2,662	199.64	63.38	4.75	13.33	6.97
48	2		W	3,994.79	3,735	259.79	77.81	5.41	14.37	6.50
33	3	E		2,837.24	2,658	179.24	80.54	5.40	14.82	6.31
13	3		W	1,338.89	1,251	87.89	96.23	0.76	14.23	6.56
13	4	E		1,154.68	1,083	71.68	83.30	5.51	15.10	6.26
9	5	E		735.93	680	45.93	75.35	5.10	14.82	6.24
10	6	E		837.84	790	47.84	79.00	4.78	16.49	5.70
200	1 to 6	121	139	20,350.63	18,957	1,393.63	72.91	5.32	13.58	6.84

TABLE 4. AVERAGE OF THE THREE YEARS.

Classified by Age and Sex, the Footing being the three Flocks collectively.

No. in Class.	Age.	Sex.	Av'age Wt. of Body.	Average Wt. of Fleece.	Pounds of Body to 1 of Wool.	Average Per Cent.
75	1	E	55.74	5.07	11.01	8.10
76	2	E	67.08	4.94	13.54	6.90
56	3	E	75.99	5.18	14.58	6.41
63	4	E	82.49	5.06	16.33	5.88
24	5	E	74.67	4.75	15.68	6.00
10	6	E	79.00	4.78	16.49	5.70
143	1	W	64.28	5.16	12.43	7.50
76	2	W	84.23	5.69	14.77	6.49
68	3	W	88.86	6.45	14.57	6.58
53	4	W	103.94	7.04	14.04	6.65
11	5	W	97.72	7.12	13.71	7.00
655	Ewes. 304	Weth. 351	79.52	5.32	14.01	6.65

TABLE 5. 1861.—CLASSIFIED BY WEIGHT,

In divisions of 10 Pounds each, except those weighing less than 50 lbs., and those more than 100 lbs.

No. in Class.	Weight of Divisions.	Sexes. Ewes.	Weth.	Gross Weight.	W't. of Bodies.	Wt. of Wool.	Aver. of Bodies.	Aver. of Fleeces.	Lbs. of Body to 1 of Wl.	Per Ct. of W. to Gr. Wt.
5	42 to 51	5		256.	234	22.	46.80	4.40	10.63	8.59
14	50 to 61	10	4	871.	803	68.	57.35	4.85	11.80	7.80
20	60 to 71	14	6	1,427.	1,320	107.	66.	5.35	12.33	7.49
34	70 to 81	21	13	2,742.	2,567	175.	75.50	5.14	14.66	6.38
39	80 to 91	19	20	3,566.	3,355	211.	86.	5.41	15.87	5.90
34	90 to 101	11	23	3,453.	3,252	201.	95.64	5.91	15.42	5.82
34	100 to 134	4	30	4,026.	3,800	226.	111.76	6.67	16.80	5.61
180	42 to 134	84	96	16,341.	15,331	1,010.	85.17	5.38	15.17	6.18

TABLE 6. 1862.—CLASSIFIED BY WEIGHT, AS BEFORE.

37	34 to 51	23	14	1,875.	1,725	150.	46.60	4.05	11.50	8.00
41	50 to 61	19	22	2,460.	2,270	190.	55.37	4.63	11.94	7.72
42	60 to 71	25	17	2,940.	2,740	200.	65.23	4.75	13.70	6.80
30	70 to 81	24	6	2,432.	2,272	160.	75.73	5.33	14.20	6.57
25	80 to 91	6	19	2,266.	2,110	156.	84.40	6.24	13.52	6.88
25	90 to 101	2	23	2,568.	2,408	160.	96.32	6.40	15.05	5.84
15	100 to 127		15	1,743.27	1,633	110.27	108.86	7.35	14.80	6.32
215	34 to 127	99	116	16,290.27	15,158	1,132.27	70.50	5.26	13.30	6.95

TABLE 7. 1863.—CLASSIFIED BY WEIGHT, AS BEFORE.

10	36 to 51	5	5	493.	455	38.	40.50	3.80	11.97	7.91
34	50 to 61	15	19	2,009.	1,850	159.	54.44	4.67	14.15	7.90
67	60 to 71	33	34	4,828.	4,480	348.	66.88	5.19	12.87	7.20
96	70 to 81	44	52	7,755.	7,230	525.	75.30	5.46	13.77	6.76
28	80 to 91	14	14	2,550.	2,390	160.	85.35	5.71	14.93	6.23
16	90 to 101	7	9	1,628.	1,532	96.	95.75	6.00	15.85	5.89
9	100 to 140	3	6	1,087.63	1,020	67.63	113.33	7.51	15.09	6.21
260	36 to 140	121	139	20,350.63	18,957	1,393.63	72.91	5.32	13.58	6.84

TABLE 8. THE AVERAGE OF TABLES 5, 6 AND 7.

No. in Class.	Weight of Divisions.	Sexes. Ewes.	Weth's.	Average Weight of Bodies.	Average Weight of Fleeces.	Pounds of Body to 1 of Wool.	Per Cent. of Wool.
52	34 to 51	33	19	44.63	4.08	11.36	8.16
89	50 to 61	44	45	55.78	4.71	11.90	7.80
129	60 to 71	72	57	66.03	5.09	12.96	7.13
160	70 to 81	89	71	75.52	5.31	14.21	6.53
92	80 to 91	39	53	85.25	5.78	14.77	6.33
75	90 to 101	20	55	95.90	6.10	15.44	5.85
58	100 to 140	7	51	111.31	7.17	15.56	6.04
655	34 to 140	304	351	79.52	5.32	14.01	6.65

The value of these tables can only be known by careful comparison and thorough study of them. What may be learned I have not now the time to determine; but from a very cursory glance at them, I learn that Mr. Livingston's proposition is true. *Small sheep do shear more in proportion to their bulk than large ones*, without regard to age or sex. I learn, also, that yearling ewes shear the largest per centage they ever will shear, and that they shear less and less per centage as they grow older, till they are four years old. They gain until five, when they are in their prime, and raising a lamb at that age does not decrease the product of wool as it has done; but at six they have passed the meridian, and for the product of wool commence going "down hill."

It can be seen at a glance that wethers shear their largest per cent. when yearlings. At two, they have lost 1 per cent., after which they commence gaining, and continue to gain till they are five years old, after which I know nothing of the facts.

The facts are just as obvious in the classification by weight. The smallest sheep shear the largest per centage, and as their weight increases the fleece decreases in proportion, till they weigh more than 100 lbs., when it increases the fifth of 1 per cent.—a smaller increase than any decrease in either of the tables. This being the exception to what before seemed to be the rule, leads me to believe that the number in the class is too small, and that I ought to have had 100 sheep at least in this class to arrive at the truth. If it could be ascertained what per cent. of lambs 100 or 1,000 ewes would raise, and the average market price of average lambs on the 1st of October, it could be very easily calculated which would be the most profitable to keep, a flock of ewes or wethers. But as there is no likelihood of this being done, and as ewes are absolutely necessary to increase the flock, perhaps no farmer will be bold enough to have a flock exclusively of wethers, though I am confident that these tables will prove that the wethers have brought to the farm the most money at the average price of wool and lambs.

If I had the time I might pursue these deductions further, with profit to myself if not to those who read; but I think enough has already been disclosed to give any inquiring mind a stimulus to pursue the investigation. Every wool raiser ought to know which of his sheep he is keeping at a *profit* and which at a loss. By weighing the fleeces as they are shorn, he thinks he knows all about it, when in reality he knows nothing, or at the best only half. At sheep shearing the careful breeder ought to know what any sheep ought to shear when it comes on the floor. For instance, next year we shall have a dozen four year old wethers, any one of which ought to weigh somewhere near ninety pounds and shear seven pounds. If any one weighs up to the average of the last three years, and shears above the average, keep him — if below, sell him. When a ewe is brought on the floor, other things have to be taken into consideration, as she is to breed, viz., the quality of the wool, the form of the body, beside the weight of the fleece and weight of the body. If she has raised a lamb, it must be examined; if a ewe lamb, particularly. In our flock we have now made a *standard* to which we can refer; our efforts of course will be to excel it. Those who keep flocks expressly for their increase, will make a standard of their own, and those who keep sheep exclusively for wool, will make their standard accordingly. Every *breeder* ought to know every fact certainly, and have his *record* to refer to.

APPENDIX II — (page 75.)

THE AMERICAN MERINOS AT THE INTERNATIONAL EXHIBITION OF 1863.

It was noticed at page 75 that Mr. George Campbell, of West Westminster, Vermont, took American Merino sheep to exhibit at the International Exhibition at Hamburg, in July, 1863. The result was not ascertained in time to be alluded to in the body of this work.

Mr. Campbell found 1,761 sheep competing in the same class with his own. They were from the Austrian, Prussian and other States of Germany, and from France. Among the French sheep competing were about sixty belonging to the Emperor Napoleon. Mr. Campbell was awarded the first prize of fifty thalers for the best ram, the second prize of twenty-five thalers for the second best ram, and the first prize of fifty thalers for the best ewes.

The Committee of Award consisted of eighteen noblemen and gentlemen. The examinations were made by sub-committees, whose preliminary reports were subject to the revision of the general committee. The American sheep had encountered a certain degree of prejudice from their first arrival. The breeders of the old world, and particularly of Germany, seemed to think it audacious that Americans, who had so often imported sheep from Germany, should now enter the lists as competitors against them. And when a rumor began to gain ground that the sub-committee were disposed to award one and then two first prizes to the American Merinos, it caused loud expressions of dissatisfaction, which were promptly re-echoed in the German newspapers. Notwithstanding, and in defiance of all of this, the general committee with manly independence ratified the action of the sub-committee by a unanimous vote. On the official promulgation of the decision, the previous censures took the form of accusations. It was asserted that the committee had been unduly influenced. Thereupon Col. Daniel Needham, Corresponding Secretary of the Vermont State Agricultural Society, who was present at the Exhibition as the Commissioner of the State of Vermont, after conferring with the U. S. Commissioner, Gov. Wright, and Mr. Campbell, published a card in the German tongue, proposing a sweepstakes open to all the previous competitors — the award to be made by *a new committee*, to be selected by the German association under whose auspices and direction the International Exhibition took place. Col. Needham's proposal was that each competitor pay an entrance fee of $10; and if there were less than ten entries he offered himself to make up the prize to $100. This offer, (substantially a challenge to a new trial,) was posted and circulated among all the competitors. Mr. Campbell immediately entered his sheep, *but his was the only entry!* This rendered the triumph of the American Merinos absolute and undeniable; and the press and public, with that hearty honesty which always marks the German national character, did ample justice to the Americans and to the American sheep. Mr. Campbell sold his prize sheep, twelve in number, to a Prussian nobleman for $5,000.

APPENDIX H.

The highest priced foreign Merino sold at the Exhibition fetched but £40, or $200. The preceding facts are stated on the personal authority of Mr. Campbell and Col. Needham.

I cannot here withhold a pleasing fact which strikingly evidences the fairness and the modesty of the victorious exhibitor at Hamburg. Col. Needham informs me that Mr. Campbell on all occasions, signified to the breeders of Germany and France, and *requested* him, (Col. Needham,) to signify that he was not the founder or leading breeder of the improved family of American Merinos, which his (Mr. Campbell's,) sheep chiefly represented—but that this honor belonged to Mr. Hammond. Mr. C.'s show sheep were, if I remember aright, all from his celebrated ram "Old Grimes," bred by Mr. Hammond and got by his "Sweepstakes." "Old Grimes" competed against his sire in the great sweepstakes at the Vermont State Fair of 1861, and stood second. He is remarkable for individual excellence and as a stock getter.

I was one of those consulted by Mr. Campbell in reference to taking American Merinos to the International Exhibition, and I strongly encouraged him to do so. I had just as little doubt of their success then as now, provided they could receive fair play; and I never for an instant doubted that among the many Germans they would receive the same fair play which our stock and products have received at all these World's Fairs. In Germany as in England, we encountered some prejudice—but when the time for official action arrived, it always gave way like a morning mist before the broad, bright sun of personal and official honor.

LIST OF ILLUSTRATIONS.

	Page.
Merino Ram "Sweepstakes,"	Frontispiece
Spanish Wool,	16
Saxon Ram,	26
Merino Ewe, (Imported Paular,)	31
Merino Ewe, (Old Fashioned,)	34
Silesian Merino Ram,	38
Group of Silesian Ewes,	41
Leicester Ram,	45
Leicester Ewe,	47
Cotswold Ram,	48
Cotswold Ewe,	50
South Down Ram,	56
South Down Ewes,	57
Shropshire Down Ram,	62
Shropshire Down Ewe,	64
Shepherd's Crook,	139
Tagging, illustrated,	141
Toe-Nippers,	169
Folding Tables,	173
Fleece Ready for Press,	173
Fleece in Press,	174
Wool Press,	174
Tattooing Instruments, (three figures,)	184
Ears Tattooed,	184
Metal Ear-Mark,	185
Dipping Box,	187
Shed of Poles,	211
Sheep Barn, with Open Sheds,	213
Ground Plan of Sheep Barn and Yards,	217
Ground Plan of a Sheep Establishment,	218
Slatted Box Rack,	230
Wall Rack and Trough,	231
End View of Wall Rack and Trough,	232
Skull of a Sheep,	265
Teeth of the Sheep,	266
Section of Sheep's Head,	273
Gad-Fly of the Sheep,	274
The "Grub" or Larva of the Gad-Fly, (three figures,)	274
The Stomachs,	294
Internal Appearance of Stomachs,	295
The Intestines and Mesentary,	303
Spanish Sheep Dog,	397
The Scotch Sheep Dog, or Colley,	400
The English Sheep, or Drover's Dog,	407

INDEX.

A

Abortion, 329.
Abscess, 332.
Adams, Seth imports Merinos into United States, 24.
Allen, A. B. describes first French Merinos imported into United States, 35.
    recommends tar, sulphur and alum for diseased sheep, 194.
Anatomy of the sheep, 264, *et seq.*
    cut of skeleton, 264.
    cut of skull, 265.
    cut of teeth, 266.
    cut of section of sheep's head, 273.
    the omentums described, 293.
    cut of external appearance of stomachs, 294.
    cut of internal appearance of stomachs, 295.
    stomachs and their functions described, 295.
    mode of introducing medicines into the stomach, 299.
    cut of the intestines, 303.
Apoplexy, 280.
Arlington long-wooled sheep, origin of, 44, 54.
Atwood, Stephen, his family of Merinos described, 28, 29.
    his family of Merinos compared with Mr. Jarvis' 28.
    their improvement in other hands, 29, 30.
    a strict in-and-in breeder, 120.
    the improved Paulars receive a cross from his flock, 417-419.

B

Baker, the Messrs., their experiments in crossing French and American Merinos, 129 note.
Bakewell, Robert, the great improver of Leicester sheep, 45.
    an in-and-in breeder, 46, 119.
    in-and-in breeding formed an element of his success, 122.
    origin of his flock not probably drawn from different breeds, 133.
    his sheep improved by Cotswold blood, 47, 133.

Bakewell, Robert, he purposely rotted sheep, 376.
Barns for sheep, construction of, 212-219.
    cuts of 213, 217, 218.
    should be cleaned out in winter, 219.
Beanes, Capt., imports Teeswater and South Down sheep, 44 note.
Bedford, Dr., on the necessity of exercise, etc., to pregnant females, 222.
Beets as sheep feed, 243.
Bement, Caleb N., his account of C. Dunn's flock, 44 note.
Biflex Canal, disease of, 354, 355.
Bigelow, Dr., account of St. Johns-wort, 276.
Black-faced Scotch sheep described, 51.
    introduced into the United States by Samuel Campbell, 52.
    weight of their fleeces, 52.
    imported by Sanford Howard, 52.
Blacklock, Mr., cited in regard to diseases of sheep, 277, 316.
Blain, 291, 292.
Blanchard, H., introduces the Wool Depot system, 177.
Bleeding, place for, 314, 315.
    mode of performing, 314, 315.
Boardman, S. P., states cost of getting wool and other products to market from Illinois, 251 note.
    his article on prairie sheep husbandry, 260.
Brain, hydatid on, 277-279.
    water on, 279, 280.
    inflammation of the, 281.
Braxy, 311.
Breeding, in-and-in, extent of among improved Infantados, 30.
    definition of the term, 101.
    like produces like, 101.
    breeding back, 101.
    causes of hereditary transmission partly controllable, 101, 102.
    likeness inherited with uniformity among full bloods, 102.
    mongrels, etc., do not transmit likeness with uniformity, 102.
    counteracting the defects of one parent by the excellencies of the other, 103.
    hereditary predispositions to be regarded, 103.
    accidental characteristics, how accounted for, 103, 104.

Breeding, accidental characteristics are sometimes vigorously reproduced and become established, 103–106.
are peculiarities acquired after birth transmissable? 103 note.
accidental characteristics less transmissable when opposed to the special ones of the breed, 105.
breeding between animals possessing the same defect to be avoided, 106.
relative influence of sire and dam on progeny, 106.
the theory that the animal organization is transmitted by halves, 107.
Mr. Walker's modification of this theory, 107, et seq.
Mr. Spooner's views on the same subject, 107, et seq.
the foregoing theories examined, 107–110.
properties transmitted by degrees, not by halves, 109.
mode of their transmission, 109.
the ram oftenest transmits his external structure to progeny, 109, 110.
the ram oftenest gives size and a part of the qualities of the fleece, 110.
influence of the ewe on the progeny, 110.
causes of the ram's superiority in this particular, 110, et seq.
influence of higher breeding among full bloods, 111, 112.
influence of pure over grade, etc., blood, 111.
why rams of same blood differ in transmitting their qualities, 111.
influence of physical and sexual vigor, 112.
indications of these in the ram, 112 note, 113.
ability of rams to procreate at different ages, 113.
period of procreation in Merino, 113.
longevity of different breeds, 113, 114.
does the male which first impregnates a female influence her subsequent offspring? 114.
Mr. Cline's theory that small males and large females should be coupled, 114, 115.
In-and-in breeding, how the term is used in this work, 116.
Sir John Sebright's views, and his use of this term, 116–118.
prejudice against breeding in-and-in in the United States, 116.
its effect where hereditary diseases prevail, 117.
it results from Divine ordination in many instances, 117, 118.
difference between men and brutes in this particular, 118.
difference between wild and domesticated brutes in this particular, 118.
under what circumstances in-and-in breeding is fatal, 118.
under what circumstances it is innocuous, 118.
eminent foreign in-and-in breeders, 119.

Breeding, great extent of their in-and-in breeding, 119 note.
it formed an important element of their success, 122.
it is almost necessary in some cases, 122.
it is not safe for ordinary breeders, 122.
more have failed than have succeeded in it, 123.
is it more dangerous among grade animals? 123.
crossing breeds and families — (For everything connected with crossing see Cross-Breeding).
expedient to adhere to one breed and family if it possesses proper elements of improvement, 131.
the most splendid successes have been won in this way, 131, and note.
great skill of English breeders in breeding mutton sheep, 132 note.
breeding lambs for butcher, 133, 134.
breeding mutton sheep on the prairies, 135.
when cross-breeding is expedient, and when inexpedient generally, 136–138.
Breeds of sheep best adapted to different situations, 82–90.
rules for determining that adaptation, 82–90.
influence of markets, 82–85.
influence of climate, 85, 86.
influence of vegetation, 86–88.
influence of soils, 88, 89.
influence of herding, 89.
influence of associated branches of husbandry, 89, 90.
comparative hardiness of English, 87.
working qualities of different breeds, 87.
crossing between different—(see Cross-Breeding.)
longevity of different, 113.
Broad-Tailed sheep introduced into the United States, 53.
bred pure in South Carolina, 53.
Bronchitis, 326.
Brugnone cited in regard to diseases of sheep, 277–302.
Bruises and strains, 382.
Buignot inoculates for small pox, 349.
Burs should be eradicated from pastures, 142.
the different kinds of, injurious to wool, 142.

C

Campbell, George, takes Merinos to World's Fair at Hamburg, 75.
length of wool on sheep taken to World's Fair, 75.
pedigrees of the sheep, 76.
his mode of tattooing sheep, 184.
his sheep victorious at the World's Fair, 438, 439.
his honorable conduct, 439.
pedigree of his stock ram, 439.
Campbell, Samuel, and James Brodie, import Leicester sheep, 47.

INDEX.   443

Campbell, Samuel, and James Brodie, cuts of a ram and ewe belonging to them, 45, 47.
  import Cheviot sheep, 52.
Canada Breeders of, 351.
Carcass the first point to be regarded in sheep, 69.
  proper form and size of the Merino, 69.
Carrots as sheep feed, 243.
Castration, 161.
Catarrh, 268, 318, 319.
  Malignant epizootic, 319-324.
Catching and handling sheep, proper mode of, 131-141.
Chamberlain, William, his account of the present Merinos in Spain, 17, 18.
  introduces Silesian Merinos into the United States, 39.
  his description of his sheep, 39-42.
  cut of a group of his ewes, 41.
  a close in-and-in breeder, 120.
  time he has his lambs yeaned, 143 note.
Cheviot sheep introduced into the United States, 52.
  character of the unimproved family, 52.
  the improved family described, 52, 53.
Chilled Lambs, how treated, 148, 149.
Chinese, or Nankin sheep in the United States, 54.
Choking, 292, 293.
Clapp, the Messrs., their experiments in crossing French and American Merinos, 129 note.
Clark, Bracy, cited in regard to diseases of sheep, 274, 275.
Clift, Leonard D., imports Lincoln sheep in 1835, 50.
  character of his sheep, 50.
Climate to be regarded in selecting a breed of sheep, 85, 86.
Cline, Mr., his views on disparity in size of sire and dam in breeding, 114.
Closed Teats, 157.
Clover, as sheep feed, 235, 237, 246.
Clumps of trees in pastures, utility of, 212.
Colic, 310.
Colley, (See Dog.)
Collins, D. C., introduces French Merinos in the United States, 34.
  description of his sheep, 35.
Coloring Sheep artificially, a fraud, 81.
Confinement, effect of on pregnant ewes, 222, 223.
Congenital Goitre, or swelled neck, 152-154.
Constipation of sheep, 221, 228, 310.
  of young lambs, 149.
Consumption, 327, 328, 379.
Corning, Erastus, with Wm. H. Sotham, imports Cotswold sheep, 48.
Cornstalks as sheep feed, 245, 258.
Cossit, Capt. Davis, his remarkable success in crossing Infantado and Saxon Merinos, 130 and note.
  pedigree of his ram, "Wrinkly 3d," 415.
Costiveness, (See Constipation.)
Cotswold Sheep introduced into the United States about 25 years since, 48.

Cotswold Sheep, imported by Mr. Dunn in 1832, 48.
  imported by Messrs. Corning & Sotham in 1840, 48.
  imported by Henry G. White, 49.
  described by Mr. Spooner, 49.
Crook, shepherd's, manner of using, 130.
  cut of, 139.
Cross-breeding, meaning of term as used in this book, 124.
  effects of between the Merinos and coarse breeds, 124.
  the Merino unimprovable by such a cross, 124.
  the Merino cross improves coarse sheep for certain purposes, 125.
  the cross between Merino and mutton sheep results in failure, 124, 125.
  the cross between the Merino and long wools, 125.
  the cross between the Merino and Downs, 125.
  permanent intermediate varieties unattainable, 125.
  peculiar tenacity of hereditary transmission in the Merino, 125.
  due probably to its great purity and antiquity of blood, 125 note.
  coarse breeds can be merged in it, 126.
  grade flocks started in Texas, 126.
  successful cross between Merino and Mexican sheep, 126.
  experience of Mr. Kendall in this particular, 126, note.
  choice rams desirable in such a cross, 127.
  grades never equal to pure Merinos, 127.
  French ideas on this subject, 127.
  German ideas on same subject, 127.
  degrees of blood in ascending crosses reckoned, 127 note.
  crossing different families of Merinos, 127-130.
  effect of in the French Merino, 128.
  effect of, in Mr. Jarvis' flock, 128.
  effect of, in the Rich or improved Paulars, 128 and note.
  effect of in the Silesian Merinos of the United States, 128, 129.
  between the American and French Merino, 129 and note.
  between the American and Saxon Merino, 129.
  remarkable result of an improved Infantado and Saxon cross, 130 and note.
  inexpediency of crossing for the sake of crossing, 130, 131.
  ordinary reasons for crossing unfounded, 131.
  bad effects of frequent and unmeaning crosses, 131.
  always better to adhere to one breed and family if it contains the elements of improvement, 131.
  the most splendid successes have been secured in this way, 131, and note.
  crossing between English breeds and families, 132.

Cross-breeding, the Hampshire, Shropshire and Oxfordshire Downs produced in this way, 132.
   but the failures in blending breeds have been far more numerous, 132.
   skill of the English breeders, 132 note.
   successful to obtain larger and earlier lambs for the butcher, 133.
   expediency of thus crossing with local families, 134.
   Mr. Thorne's experience in this particular, 134, 135 note.
   an analogous cross for mutton raising expedient in Western States, 135.
   the English family which should be selected for this purpose, 135, 136.
   the cross should stop with the first one, 134.
   recapitulation, showing when crossing is expedient, and when inexpedient, 136–138.
Crossing, (See Cross-breeding.)
Cutaneous Diseases, unnamed ones, 344, 345.
Cuts, 380.
Cutting teeth, 150.
Cystitis, 337.

D

D'Arboval Hurtel cited in regard to diseases of Sheep, 314, 349, 350.
Darlington, Dr. his account of St. John's-wort, 269.
Darwin, M., his account of South American sheep-dogs, 405.
Daubenton's directions for bleeding sheep, 314.
Delafond, Mr., on history of small pox, 349.
Delessert, M., imports Merinos into United States, 22.
Dewees, Dr., on proper treatment of pregnant females, 336.
Diarrhea, 306–308, 380.
   in young lambs, 151.
Dickens, Mr., cited in regard to diseases of sheep, 377.
Dick, Professor, on hoof-rot, 358 note.
Diseases and wounds of Sheep, 261, et seq.
   comparatively small number of in United States, 261, 262.
   low type of American sheep diseases, 262.
   Abortion, 329.
   Abscess, 382.
   Apoplexy, 280.
   Biflex Canal, disease of 354.
   Blain, 291, 292.
   Braxy, or inflammation of the bowels, 311.
   Bronchitis, 326.
   Bruises and Strains, 382.
   Catarrh, 268, 318, 319.
   Catarrh, malignant epizootic, 319–324.
   Choking, 292, 293.
   Cold (see Catarrh.)
   Colic, 310.
   Constipation, 221, 228, 310.
   Constipation in young lambs, 149, 150.
   Consumption, 327, 328, 379.

Diseases and wounds of Sheep, Costiveness, (see Constipation.)
   Cutaneous diseases, unnamed ones, 344, 345.
   Cuts, 380.
   Cystitis, (see Inflammation of the bladder.)
   Diarrhea, 306–308, 380.
   Diarrhea in young lambs, 151.
   Distemper, the, 324.
   Dog Bites, 381.
   Dropsy, acute, or Red Water, 304.
   Dysentery, 308–310, 379, 380.
   Enteritis, 306.
   Epilepsy, 282, 283, 380.
   Epizootic of 1846–47, 319 et seq.
   Eye, inflammation of, 272.
   Fever, 316.
   Fever, inflammatory, 316, 317.
   Fever, malignant inflammatory, 317, 318.
   Fever, parturient, 331–337.
   Fever, puerperal, 331–337.
   Fever, typhus, 318.
   Foot-rot—(see Hoof-Rot.)
   Fouls, 356.
   Fractures, 354.
   Garget, 157, 330.
   Gravel, 355.
   Grub in the head, 273, 277.
   Goitre, congenital, 152, 154.
   Head, Grub in, 273–277.
   Hereditary diseases, 379, 380.
   Hoof-Rot, 356–371, 381.
   Hoove, 299–301.
   Hydatid on the Brain, 277–279, 380.
   Ignis Sacer, 344.
   Inflammation of cellular tissue under the tongue—(see Blain.)
   Inflammation of the bladder, 337.
   Inflammation of the brain, 281.
   Inflammation of the coats of the Intestines, 306.
   Inflammation of the Eye 272.
   Inflammation of the lungs, (see Pneumonia.)
   Inflammation of the udder, (see Garget.)
   Inversion of the womb, 145, 330.
   La Clavelee, (see Small-pox.)
   Lameness, 355, 356.
   Madness, (see Rabies.)
   Obstructions of the gullet, 292, 293.
   Opthalmia, 272, 279.
   Palsy, 283.
   Parturient fever, 331–337.
   Phthisis, (see Consumption.)
   Pining, 312.
   Pinning, 151.
   Pleurisy, (see Pleuritis.)
   Pleuritis, 326, 327.
   Pneumonia, 325, 379.
   Poisons, 301, 302.
   Puerperal fever, 331–337.
   Rabies, 283–290.
   Rheumatism, 155, 156, 379.
   Rot, the 372–378.
   Rot, cut of the Fluke, 374.
   Scab, erysipelatous 344.
   Scab, the 338, 343.

INDEX.   445

Diseases and wounds of Sheep, Scours (see Diarrhea.)
  Scrofula, 378, 380.
  Small-pox, 345–353.
  Sore Face, 269–271.
  Sprains, 382.
  Stretches, 310.
  Swelled Head, 268.
  Swelled Lips, 271.
  Swelled Neck, 152, 154, 380.
  Teeth, cutting of the 150.
  Tetanus, or Locked-Jaw, 281, 282.
  Variola Ovina—(see Small-pox,)
  Water on the brain, 279, 280.
  Wild Fire, 344.
  Worms, 312.
  Wounds, 380–382
  Wounds, lacerated and contused, 381.
  Wounds, poisoned, 381, 382.
  Wounds, punctured, 381.
Disowning Lambs, 158, 159.
Distemper, the, 324.
Docking Lambs, 160, 161.
Dog, bites of the, 381.
  the dog, in connection with sheep, 393, et seq.
  injuries inflicted by, on sheep, 393–396.
  sheep dog described by Buffon, 396.
  Spanish, 397.
  Hungarian, 400.
  French, 401.
  Mexican, 401–405.
  South American, 405, 406.
  other large races, 406.
  English, or drover's, 407.
  Scotch, or Colley, 408–410.
  mongrel Colley, a sheep killer, 410.
  accustoming the sheep to the dog, 411.
Down Sheep, (see South Downs, Hampshire Downs, Shropshire Downs and Oxfordshire Downs.)
Drafting and selection, in flocks, 179.
Dropsy, acute, 304.
Drying off ewes, 158.
Dun, Finlay, on hereditary diseases, 379, 380.
Dunn, Christopher, origin of his Leicester flock, 44.
  character of his flock, 44 note.
  crosses it with Cotswold rams, 48.
Dunglison's Medical Dictionary, cited passim.
Dupont de Nemours, imports Merinos into United States, 22.
Dysentery, 308–310, 379, 380.

E

Elithorp, Prosper, length of his Merino wool, 76.
  crosses the Paular and Infantado sheep, 128 note.
  his remedy for stretches, 310.
  his connection with the origin of the improved Paular family, 417–419.
  furnishes an account of origin of, 419.
Ellman, Mr., his success in breeding South Down sheep, 55 et seq.
  followed in-and-in breeding, 119.
  it was an element of his success, 122.

Ely, David, his "little-cared" sheep, 104.
English Breeders, their great skill in breeding mutton sheep, 132 note.
Enteritis, 306.
Epilepsy, 282, 283, 380.
Epizootic among sheep in 1846–47, 319, et seq.
  the lamb epizootic of 1862, 154, 226.
  the term defined, 226 note.
Escurial Merino, 14.
Ewe, influence of on progeny, 110.
  fall feed and shelter necessary for, 202–205.
  effect of neglect in this particular, 203, 204.
  "hunger rot" described, 203, 204.
  subject to other diseases when in low condition, 204.
  does not take the ram uniformly when poor, 205.
  selection of for the ram, 205, 206.
  coupling with the ram, modes of, 206, 207.
  period of gestation in, 207.
  want of sagacity in protecting its young, 213.
  injurious effects of close confinement on, 222, et seq.
  should not be confined to dry feed in winter, 222, et seq.
  its prolificacy affected thereby, 222 et seq.
Exercise important for pregnant ewes, 223, 226.
Experiments in fattening sheep, 418–425.
Eye, inflammation of, 272.

F

Face, sore, 269–271.
Fall management of sheep, (see management of sheep in fall).
Fat-Rumped sheep introduced into the United States, 53.
Fattening Sheep, 418–425.
Fay, Richard S., imports Shropshire sheep into United States, 66.
  character of his sheep, 66, 67.
Feed, different values of, for fattening, 420–426.
  experiments in mixing, 419 et seq.
Feeds for sheep—(see Fodder.)
Feeding sheep, Mr. Pawlett's experiments in, 418–425.
Felting property of wool, how produced, 16.
Fences for sheep, value of different, 233, 245.
Fever, 316.
  inflammatory, 316, 317.
  malignant inflammatory, 317, 318.
  typhus, 318.
  parturient, 331–337.
  puerperal, (see fever parturient).
Fischer, Ferdinand, established the family of Merinos, now termed Silesian in the United States, 39.
Fischer Louis, son of preceding, continues the flock, 39.

G

Fischer, Louis, effect of his cross between the Negretti and Infantado, 128, 129.
Fleece, proper characteristics of in a Merino, 71, 72.
Fleischmann, Charles L., his drawing of a Saxon ram, 26.
  his statements about German cross-bred sheep, 127.
  his drawings of marking instruments, 184.
Fodder for sheep, value of different, 233-245.
Folds, or wrinkles, proper amount in the skin of the Merino, 70, 71.
Forshey, Caleb G., on the climatology of Texas, 428 *et seq.*
Foster, William, first introduced Merinos into United States, 22.
Foot-Rot, (see Hoof-Rot.)
Fouls, 356.
Fractures, 254.

G

Gad-fly of the sheep, cut of, 274.
  cut of Larvæ of, 274.
  their effect on sheep, (see Grub in the Head.)
Garget, 157.
Gayot, inoculates for small-pox, 350.
Gasparin, cited in regard to sheep diseases, 283, 314.
Gandeloupe Merino, 14.
Geddes, James, cut of his Silesian Merino ram "Carl," 38.
  cut of his improved wool-press, 174.
Geddes, Hon. George, experiments in feeding beets to sheep, 243.
Germany, Breeders of, at World's Fair, 438, 439.
Gestation, period of in the ewe, 207.
Gilbert, his description of the origin of the Rambouillet flock, 19.
Girard, inoculates for small-pox, 349.
Goitre, congenital, 152, 154.
Gold Drop, Mr. Hammond's ram, pedigree of, 121, 122.
Goodale, S. L., his work on the principles of breeding, 114 note, 123.
Gossip, George H. and Brother, import Lincoln sheep into the U. States, 50.
Gragnier inoculates for small pox, 349.
Grasses, most valuable ones for sheep, 233, 234, 235, 237.
Gravel, 355.
Grease in wool—(see Yolk.)
Greaves, Mr. W., cited in regard to sheep diseases, 305.
Greer, W. F., in regard to hoof-rot, 371.
Grennel, James S., his report on sheep husbandry to the Massachusetts State Board of Agriculture, 51.
  his account of New Oxfordshire sheep, 51.
  his statement of comparative waste in cooking beef and mutton, 83.
  his statement of increase of sheep bought in Boston market between 1839 and 1859, 84.
  his account of sheep poisons, 301

Grinnel, J. B., his statement of cost of getting wool and other products to market from Iowa, 251 note.
  his article on prairie sheep husbandry, 260.
Grove, Henry D., his account of importations of Saxon sheep, 25.
  weight of fleeces of his Saxon flock, 25 note.
  his account of origin of the "little eared" sheep, 104.
Grognier, Prof., his account of French sheep dogs, 401.
Grub in the head, 273-277.
Guillaume inoculates for small-pox, 349.
Gullet, obstructions of, 292, 293.
Gum on wool—(see Yolk.)

H

Hammond, Edwin, commences his flock with Infantado or Atwood sheep, 29, 30.
  the great improver of the Infantados, 29.
  present character of his flock, 29, 30.
  his ram Sweepstakes—(the frontispiece of this volume,) 29.
  length of Sweepstakes' wool, 76.
  pedigree of Sweepstakes, 121.
  description of Sweepstakes, 413.
  the points which Mr. H. has bred for, 30.
  the extent of his in-and-in breeding, 30, 120.
  pedigrees of his leading stock rams and ewes, 14, 122.
  in-and-in breeding a lever of his success, 122.
  plan of his sheep establishment, 218.
  description of his leading animals, and course of breeding, 412-416.
Hampshire Downs described by Professor Wilson, 59, 60.
  Mr. Spooner's account of their origin and blood, 60, 61.
Handling Sheep—(see Catching and Handling.)
Harrison, Dr., on symptoms of rot, 372.
Head, grub in, 273-277.
  swelled, 268.
Herding, capacity for in different breeds of sheep, 89.
Hereditary Diseases, 379, 380.
Hogg, James, cited in regard to diseases of sheep, 268, 278, 291, 312, 364.
Hoof-Rot, 356-371, 380.
Hoofs, shortening of the, 168, 169.
  cut of toe-nippers, 169.
Hoove, 299-301.
Horns on sheep, shortening, etc., 189.
Howard, Charles, describes origin of Shropshire Downs, 63, 64.
Howard, Sanford, imports Cheviot sheep, 52.
Huard inoculates for small-pox, 349.
Humphreys, David, imports Merinos into the United States, 23.
  breeds in-and-in, 120.

INDEX. 447

Humrickhouse, T. S., his inquiries as to present flocks of Spain, 16.
"Hunger-Rot," how produced, 203, 204.
Hydatid on the brain, 277-279, 380.
Hyde, Professor, his dissections of sheep, 290, 321.

I

In-and-in breeding — (see breeding in-and-in.)
Ignis Sacer, 344.
Illinois, sheep husbandry in, 248, *et seq.*
Infantado Merinos in Spain, 14.
  the improved Infantados of the United States, 28, *et seq.*
  closely bred in-and-in in the United States, 120.
  one of the families on which the American Silesian are based, 129.
  leading animals of the improved family, 412-416.
Inflammation of the eye, 272.
  of the brain, 281.
  of cellular tissue under the tongue, (see Blain.)
  of coats of intestines, 306.
  of the bowels, 311.
  of the lungs, 325.
  of the bronchial tubes, 326.
  of the udder, 157, 330.
  of the bladder, 337.
Injections, 150.
Inoculation for small-pox, 349, *et seq.*
Iowa, starting a sheep establishment in, 427, 428.
International Exhibition at Hamburg, 438.
  triumph of American Merinos at, 438, 439.
Inverted womb, how treated, 145.

J

Jarvis, William, imports Merinos into the United States, 23, 24.
  crosses them with the Saxons, 24.
  breeds back, but crosses his Merino families, 24.
  weight of his fleeces and prices of his wool, 24.
  his Merinos established as a family, 27.
  his sheep described, 27.
  effect of his crossing different families, 128.
  his remedy for hoof-rot, 363.
  his family crossed with the Improved Paulars, 417, 418.
John's-wort — (see St. John's-wort.)

K

Kendall, George Wilkins, the wintering of his sheep in 1860, 89.
  his successful cross between Merinos and Mexican sheep, 126 note.
  mean temperature near his residence, 249 note.
  his account of Mexican sheep dogs, 404.

Klippart, John H., his statement of the number of sheep killed by dogs in Ohio, 393-396.

L

La Clavelee — (see small-pox.)
Lambs, management and diseases of in spring — (see Spring Management.)
  management of in fall, after weaning, 198-201.
  importance of fall shelter for, 201.
Lambing, proper time for, 142.
  proper place for, 143.
  mechanical assistance in, 144.
  administering cordials, etc., during, 145.
Lameness from traveling — (see Travel Sore.)
Langlois inoculates for small-pox, 349.
Lasteyrie, his description of the Merino families, 14.
  his account of the weight of French Merino fleeces, 19.
Lax, Mr., imports Leicester sheep into the United States, 44.
Leicester sheep, 43.
  probably introduced into United States by Gen. Washington, 44.
  imported by Mr. Lax, 44.
  imported by Capt. Besnes, 44.
  cut of Messrs. Campbell & Brodie's ram, 45.
  cut of one of their ewes, 47.
  Prof. Wilson's description of the Leicesters, 45-47.
  their origin, 45.
  Mr. Bakewell selected from different families, 46.
  he then bred in-and-in, 46.
  not so hardy as the other large breeds, 46.
  their early maturity, 46.
  now improved by a dip of Cotswold blood, 47, 133.
Leonesa, the best Spanish families of the Merino, so called, 14.
Lewis, Dr., statement regarding Spanish sheep dogs, 399.
Lincolnshire sheep imported into the United States by Leonard D. Clift, 50.
  imported by Geo. H. Gossip & Brother, 50.
  character of the imported sheep, 50.
Lips, swelled, 271.
Livermore, George, table of wool prices furnished by him, 92-94.
Livingston, Robert R., states weight of Spanish fleeces, 16.
  imports Merinos into United States, 22.
  character of their descendants, 23.
  weight of his Merino fleeces, 23.
  cited in regard to diseases, 340, 341.
  on proportion of wool to surface, 433.
Locked-jaw, 281, 282.
Longevity of different breeds, 113.
Loveland, R. A., his account of starting a sheep establishment in the new Western States, 427, 428.

INDEX.

Lyman. J. H., his account of Mexican sheep dogs, 401-404.

M

Madness—(see Rabies.)
Maggots on sheep, how destroyed, 189, 190.
Management of sheep in spring, 139.
   catching and handling, 139-141.
   tagging, 141, 142.
   burs in pastures to be eradicated, 142.
   lambing, 142, 143.
   proper place for lambing, 143, 144.
   mechanical assistance in lambing, 144, 145.
   inverted womb, how treated, 145, 146.
   management of new-born lambs, 146.
   artificial feeding of lambs, 146-148.
   chilled lambs, 148, 149.
   constipation or costiveness of lambs, how treated, 149, 150.
   cutting teeth, 150.
   pinning, how treated, 151.
   diarrhea or purging of lambs, how treated, 151.
   congenital goitre, or swelled neck, 152-154.
   imperfectly developed lambs, 154, 155.
   rheumatism in lambs, 155, 156.
   treatment of ewe after lambing, 156, 157.
   closed teats, 157.
   inflamed udder, 157.
   drying off ewes, 158.
   disowning lambs, 158, 159.
   pens, 159.
   foster lambs, 159, 160.
   docking lambs, 160, 161.
   castration of lambs, 161.
Management of sheep in summer, 163-197.
   modes of washing sheep, 163, 164.
   utility of washing sheep, 163, 168.
   shortening the hoof, 168, 169.
   cut of toe-nippers, 169.
   time between washing and shearing, 170.
   shearing, 170-172.
   stubble shearing and trimming, 172.
   shearing lambs and shearing sheep semi-annually, 172.
   doing up wool, 173-175.
   cut of folding table, 173.
   cut of fleece ready for press, 173.
   cut of fleece in press, 174.
   cut of wool-press, 174.
   frauds in doing up wool, 175.
   storing wool, 176.
   place for selling wool, 177.
   wool depots and commission stores, 177.
   sacking wool, 177.
   drafting and selection of flock, 179.
   registration, 180.
   marking and numbering, 182-186.
   Von Thaer's mode of, 183.
   German mode of tattooing, 183.
   a third mode of marking, 184.
   a fourth mode of marking, 185.
   cut of instruments for tattooing, 184.

Management of sheep in summer, 163-197.
   cut of ears tattooed, 184.
   cut of copper ear marks, 185.
   storms after shearing, 186.
   sun-scald, 186.
   ticks, how destroyed, 187-189.
   cut of dipping box, 187.
   shortening horns, etc., 189.
   maggots, 189, 190.
   confining rams, 190, 191.
   training rams, 191.
   fences, care of, 192.
   salt necessary for sheep, 192.
   tar, sulphur, alum, etc., for sheep, 193.
   water in pastures, 194.
   shade in pastures, 195.
   housing sheep in summer, 195.
   pampering sheep, 196, 197.
Management of sheep in the fall, 197-210.
   weaning and fall feeding lambs, 197-201.
   sheltering lambs in fall, 201.
   fall feeding and sheltering breeding ewes, 202-205.
   selecting ewes for the ram, 205, 206.
   coupling, 206, 207.
   period of gestation, 207.
   management of rams during coupling, 207, 209.
   dividing flocks for winter, 209, 210.
Management of sheep in winter, 210-247.
   winter shelter, 211.
   temporary sheds, 211.
   cut of shed of poles, 211.
   clumps of trees and stalls, 212.
   hay barns with open sheds, 212.
   sheep barns or stables, 214, 219.
   cut of sheep barn and yards, 217.
   cut of a sheep establishment, 218.
   cleaning out stables in winter, 219.
   yards, how arranged, etc., 220.
   littering yards, 220.
   confining sheep in yards and to dry feed, 221-228.
   hay racks, 229.
   cut of slatted box rack, 229.
   cut of wall rack and trough, 230.
   cut of end view of wall rack and trough, 231.
   water for sheep in winter, 232.
   amount of food consumed by sheep in winter, 233.
   value of different fodders, 233, 243.
   nutritive equivalents, 234.
   table of nutritive equivalents, 235.
   proportion in which different nutriment increases live weight, wool and tallow, 236, 238.
   cost and economy of the different kinds of, 238, *et seq.*
   experiments in feeding, 239-242.
   mixed feeds, 243-245.
   fattening sheep in winter, 245, 246.
   regularity in feeding, 246, 247.
   Salt in winter, 247.
Management of Sheep on Prairies—(see Prairie Sheep Husbandry.)
Markets, influence of, in determining the selection of a breed, 82.

INDEX. 449

Marking and numbering sheep, different modes of, 182-186.
Marshall, Gen. O. F., his mode of salting sheep in winter, 247.
Marshes, access to not dangerous to sheep in Northern States, 88, — (see Salt Marshes.)
Mauchamp Merinos in France, 104.
Meat and Wool, proportion of, between sheep of different ages and sexes, 433 *et seq.*
Medicines, mode of introducing into the stomach of sheep, 299.
    explanation of medical terms used, 343, 344.
    list of medicines used in diseases of sheep, 383, 392.
Merino, American, introduced into United States, 22.
    little noticed before 1807, 24.
    prices of wool from 1807 to 1824, 24.
    prices of sheep from 1807 to 1815, 24.
    circumstances affecting prices of wool, 24.
    established as a variety in United States, 27.
    the mixed Leonese or Jarvis family, 27, 28.
    the Infantado or Atwood family, 28.
    Mr. Hammond, founder of the improved Infantados, 29, 30.
    the improved Paular or Rich family, 30-33.
    other American Merino families, 33.
    prices of, in winter of 1862-63, 69 note.
    proper form and size of, 69.
    the different families should not be merged, 69, 70.
    proper qualities of skin of, 70.
    proper amount of folds or wrinkles, 70.
    characteristics to be sought in the fleece, 71, 72.
    spotted and black Merinos, etc., 72 note.
    the most profitable quality of wool and breed of sheep to propagate, 72, 73.
    evenness of the fleece, 73.
    trueness and soundness of wool, 74.
    pliancy and softness of wool, 74.
    style of wool, 75.
    length of wool, 75, 76.
    endures extremes of weather better than any other valuable breed, 86.
    is a better working sheep than the English, 87.
    effect of abundant food on, 88.
    will not endure wet soils, 88.
    the great capacity of, for herding, 80.
    average production of wool per head in large flocks, 98.
    annual value of manure of, 99.
    its manure far more valuable than that of the horse or cow, 99 note.
    annual value of lambs, 99.
    comparative profits of, in different parts of the United States, 99.
    full bloods as cheaply raised as grades, 99.
    profits of growing on lands worth $50 per acre, 100.

Merino, American, breeding in-and-in of the improved Infantados, 120.
    pedigrees of celebrated improved Infantados, 121, 122.
    origin of the improved Paulars, 128 note.
    effect of crossing American Merinos with coarse breeds — (see Cross-Breeding.)
    effect of crossing different families of Merinos—(see Cross-Breeding.)
    origin of improved Infantados, 412-416.
    leading early animals of Mr. Hammond's flock, 412-416.
    origin of improved Paulars, 416-418.
    leading early animals of the family, 416-418.
    victorious at World's Fair at Hamburg, 438.
Merino, French, origin of, 18, 19.
    stock from which the Rambouillet flock sprung, 19.
    weight of fleece given by Lasteyrie, etc., 19, 20.
    general description of, by Trimmer, in 1827, 19.
    introduced into the United States by D. C. Collins, 35.
    A. B. Allen, description of them, 35.
    imported by John A. Taintor, 36.
    weight of fleeces of this family, given by J. D. Patterson, 36.
    character of the variety, 36, 37.
    crossed with American Merinos, 129.
Merino, Saxon, origin of, 20.
    management of, in Germany, 20.
    its characteristics of carcass and fleece, 20.
    introduced into United States in 1824, 25.
    circumstances affecting its success in United States, 25, 26.
    supercedes the Spanish, and in turn superceded by them, 25.
    cut of Von Thaer's Saxon ram, 26.
Merino, Silesian, introduced into the United States, 39.
    description of them by Mr. Chamberlain, 39-42.
    cut of a group of Mr. Chamberlain's ewes, 41.
    have been closely bred in-and-in, 120.
    effect of the original cross from which the family was established, 128, 129.
Merino, Spanish, origin of, 13, 125 note.
    provincial varieties of, in Spain, 13.
    cabanas, or families of, in Spain, 13, 14.
    migrations of, in Spain, 13.
    general treatment of, in Spain, and effects, 13, 14.
    its wool, character and color, 15, 16.
    its wool, compared with that of American Merino, 15.
    fineness and felting properties of its wool, 16.
    cut illustrating appearance of wool, 16.
    best families of, lost to Spain, 16, 17.
    the character of the present flocks of Spain, 17, 18.

450 INDEX.

Merino, Spanish, the earlier families introduced into the United States by different persons, 22, 23.
    black ones imported, 23.
    the different families bred in-and-in, 119, 120.
Messenger, Thos., imports Hampshire Downs into the United States, 61.
Miguel inoculates for small pox, 350.
Miller, George, imports Shropshires into Canada West, 65.
Mississippi, sheep husbandry in, 248 *et seq.*
Mixed feeds for sheep, 243–245.
Morrell, L. A., author of American Shepherd, 269.
    cited in regard to sheep diseases, 269, 301, 311.
Myrtle & Ackerson, length of their Merino wool, 76.

N

Nankin sheep in the United States, 54.
Native sheep of the United States, 43.
Neck, swellings of, 152, 154, 380.
Needham, Col. Daniel, attends World's Fair as Commissioner of Vermont, 438.
    challenges the breeders of Europe, 438.
Negretti Merinos, 14, 129.
    weight of fleeces of flock of King of England, 16.
Nelson, Capt. Allison, his account of Mexican sheep dogs, 405.
New Oxfordshire sheep imported into the United States, 51.
    described by L. Smith, a breeder of them, 51.
Nomadic shepherds on the prairies, 250.
Numann, Prof. A., on treatment of small pox, 348, 349.
Nutritive equivalents in sheep feed, 334 *et seq.*
    table of nutritive equivalents, 235.

O

Ohio, destruction of sheep in, by dogs, 393, 396.
Oil in wool—(see Yolk.)
Old Robinson Ram, his history and qualities, 113.
    his pedigree, 128 note.
    his pedigree and qualities, 416–418.
Opthalmia, 272, 379.
Orton, Mr., his theory of breeding, 107 *et seq.*
Otter sheep, 43.
Oxfordshire Downs, described by Mr. Howard, 65.
    introduced into United States, 66.
    description of Mr. Fay's sheep, 66, 67.

P

Paget, Mr., his account of Hungarian sheep dogs, 400.
Palsy, 283.
Pampering sheep, effects of, 196, 197.

Parturient fever, 331–337.
Patterson, John D., describes French Merinos, 36.
Paular Merinos, 14.
    improved in United States, 32, 33, 119.
Pawlett, T. E., his essay on management of sheep, 190.
    his views on fall feeding of lambs, 199.
    his experiments in winter feeding, 418–425.
Pea-haulm as sheep feed, 235, 245.
Pedigree, mode of keeping, 121.
Persian sheep in United States, 54.
Peters, Theodore C., opens a Wool Depot in 1847, 177.
    his letter in regard to sheep diseases, 262.
    his account of sheep dogs, 407, 409.
Petri, his measurements, etc., of Spanish sheep, 14.
Pining, 312.
Pinning, 151.
    of young lambs, how treated, 151.
Pleurisy, 326, 327.
Pleuritis, 326, 327.
Pneumonia, 325, 379.
Poisons, 301, 302.
Porter, Commodore, imports Broad-Tailed sheep into United States, 53.
Powell, John Hare, breeds Tunisian Mountain sheep, 53.
    imports South Downs into United States, 57.
    his account of Spanish sheep dogs, 400.
Prairie Sheep Husbandry, 248–260.
    comparative climate of Prairie States, 248.
    great advantages for wool growing in, 249.
    nomadic shepherds in, 250.
    acclimation of sheep in, 250.
    profits of wool growing in, over Eastern States, 251.
    wool the most profitable staple in, 251 and note.
    management of sheep in summer in, 252
    lambing in prairie flocks, 252, 253.
    folds and dogs, 253.
    stables, 253.
    herding, 254.
    washing, 254.
    shearing, 254.
    storing and selling wool, 254.
    ticks on sheep, 255.
    prairie diseases, 255, 256.
    feeding salt, 256.
    weaning lambs, 256.
    prairie management in winter, 256.
    winter feed, 258, 259.
    sheds or stables, 259.
    water, 260.
    location of sheep establishment, 260.
Pregnancy, proper treatment during, 221–228, 336.
Price, Mr., cited in regard to sheep diseases, 262.
Puerperal Fever—(see Parturient Fever.)
Pulse, its frequency in healthy sheep, 314.
    where it is felt, 314.
Purging—(see Diarrhea, Dysentery.)

INDEX. 451

R

Rabies, 283, 290.
Racks for feeding sheep, 229-231.
  cut of slatted box rack, 229.
  cut of wall racks, 230.
  cut of end view of same, 231.
Ram, influence of, in breeding, 108-115.
  oftenest gives the form to progeny, 109, 110.
  points to be regarded in, 111, 112.
  capacity of, to procreate, 113, 209.
  proper size of, 114.
  horns of, require attention, 189.
  confinement of, 190.
  training of, 191.
  treatment of, when vicious, 191.
  selecting ewes for, 205.
  modes of coupling, 206, 207.
  management of, during coupling, 207-209.
  causes which sometimes render them unsure stock-getters, 207 and note.
  when they require mechanical assistance, 207 note.
  preparation of, for coupling season, 208.
  feed inclosures, etc., 208.
Reaumur's experiments, showing how feeds increase animal products, 236, 238.
Red Water, 304.
Registration of sheep, 180-182.
  form of a register, 181.
Regularity in feeding, importance of, 246, 247.
Remelee, Loyal C., crosses the Paular and Infantado sheep, 128 note.
  his connection with the origin of the improved Paular, 417.
Rheumatism, 155, 156, 379.
  in lambs, 155, 156.
Rich, Charles, origin of his Paular flock of Merinos, 30-33.
  John T. succeeds to the flock of his father, 31.
  Messrs. John T. and Virtulan, succeed to the flock of John T. Rich, Sen., 31, 32.
  the course of breeding and character of the Rich flock, 32, 33, 119.
  cut of a ewe bred by the Messrs. Rich, 31.
  effect of a dip of other blood on the flock, 128 and note.
Rickets, the 380.
Rives, William C., imports Shropshire sheep into United States, 66.
Robinson, Erastus, breeds the "Old Robinson Ram," 128 note.
  originates the "Robinson Sheep" of Vermont, 128 note.
  his connection with the origin of the improved Paulars, 416, 418.
Robinson Ram, the old, his pedigree, 416-418.
Roots, value of, for fattening sheep, 418, et seq.
Rot, the, 372-378.

Rotch, Francis, his flock of early American Merinos, 33.
  cut of one of his ewes, illustrating those early Merinos, 34.
  imports South Downs into the United States, 57.
  his account of a Spanish Sheep Dog, 398.

S

Sacking wool, 177.
Salt marshes healthy for sheep, 88.
Salt necessary to sheep in summer, 192.
  necessary in winter, 247.
Sanford, William R., his account of the present Merinos in Spain, 18.
  his remedy for stretches, 310.
  his purchases of sheep, 412, 414.
Saxton, Nelson A., his remedy for stretches, 310.
Scab, the, 338-343.
  cut of the acarus, 339.
  erysipelatous, 344.
Scotch Black-faced sheep — (see Black-Scotch sheep)
Scours — (see Diarrhea.)
Scrofula, 378, 380.
Seaman, Isaac, his prize essay on parturient fever, 331, 335.
Selection — (see Drafting and Selection.)
Shade in pastures of much utility, 212.
Shearing sheep, mode of performing, 170-172.
  stubble shearing and trimming, 172.
  shearing lambs and shearing sheep semi-annually, 172.
Sheds temporary and permanent, for sheep, 211-214.
Sheep, the most profitable animals to depasture our cheap lands, 96.
  necessary to good farming on grain farms, 96.
  more profitable than dairy cows in portions of New York, 97.
  the best cleaners of new lands, 97.
  best adapted to the pecuniary means of a portion of our rural population, 97.
  their management simple and easily learned, 97.
  they never die in debt to man, 97.
  catching and handling, mode of, 139-140.
  turning out to grass, 141.
  tagging, how performed, 141.
  cut illustrative of tagging, 141.
  necessity of eradicating burs from pastures, 142.
  lambing time, place for and assistance in, 142-144.
  spring management of, 139-162.
  summer management of, 163-197.
  administering medicines to when in health, 193.
  housing of in summer, 195.
  pampering of, 196.
  fall management of, 198-210.
  former mode of fall feeding, 202, 203.

452　INDEX.

Sheep, dividing flocks for winter, 209.
　its want of providence in protecting its young, 213 and note.
　winter management of, 211–247.
　confinement to yards and dry feed, 221–228.
　consumption of food by, in winter, 233.
　comparative value of different fodders for, 233–245, 418–425.
　the fattening of in winter, 245, 246.
　management of, on the prairies — (see Prairie Sheep Husbandry.)
　their ready acclimation on the prairies, 250.
　their non-deterioration on prairies, 251.
　diseases of—(see Diseases of Sheep.)
　diseases of, comparatively few in the United States, 261, 262.
　diseases of a low type in the United States, 262, 263.
　anatomy—(see Anatomy of Sheep.)
　longevity of, 268.
　mode of administering medicines to, 299.
　medicines used in diseases of, 384–392.
　destruction by dogs, 393–396.
　amount of food consumed by 418 et seq.
　Mr. Pawlet's experiments in fattening, 418–425.
　number of in United States, 426.
　proportion of wool to meat in, 433.
Sheep Husbandry on the Prairies — (see Prairie Sheep Husbandry.)
Shelters for sheep, 211, 219.
Shropshire Downs, described by Professor Wilson, 61–63.
　Mr. Spooner's account of their origin, 63.
　Mr. Howard describes their origin and character, 63, 64.
　cut of Judge Chaffee's Shropshire ram Lion, 62.
　cut of Judge Chaffee's Shropshire ewe Nancy, 65.
　Judge Chaffee's description of his sheep, 65.
Sibbald, W. C., on parturient fever, 337.
Silesian Merinos—(see Merinos Silesian.)
Simonds, Prof., his remedy for scab, 343.
Skin, proper qualities of, in the Merino, 70.
　diseases of, unnamed ones, 344, 345.
Small-pox, 345, 353.
　its introduction into America to be guarded against, 351, 352.
Smith, Robert, his prize essay on Management of sheep, 198.
　his views in respect to fall feeding lambs, 198, 199.
　his experiments in feeding sheep, 239–262.
　his remedy for diarrhea and dysentery, 308, 309.
　his remedy for scab, 342, 343.
　his remedy for hoof-rot, 364, 365.
Smith's Island sheep, 43.
Soils to be regarded in selecting a breed of sheep, 88.
　the long-wooled sheep preferable on wet soils, 88.
　the Merino cannot endure wet soils, 88.

Soils, effect of low, flat, moist and very rich soils on sheep, 88.
　effect of light, sandy soils, 89.
　kind of, adapted to Merino and Down sheep, 89.
Sore face, 269–271.
Sotham, William H., his account of Mr. Dunn's wethers, 44.
　imports Cotswold sheep in 1840 with Mr. Corning, 48.
South Downs, described by Professor Wilson, 55–57.
　imported into the United States by Mr. Powell, 57.
　imported by Rotch, 57.
　imported by Mr. Thorne, of New York, Mr. Alexander, of Kentucky, and Mr. Taylor, of New Jersey, 58.
　Mr. Thorne describes his mode of managing them, 58, 59.
　cut of Mr. Thorne's ram Archbishop, 56.
　cut of two of his ewes, 57.
　annual value of manure in England, 98.
Spooner, William, describes the Cotswold sheep, 49.
　describes the improved Cheviots, 52, 53.
　describes origin and blood of Hampshire Downs, 60, 61.
　his account of the origin of the Shropshire sheep, 63.
　his estimate of the value of sheep manure, 98, 99.
　his theory of hereditary transmission, 107.
　cited in regard to diseases of sheep, 277, 280, 281, 300, 302, 304, 307, 311, 312, 326, 329, 330, 342, 347, 364, 370, 372, 378, 381, 382, 387, 390.
Sprains, 382.
Spring management of sheep, — (see Management of sheep in spring.)
Stables for sheep, — (see Barns.)
Stells for sheep, 212.
Stevenson, Mr., cited in regard to diseases of sheep, 344.
Stickney, Tyler, his connection with the improved Paulars, 128 note, 417, 418.
St. John's-Wort injurious to sheep, 269–271.
　popular opinions respecting, 270.
Stone, Frederick William, of Canada West, a distinguished breeder of Cotswold sheep, 48, 49.
Storms after shearing, effect on sheep, 186.
Strains, — (see Bruises and Strains.)
Stretches, 310— (see Colic.)
Straw, as sheep feed, 235, 236, 245.
Summer management of sheep, — (see Management of sheep in summer.)
Sun-Scald, how produced, 186.
Swamps, effect of on sheep, — (see Marshes.)
Sweepstakes, Mr. Hammond's ram, — (see Hammond, Edwin.)
Sweet, H. D. L., on comparative weight of wool and bodies of sheep, 433.
Swelled Head, 368.

INDEX. 453

Swelled Lips, 271.
Swelled Neck, 152, 154, 380.

T

Tagging, how performed, 141.
    cut, illustrative of, 141.
Taintor, John A., his account of present Merinos of Spain, 17.
    imports French Merinos into United States, 36.
    description of his Merinos, 36.
Tariffs of the United States, effects of different ones on production, price, etc., of wool, 25, 26.
    those in force from 1824 to 1861, 92–94.
Teats, closed ones, how opened, 157.
Teeth, cutting of the, 150.
    described, 266.
    the most reliable test of age, 266, 267.
    to be extracted sometimes, 267.
Tessier, cited in regard to sheep diseases, 238, 318.
Tetanus, 281, 282.
Texas, adaptation of to wool growing, 248, et seq.
    climate of, 248, 249.
    mean temperature at New Braunfels, 249.
    mean temperature at Austin, 249.
    climate of, 428, et seq.
    northers of, 429, 430.
    seasons and crops, and their vicissitudes, 431, 432.
Thomiere, inoculates for small-pox, 350.
Thorne, Samuel, imports South Down sheep, 58.
    describes his mode of managing them, 58, 59.
    his crosses to procure lambs for the butcher, 134, 135 note.
    his account of parturient fever in his flock, 334, 335.
Ticks, effects of on sheep, 187.
    how exterminated from flocks, 187–189.
    cut of dipping box, 187.
Toe-nippers, cut of, 169.
Torry, Dr., his account of St. John's-Wort, 269.
Travel-sore, 355.
Treatment of ewe after lambing, 156, 157.
Trees in pastures, 212.
Trimmer, Mr., his description of French Merinos in 1827, 19.
    his description of Spanish sheep dogs, 399, 400.
Tunisian Mountain sheep introduced into Pennsylvania, 53.
    bred and commended by John Hare Powell, 53.
Turnips as sheep feed, 221, 235, 239–243.
"21 per cent.," the ram so called, 15.
    length of his wool, 76.
    his qualities as a sire, 109.
    remarkable cross between him and Saxon ewes, 130, and note.
    his pedigree, 415.

U

Udder, inflamed, 157, 330.
    opening closed teats, 157.

V

Vaccination for small-pox, 350.
Valois inoculates for small-pox, 349.
Variola Ovina—(see Small Pox.)
Vegetation, kind of, required by different breeds of sheep, 86, 87.
Vermont, Merino sheep breeders of, 27–30.
Von Thaer, Albert, cut of his Saxon ram, 26.
    his mode of numbering sheep, 183.

W

Walker, Mr., his theory of hereditary transmission, 107 et seq.
Walz M., his description of scab, 388.
Washing sheep, 163, 164.
    its utility considered, 164–168.
Water for sheep, its utility in summer, 194.
    its necessity in winter, 231.
    modes of watering in winter, 231–233.
Water in pastures highly beneficial, 194.
    indispensable in winter, 232.
Weaning lambs, age and mode, 198.
    feeding after weaning, 198–201.
    English mode of fall-feeding, 198, 199.
Webb, Jonas, his success in breeding South Down Sheep, 57 et seq.
Wells, Thomas, describes symptoms of small-pox, 347.
White, Henry G., imports Cotswold Sheep into United States, 49.
    cut of his Cotswold ram Pilgrim, 48.
    cut of his Cotswold ewe Lady Gay, 50.
    an account of his sheep, 49.
Wilcox, Asahel F., pedigree of his "Thousand Dollar Ram," 415.
Wild-fire, 344.
Wilson, Professor John, his description of Leicester sheep, 45–47.
    his description of South Down Sheep, 55.
    his description of the Hampshire Downs, 59, 60, 61, 63.
Womb, inversion of, 145, 330.
Wool, characteristics of Spanish, 15, 16.
    fineness and felting property of Spanish, 16.
    felting property of Saxon, 16.
    characteristics of Saxon fleeces, 20.
    proper degree of fineness of in the American Merino, 72, 73.
    that of the Merino sometimes black, 72 note.
    evenness of, the term defined, 73.
    trueness and soundness of, 74.
    pliancy and softness of, 74.
    style of, 75.
    length of, 75.
    yolk in (see Yolk.)
    oil, grease, and gum in, (see Yolk.)
    prices of in United States from 1800 to 1861, 91–94.

Wool, table of average quarterly prices from 1824 to 1861, 92–94.
  prices medium have never sunk below cost of production, 94.
  prices have been generally remunerative, 94.
  annual exports and imports of from 1840 to 1861, 95, 96.
  the domestic supply has never met the demand, 96.
  cost of producing in New York and New England, 97.
  cost of producing in the South and South-west, 98.
  cost of producing in the Western and North-western States, 98.
  cost of producing in intermediate situations, 98.
  average production of per head by Merinos in large flocks, 98.
  comparative profit of producing in different parts of the United States, 99.
  profits of producing on land worth $50 per acre, 100.
  washing of on the back, 163, 164.
  shearing, mode of, 170–172.
  doing up, mode of, 173–175.
  frauds in doing up, 175.
  storing wool, 176.
  place for selling wool, 177.
  wool depots and commission stores, 177.
  sacking wool, 177.
  cost of getting to market, 251.
  product of, in the United States in 1860, 426.
  proportion to meat in sheep of different ages, sexes and sizes, 433 *et seq.*
Woolens, exports and imports of, from 1840 to 1861, 95.
Wooster, Abel J., describes the "Wooster Ram," 113 note.
Wooster Ram described, 113 and note.
Worms, 312.
Wounds, (see Diseases and Wounds.)
  cuts, 380.

Wounds, lacerated and contused wounds, 381.
  punctured wounds, 381.
  dog bites, 381.
  poisoned wounds, 381.
Wright, Loyal C., his ram, 113.
Wright, M. W. C., first crosses the Paular and Infantado Sheep in Vermont, 128 note.
  originates the Paular and Infantado cross, 416.
  his statements, 418.
Wright, Gov., of Indiana, at World's Fair, 438.
Wrinkles, (see Folds.)

Y

Yards for sheep (see Barns.)
  size, situation of, etc., 220.
  littering yards, 220.
  confining sheep to them in winter, 221 *et seq.*
Yolk described, 77.
  chemical analysis of, 77.
  uses of, in wool, 77.
  proper amount and consistency of, 78, 79.
  proper color of, 80, 81.
  artificial imitation of its color externally, 81.
  artificial propagation and preservation of in fleece, 81.
Youatt, William, discovers conformation of wool, 16.
  his testimony in favor of pure blood, 131 note.
  in regard to sagacity and affection of sheep, 213.
  in regard to defects of the Merino, 223 note.
  cited in regard to diseases of sheep, 268, 274, 275, 278, 279, 280, 282, 283, 291, 300, 301, 306, 309, 314, 315, 317, 318, 326, 327, 329, 330, 336, 339, 340, 342, 344, 345, 347, 350, 354, 356, 357, 363, 364, 373, 385, 389.

MANUAL OF FLAX AND HEMP CULTURE.

JUST PUBLISHED, A NEW EDITION OF

A MANUAL OF FLAX CULTURE AND MANUFACTURE: embracing Full Directions for Preparing the Ground, Sowing the Seed, Harvesting the Crop, Etc. Also comprising an Essay, by a Western Man, on HEMP AND FLAX IN THE WEST:—Amount Grown, Modes of Culture, Preparation for Market, &c., &c. With Botanical Descriptions and Illustrations.

THIS work is composed of Nine Essays from the pens of Practical and Scientific Men who are well advised on the various branches of the subject discussed. It comprises, in a neat and compact form, a large amount of valuable information, and is designed to enable new beginners to cultivate Flax and Hemp successfully. The leading Essay is by a gentleman who has had over thirty years experience in Flax Growing, and thoroughly understands the whole business.

The Manual is published in handsome style, pamphlet form. Price only 25 cents — for which a copy will be sent to any point reached by the United States or Canada mails. Liberal discount to Agents and the Trade. Address

D. D. T. MOORE,

September, 1863. EDITOR RURAL NEW-YORKER, ROCHESTER, N. Y.

OPINIONS OF THE PRESS.

FLAX AND HEMP.—A Manual of Flax Culture and Manufacture, embracing full directions for preparing the ground, sowing, harvesting, dressing, and manufacturing, with the process of making flax cotton, and also an essay upon hemp culture, has been published by D. D. T. MOORE, editor of the *Rural New-Yorker*, Rochester, in pamphlet form at 25 cents, and is well worthy the attention of all who are embarking in flax culture.—*New York Daily Tribune.*

THE CULTIVATION OF FLAX.—A Manual of Flax Culture and Manufacture, has been published in neat pamphlet form, at the office of *Moore's Rural New-Yorker*, and is on sale at the Bookstores generally,—price 25 cents. It is a work pretty exhaustive on the subject. The production of flax is a matter of increasing importance, and our agricultural friends should consult the new Manual.—*Syracuse Daily Journal.*

MANUAL OF FLAX CULTURE.— * * * Those who wish to know all about Flax and Hemp Culture, and to aid in killing "King Cotton" and *suspending* traitors should remit the cost of the Manual—25 cents—to D. D. T. MOORE, Rochester, N. Y.—*Rochester Daily Democrat and American.*

MANUAL OF FLAX CULTURE.—We have received from the publisher, D. D. T. MOORE, Rochester, N. Y., Rural Manual, No. 1, being a collection of valuable information on the culture and manufacture of Flax and Hemp; with illustrations. The wants of a large number of persons who are experimenting with these crops for the first time will be filled with this book. It can be had by addressing the publisher, inclosing 25 cents.—*Prairie Farmer.*

MANUAL OF FLAX AND HEMP CULTURE.—We are pleased to learn that this valuable little work is selling rapidly and widely. The publisher is daily receiving orders from various parts of the Loyal States and Canadas. Three editions have been published within as many weeks, and the demand is such that a fourth is now in press. Those desirous of obtaining reliable information on the culture of Flax and Hemp, and the preparation of their staples for market, should send 25 cents to D. D. T. Moore, Rochester, for his Manual on the subject.—*Rochester Daily Union & Adv.*

A MANUAL OF FLAX CULTURE.— * * * Our farmers have had their attention frequently called this season to the importance of flax-growing, and will probably sow twice or three times the usual amount of seed. But many, and perhaps most of them, are ignorant of the best methods of culture, the improved methods of preparing the fiber, etc. They will find just the information they need in Mr. MOORE's seasonable little Manual.—*Utica Morning Herald and Daily Gazette.*

"Excelsior" its Motto—"Progress and Improvement" its Objects.

MOORE'S RURAL NEW-YORKER,

THE MOST COMPLETE AND POPULAR WEEKLY

AGRICULTURAL, LITERARY AND FAMILY JOURNAL.

This Standard and Unrivaled AGRICULTURAL, HORTICULTURAL, LITERARY AND FAMILY NEWSPAPER, is now in its Fourteenth Year and Volume. The RURAL NEW-YORKER is well known as the *Best, Cheapest and Largest Circulated Journal of its Class* on the Continent—as the Favorite HOME WEEKLY of America—and the Volume for 1864 will at least equal either of its predecessors in CONTENTS, STYLE AND APPEARANCE. Its ample pages comprise various Departments, such as

Agriculture,	Rural Architecture,	Education,
Horticulture,	Choice Miscellany,	Arts and Sciences,
Domestic Economy,	Sabbath Musings,	General News,
Ladies Reading,	Reading for the Young,	Market Reports, &c'

Including Numerous Illustrations, Tales, Sketches, Music, Poetry, Enigmas and Rebuses, &c., &c.

The RURAL NEW-YORKER is and will continue to be THE PAPER FOR THE TIMES, furnishing a weekly variety of appropriate and interesting reading for the various members of the Family Circle. It is National, Patriotic and Progressive — earnest in its support of the Union, Constitution and Laws — ardently advocates the RIGHT condemns WRONG, and constantly endeavors to promote the Best Interests of the People and Country.

FORM, STYLE AND TERMS:

THE RURAL NEW-YORKER is published in Quarto Form, each No. comprising Eight Double Quarto Pages, [forty columns,] printed in Superior Style. An Index, Title Page, &c., given at the close of each Volume.

TERMS, IN ADVANCE:—$2 a Year; Three Copies, $5; Six for $10; Ten for $15, and any greater number at same rate — only $1.50 per copy. Club papers sent to different Post-offices, if desired. As we pre-pay American postage, $1.70 is the lowest Club rate to Canada, and $2.50 to Europe. Subscriptions can begin with the volume, (Jan. 1st,) or any number. Specimens sent free. Address

September, 1863.   **D. D. T. MOORE, Rochester, N. Y.**

OPINIONS OF THE PRESS.

MOORE'S RURAL is full of variety, original and select. No paper on our exchange list comes so near our ideas of perfection, for a secular family paper. It maintains a high moral standard.—*New York Observer.*

THE frequency with which we publish extracts from the RURAL shows our own appreciation of it.—*N. Y. Evening Post.*

THE RURAL is not only a favorite in the rural districts, but deservedly popular in the cities. No newspaper in this or any other country has ever run a more prosperous career.—*Louisville Journal.*

THE RURAL is a very valuable paper, eminently practical in its character, and pure in its tone. Deserves and is achieving abundant success.—*N. Y. Times.*

Mr. MOORE *ought* to make a fortune out of his journal, and we trust he will, for *he is helping to make the fortune of the country.*—*Ohio Statesman.*

THE RURAL is the *best Farm and Fireside Journal in America*, and has justly earned all its devoted editor claims for it. —*Chicago Daily Democrat.*

No one can possibly regret subscribing for the RURAL, as it will be read with profit by every family. It has excellent illustrations.—*Ind. State Sentinel.*

WITHOUT exception, the best Agricultural and Family Newspaper. Mr. MOORE lately received a $1,000 draft for one club of new subscribers!—*Minnesota Statesman.*

THE RURAL is a perfect typographical luxury, teeming with originality, pure morals, and useful reading.—*Vt. Citizen.*

THE RURAL is *the best* Agricultural, Horticultural and Family paper published on the Continent. —*Recorder, Newcastle, C. W.*

WE wonder not at the RURAL's great success; it richly merits it.—*Gospel Banner, Maine.*

www.ingramcontent.com/pod-product-compliance
Lightning Source LLC
Chambersburg PA
CBHW020738020526
44115CB00030B/163